Thomas E. Barman

Enzyme Handbook

Supplement I

Springer-Verlag
New York Berlin Heidelberg Tokyo

Dr. Thomas E. Barman
Institut National de la Sante
et de la Recherche Medicale
INSERM U 128
F-34033 Montpellier Cedex
France

Printed and bound by R.R. Donnelley & Sons, Harrisonburg, Virginia.
Printed in the United States of America.

9 8 7 6 5 4 3 2 (Second printing, 1985)

ISBN 0-387-06761-2 Springer-Verlag New York Berlin Heidelberg Tokyo
ISBN 3-540-06761-2 Springer-Verlag Berlin Heidelberg New York Tokyo

Preface

In the five years since the appearance of the *Enzyme Handbook* several hundred new enzymes have been described. The *Supplement* includes molecular and kinetic data on about half of these and also on several enzymes omitted from the *Handbook;* information on many of the remainder will be found in *Enzyme Nomenclature (1972).* An incomplete search of the literature may have resulted in the omission of a number of enzymes from the *Supplement* and I apologize to the authors whose contributions I have missed.

The labour involved in compiling the data for the *Supplement* was greatly reduced by the kindness of Professor E. C. Webb who sent me a draft copy of *Enzyme Nomenclature (1972)* prior to publication. Dr. E. A. Jones and Dr. G. S. Knaggs reviewed the entire text of the *Supplement* and I am grateful for their criticisms and suggestions. I also thank the many people who offered constructive criticisms of the earlier *Enzyme Handbook.* Several of these suggestions have been incorporated into the *Supplement.* I am grateful to Mrs. Anna Knaggs for her expert typing and for making the manuscript ready for photomechanical reproduction, to Mrs. Lucia Barman for preparing the lists of references and the index and to the staff of Springer-Verlag for giving me every possible assistance in seeing the *Supplement* to completion. I am indebted to Professor E. C. Webb, the IUPAC-IUB and the Elsevier Publishing Company for permission to reproduce the Key to the Numbering and Classification of Enzymes which appears on pages 7–12.

Shinfield, Reading
January 1974

Thomas E. Barman

Contents

SYMBOLS AND ABBREVIATIONS

General

ADP, CDP, GDP, IDP, dTDP, UDP	the 5'-diphosphates of adenosine, cytidine, guanosine, inosine, thymidine and uridine
Ammediol	2-amino-2-methyl-1,3-propanediol
AMP, CMP, GMP, IMP, dTMP, UMP	the 5'-monophosphates of adenosine, cytidine, guanosine, inosine, thymidine and uridine
Cyclic AMP	adenosine 3':5'-cyclic phosphate
ATP, CTP, GTP, ITP, dTTP, UTP	the 5'-triphosphates of adenosine, cytidine, guanosine, inosine, thymidine and uridine
dATP, dCTP, etc.	deoxy ATP, deoxy CTP etc.
ATPP	adenosine tetraphosphate
Bicine	*N*,*N*-bis(2-hydroxyethyl)glycine
C	competitive
CoA	coenzyme A
Coenzyme I	NAD
Coenzyme II	NADP
Cozymase	NAD
DNA	deoxyribonucleic acid
DPN	NAD
EDTA	ethylenediaminetetraacetate
FAD	flavin - adenine dinucleotide
FMN	flavin mononucleotide
Glc	glucose
GlyGly	glycylglycine
HEPES	*N*-2-hydroxyethylpiperazine-*N*'-2-ethane-sulphonic acid
I	inhibitor
LAD	light absorption data
M	gm molecule (1 mole) per litre
mM	10^{-3}M
µM	10^{-6}M
nM	10^{-9}M
MES	2-(*N*-morpholino)ethanesulphonic acid
NAD	nicotinamide-adenine dinucleotide

NADP	nicotinamide-adenine dinucleotide phosphate
NC	non-competitive
NDP	nucleoside diphosphate
NMN	nicotinamide mononucleotide
NTP	nucleoside triphosphate
P_i	inorganic orthophosphate
PP_{ii}	inorganic pyrophosphate
RNA	ribonucleic acid
tRNA	transfer ribonucleic acid
S	substrate
$S_{20,w}$	sedimentation coefficient in Svedbergs (in water at 20°). One Svedberg = 1×10^{-13} sec.
SDS	sodium dodecyl sulphate (≡sodium lauryl sulphate)
TAES	tris (hydroxymethyl)-methyl-2-aminoethane sulphonic acid
TEA	triethanolamine
TPN	NADP
Tris	2-amino-2-hydroxymethylpropane-1,3-diol
TPP	thiamine pyrophosphate
Tricine	2-N-glycine-2-hydroxymethylpropane-1,3-diol
UC	uncompetitive
ε	molar extinction coefficient (units = $M^{-1}cm^{-1}$)
λ	wavelength
nm	millimicrons(1nm = 10Å = 10^{-9}m)
°	degree centigrade

Symbols for amino acids

Unless otherwise stated all amino acids are of the L-configuration.

Ala	alanine	Leu	leucine
Arg	arginine	Lys	lysine
Asn	asparagine	Met	methionine
Asp	aspartic acid	Phe	phenylalanine
Cys	cysteine	Pro	proline
Gln	glutamine	Ser	serine
Glu	glutamic acid	Thr	threonine
Gly	glycine	Trp	tryptophan
His	histidine	Tyr	tyrosine
Ile	isoleucine	Val	valine

Kinetic symbols

Symbol	Definition	Unit
v	velocity of reaction catalyzed by an enzyme	-
V	maximum velocity: the value of v when the enzyme is saturated with substrate	-
K_D	dissociation constant	-
K_m	Michaelis constant; concentration of substrate at which $v = V/2$.	M
K_s	substrate constant; equilibrium (dissociation) constant of the reaction $E + S = ES$	M
K_i	inhibitor constant; equilibrium (dissociation) constant of the reaction $E + I = EI$	M
k_o	overall or observed rate constant - the number of molecules of substrate transformed per second per molecule of enzyme, sometimes called turnover number	sec^{-1}

Literature abbreviations

Other abbreviations such as those accepted without definition by the Journal of Biological Chemistry appear in the Handbook without further identification.

AB	Archives of Biochemistry
ABB	Archives of Biochemistry & Biophysics
ACS	American Chemical Society
Advances in Enzymology	ed. Nord, F. F. Interscience (John Wiley & Sons): New York
Ann. Rev. Biochem.	Annual Review of Biochemistry. Annual Reviews Inc: Palo Alto, California
B	Biochemistry
BBA	Biochimica & Biophysica Acta
BBRC	Biochemical & Biophysical Research Communications
BJ	Biochemical Journal
Brookhaven Symposia in Biology	Biology Department, Brookhaven National Laboratory, Upton, New York
BZ	Biochemische Zeitschrift
EJB	European Journal of Biochemistry
The Enzymes (second edition, 1959-1963)	eds. Boyer, P.D., Lardy, H.A. & Myrbäck, K. Academic Press Inc.: New York
The Enzymes (third edition, 1970-)	ed. Boyer, P.D. Academic Press Inc.: New York
FEBS lett.	the Federation of European Biochemical Societies letters
FP	Federation Proceedings
JACS	Journal of the American Chemical Society
JBC	Journal of Biological Chemistry
JMB	Journal of Molecular Biology
Methods in Enzymology	eds. Colowick, S.P. & Kaplan, N.O. Academic Press Inc.: New York
PNAS	Proceedings of the National Academy of Sciences (U.S.)
The Proteins (first edition)	eds. Neurath, H. & Bailey, K. Academic Press Inc.: New York
The Proteins (second edition)	ed. Neurath, H. Academic Press Inc.: New York

EXPLANATORY NOTES

The data in the Supplement are arranged in the same way as in the Enzyme Handbook and below is a summary of the way in which this has been done. The reader is referred to the Handbook for a more detailed discussion of the enzyme properties considered in the Supplement.

Enzyme classification

The enzyme list in Enzyme Nomenclature (1972) differs in several important respects from the 1964 list (especially subgroups 1.13, 1.14 and 3.4). The new key to the numbering and classification of enzymes will be found on pages 7 – 12 . The enzymes in the Supplement have been arranged according to the 1972 list.

Equilibrium constant

When a reaction is described as "essentially irreversible" the direction catalyzed by the enzyme in question is indicated by [F] (forward reaction) or [R] (reverse reaction).

Molecular properties

The Supplement includes more molecular data than the Enzyme Handbook and wherever possible details of the following are given: molecular weight [number of subunits]; carbohydrate or lipid content; multiplicity; prosthetic group (including metal ion content); stable enzyme-substrate complexes; active site directed irreversible inhibitors (e.g. diisopropylfluorophosphate. These are occasionally included under *Specificity and kinetic properties* or *Inhibitors*) and references to amino acid compositions. Details of the way in which a particular molecular weight was obtained are given under conditions. This information is given in a concise form and when several methods were used, these are separated by semicolons. For example, the entry

"pH 8.2; Sephadex G 100, pH 7.5; amino acid composition"

is an abbreviation of

"the molecular weight was obtained by an ultracentrifugation method at pH 8.2 and by gel filtration on a column of Sephadex G 100 at pH 7.5. The amino acid composition of the enzyme has been determined."

Specific activity

The specific activity of an enzyme is defined as units per mg of enzyme protein where one unit is that amount of enzyme which will catalyze the transformation of 1 μmole of substrate per minute under specified conditions.

Specific activities are only given for enzymes which have been highly purified and which are thought to be homogeneous. The figure in brackets immediately following the source of the enzyme of interest is the purification factor - i.e. the ratio of the specific activity of the purified enzyme to the specific activity of the source material. Specific activities are occasionally included under *Specificity and kinetic properties*.

Specificity and kinetic properties

The Supplement includes more information on cofactor and activator requirements than the Enzyme Handbook. In most cases purification factors are included, regardless of the purity of the enzyme. Relative activities are included wherever possible - usually this information is tabulated but often relative activities are indicated in brackets immediately following the compound listed in the text. In both cases the reference compound (i.e. that of relative activity 1.00) is usually thought to be physiologically the most important.

Inhibitors

Information on inhibitors is in many cases included under *Specificity and kinetic properties*. The data are condensed. For instance, the entry

"a number of amino acids [C(L-lysine); UC(reduced NAD)] are inhibitors"

is an abbreviation of

"a number of amino acids are inhibitory. The type of inhibition is competitive (with respect to the substrate L-lysine) or uncompetitive (with respect to the substrate reduced NAD)".

Active site irreversible inhibitors are usually included under *Molecular properties*.

References

The literature survey for the Supplement was concluded September 30 1973. Certain references were inserted after the typing of the manuscript; these are not numbered. References to review articles are underlined.

KEY TO NUMBERING AND CLASSIFICATION OF ENZYMES

From ENZYME NOMENCLATURE: *Recommendations (1972) of the Commission on Biochemical Nomenclature on the Nomenclature and Classification of Enzymes together with their Units and the Symbols of Enzyme Kinetics.* Published by Elsevier, Amsterdam; pages 17-22.

1. OXIDOREDUCTASES

1.1 Acting on the CH-OH group of donors

1.1.1 With NAD or NADP as acceptor
1.1.2 With a cytochrome as acceptor
1.1.3 With oxygen as acceptor
1.1.99 With other acceptors

1.2 Acting on the aldehyde or keto group of donors

1.2.1 With NAD or NADP as acceptor
1.2.2 With a cytochrome as acceptor
1.2.3 With oxygen as acceptor
1.2.4 With a disulphide compound as acceptor
1.2.7 With an iron-sulphur protein as acceptor
1.2.99 With other acceptors

1.3 Acting on the CH-CH group of donors

1.3.1 With NAD or NADP as acceptor
1.3.2 With a cytochrome as acceptor
1.3.3 With oxygen as acceptor
1.3.7 With an iron-sulphur protein as acceptor
1.3.99 With other acceptors

1.4 Acting on the CH-NH$_2$ group of donors

1.4.1 With NAD or NADP as acceptor
1.4.3 With oxygen as acceptor
1.4.4 With a disulphide compound as acceptor
1.4.99 With other acceptors

1.5 Acting on the CH-NH group of donors

1.5.1 With NAD or NADP as acceptor
1.5.3 With oxygen as acceptor
1.5.99 With other acceptors

1.6 Acting on reduced NAD or reduced NADP

1.6.1 With NAD or NADP as acceptor
1.6.2 With a cytochrome as acceptor
1.6.4 With a disulphide compound as acceptor
1.6.5 With a quinone or related compound as acceptor
1.6.6 With a nitrogenous group as acceptor
1.6.7 With an iron-sulphur protein as acceptor
1.6.99 With other acceptors

1.7 Acting on other nitrogenous compounds as donors

1.7.2 With a cytochrome as acceptor
1.7.3 With oxygen as acceptor
1.7.7 With an iron-sulphur protein as acceptor
1.7.99 With other acceptors

1.8 Acting on a sulphur group of donors

1.8.1 With NAD or NADP as acceptor
1.8.2 With a cytochrome as acceptor
1.8.3 With oxygen as acceptor
1.8.4 With a disulphide compound as acceptor
1.8.5 With a quinone or related compound as acceptor
1.8.6 With a nitrogenous group as acceptor
1.8.7 With an iron-sulphur protein as acceptor
1.8.99 With other acceptors

1.9 Acting on a haem group of donors

1.9.3 With oxygen as acceptor
1.9.6 With a nitrogenous group as acceptor
1.9.99 With other acceptors

1.10 Acting on diphenols and related substances as donors

1.10.2 With a cytochrome as acceptor
1.10.3 With oxygen as acceptor

1.11 Acting on hydrogen peroxide as acceptor

1.12 Acting on hydrogen as donor

1.12.1 With NAD or NADP as acceptor
1.12.2 With a cytochrome as acceptor
1.12.7 With an iron-sulphur protein as acceptor

1.13 Acting on single donors with incorporation of molecular oxygen (oxygenases)

1.13.11 With incorporation of two atoms of oxygen
1.13.12 With incorporation of one atom of oxygen (internal monooxygenases or internal mixed function oxidases)
1.13.99 Miscellaneous (requires further characterization)

1.14 Acting on paired donors with incorporation of molecular oxygen

1.14.11 With 2-oxoglutarate as one donor, and incorporation of one atom each of oxygen into both donors
1.14.12 With reduced NAD or reduced NADPH as one donor, and incorporation of two atoms of oxygen into one donor
1.14.13 With reduced NAD or reduced NADP as one donor, and incorporation of one atom of oxygen
1.14.14 With reduced flavin or flavoprotein as one donor, and incorporation of one atom of oxygen
1.14.15 With a reduced iron-sulphur protein as one donor, and incorporation of one atom of oxygen
1.14.16 With reduced pteridine as one donor, and incorporation of one atom of oxygen

1.14.17 With ascorbate as one donor, and incorporation of one atom of oxygen
1.14.18 With another compound as one donor, and incorporation of one atom of oxygen
1.14.99 Miscellaneous (requires further characterization)

1.15 Acting on superoxide radicals as acceptor

1.16 Oxidizing metal ions

1.16.3 With oxygen as acceptor

1.17 Acting on -CH_2-groups

1.17.1 With NAD or NADP as acceptor
1.17.4 With a disulphide compound as acceptor

2. TRANSFERASES

2.1 Transferring one-carbon groups

2.1.1 Methyltransferases
2.1.2 Hydroxymethyl-, formyl- and related transferases
2.1.3 Carboxyl- and carbamoyltransferases
2.1.4 Amidinotransferases

2.2 Transferring aldehyde or ketonic residues

2.3 Acyltransferases

2.3.1 Acyltransferases
2.3.2 Aminoacyltransferases

2.4 Glycosyltransferases

2.4.1 Hexosyltransferases
2.4.2 Pentosyltransferases
2.4.99 Transferring other glycosyl groups

2.5 Transferring alkyl or aryl groups, other than methyl groups

2.6 Transferring nitrogenous groups

2.6.1 Aminotransferases
2.6.3 Oximinotransferases

2.7 Transferring phosphorus-containing groups

2.7.1 Phosphotransferases with an alcohol group as acceptor
2.7.2 Phosphotransferases with a carboxyl group as acceptor
2.7.3 Phosphotransferases with a nitrogenous group as acceptor
2.7.4 Phosphotransferases with a phospho-group as acceptor
2.7.5 Phosphotransferases with regeneration of donors (apparently catalysing intramolecular transfers)
2.7.6 Diphosphotransferases
2.7.7 Nucleotidyltransferases

2.7.8 Transferases for other substituted phospho-groups
2.7.9 Phosphotransferases with paired acceptors

2.8 Transferring sulphur-containing groups

2.8.1 Sulphurtransferases
2.8.2 Sulphotransferases
2.8.3 CoA-transferases

3. HYDROLASES

3.1 Acting on ester bonds

3.1.1 Carboxylic ester hydrolases
3.1.2 Thiolester hydrolases
3.1.3 Phosphoric monoester hydrolases
3.1.4 Phosphoric diester hydrolases
3.1.5 Triphosphoric monoester hydrolases
3.1.6 Sulphuric ester hydrolases
3.1.7 Diphosphoric monoester hydrolases

3.2 Acting on glycosyl compounds

3.2.1 Hydrolysing *O*-glycosyl compounds
3.2.2 Hydrolysing *N*-glycosyl compounds
3.2.3 Hydrolysing *S*-glycosyl compounds

3.3 Acting on ether bonds

3.3.1 Thioether hydrolases
3.3.2 Ether hydrolases

3.4 Acting on peptide bonds (peptide hydrolases)

3.4.11 α-Aminoacylpeptide hydrolases
3.4.12 Peptidylamino-acid or acylamino-acid hydrolases
3.4.13 Dipeptide hydrolases
3.4.14 Dipeptidylpeptide hydrolases
3.4.15 Peptidyldipeptide hydrolases
3.4.21 Serine proteinases
3.4.22 SH-proteinases
3.4.23 Acid proteinases
3.4.24 Metalloproteinases
3.4.99 Proteinases of unknown catalytic mechanism

3.5 Acting on carbon-nitrogen bonds, other than peptide bonds

3.5.1 In linear amides
3.5.2 In cyclic amides
3.5.3 In linear amidines
3.5.4 In cyclic amidines
3.5.5 In nitriles
3.5.99 In other compounds

3.6 Acting on acid anhydrides

3.6.1 In phosphoryl-containing anhydrides
3.6.2 In sulphonyl-containing anhydrides

3.7 Acting on carbon-carbon bonds

3.7.1 In ketonic substances

3.8 Acting on halide bonds

3.8.1 In C-halide compounds
3.8.2 In P-halide compounds

3.9 Acting on phosphorus-nitrogen bonds

3.10 Acting on sulphur-nitrogen bonds

3.11 Acting on carbon-phosphorus bonds

4. LYASES

4.1 Carbon-carbon lyases

4.1.1 Carboxy-lyases
4.1.2 Aldehyde-lyases
4.1.3 Oxo-acid-lyases
4.1.99 Other carbon-carbon lyases

4.2 Carbon-oxygen lyases

4.2.1 Hydro-lyases
4.2.2 Acting on polysaccharides
4.2.99 Other carbon-oxygen lyases

4.3 Carbon-nitrogen lyases

4.3.1 Ammonia-lyases
4.3.2 Amidine-lyases

4.4 Carbon-sulphur lyases

4.5 Carbon-halide lyases

4.6 Phosphorus-oxygen lyases

4.99 Other lyases

5. ISOMERASES

5.1 Racemases and epimerases

5.1.1 Acting on amino acids and derivatives
5.1.2 Acting on hydroxy acids and derivatives
5.1.3 Acting on carbohydrates and derivatives
5.1.99 Acting on other compounds

5.2 Cis-trans *isomerases*

5.3 Intramolecular oxidoreductases

5.3.1 Interconverting aldoses and ketoses
5.3.2 Interconverting keto- and enol-groups
5.3.3 Transposing C=C bonds
5.3.4 Transposing S-S bonds
5.3.99 Other intramolecular oxidoreductases

5.4 Intramolecular transferases

5.4.1 Transferring acyl groups
5.4.2 Transferring phosphoryl groups
5.4.3 Transferring amino groups
5.4.99 Transferring other groups

5.5 Intramolecular lyases

5.99 Other isomerases

6. LIGASES (SYNTHETASES)

6.1 Forming carbon-oxygen bonds

6.1.1 Ligases forming aminoacyl-tRNA and related compounds

6.2 Forming carbon-sulphur bonds

6.2.1 Acid-thiol ligases

6.3 Forming carbon-nitrogen bonds

6.3.1 Acid-ammonia ligases (amide synthetases)
6.3.2 Acid-amino-acid ligases (peptide synthetases)
6.3.3 Cyclo-ligases
6.3.4 Other carbon-nitrogen ligases
6.3.5 Carbon-nitrogen ligases with glutamine as amido-*N*-donor

6.4 Forming carbon-carbon bonds

6.5 Forming phosphate ester bonds

ENZYME COMMISSION NUMBERS (1972) OF INCOMPLETELY NUMBERED ENZYMES IN THE ENZYME HANDBOOK (1969)

Numbers in brackets are tentative. The numbers of several enzymes included in the *Handbook (1969)* have recently been revised (especially subgroups 1.13, 1.14 and 3.4) and the reader is referred to *Enzyme Nomenclature (1972)* for details of these changes.

Enzyme Handbook	EC Number
1.1.1.a	1.1.1.77
1.1.1.b	1.1.1.88
1.1.1.c	1.1.1.140
1.1.1.d	1.1.1.103
1.1.1.e	1.1.1.129
1.1.1.f	1.1.1.138
1.1.1.g	1.1.1.130
1.1.1.h	1.1.1.126
1.1.1.k	1.1.1.127
1.1.1.m	1.1.1.125
1.1.1.n	1.1.1.150
1.1.1.p	1.1.1.85
1.1.1.q	1.1.1.137
1.1.1.r	1.1.1.132
1.1.1.s	1.1.1.95
1.1.3.a	1.1.3.15
1.1.99.a	(1.1.99.10)
1.2.1.a	1.5.1.12
1.3.1.a	1.3.1.12
1.3.1.b	1.3.1.7
1.3.99.a	1.3.99.6
1.4.1.a	1.4.1.10
1.4.3.a	1.4.3.8
1.4.3.b	1.13.12.1
1.5.1.a	1.5.1.11
1.5.3.a	1.5.3.4
1.6.2.a	1.6.2.4
1.6.4.a	1.6.4.5
1.7.99.a	(1.7.3.4)
1.11.1.a	1.11.1.10
1.14.1.a	1.14.13.1
1.14.3.a	1.14.16.2
2.1.1.a	2.1.1.27
2.1.1.b	-
2.1.1.c	2.1.1.17
2.1.2.a	2.1.2.7
2.1.2.b	2.1.2.8
2.1.3.a	-
2.3.1.a	2.3.1.21
2.3.1.b	2.3.1.32
2.3.1.c	4.1.3.12
2.3.1.d	2.3.1.36
2.4.1.a	2.4.1.21
2.4.1.b	2.4.1.21
2.4.2.a	2.4.2.19
2.4.2.b	2.4.2.22
2.4.2.c	2.4.2.17
2.6.1.a	2.6.1.30
2.6.1.b	2.6.1.21
2.6.1.c	2.6.1.33
2.6.1.d	2.6.1.31
2.6.1.e	2.6.1.6 or 42
2.7.1.a	2.7.1.63
2.7.1.b	2.7.1.60
2.7.1.c	2.7.1.54
2.7.1.d	2.7.1.53
2.7.1.e	2.7.1.62
2.7.1.f	2.7.1.59
2.7.1.g	2.7.1.61
2.7.1.h	2.7.5.5
2.7.1.k	2.7.1.7
2.7.2.a	2.7.2.7
2.7.3.a	2.7.3.7
2.7.5.a	-
2.7.7.a	2.7.7.29
2.7.7.b	2.7.7.27
2.7.7.c	2.7.7.43
2.7.7.d	2.7.7.37
2.7.7.e	2.7.7.32
2.7.7.f	2.7.7.33
3.1.3.a	3.1.3.29
3.1.4.a	3.1.4.13
3.1.4.b	3.1.4.26
3.1.4.c	3.1.4.17
3.1.4.d	3.1.4.16
3.2.2.a	3.2.2.7
3.4.4.a	3.4.99.26
3.4.4.b	3.4.21.2
3.4.4.c	3.4.21.15
3.5.1.a	3.5.1.19

Enzyme Handbook	EC number
3.5.5.a	3.5.5.2
3.8.1.a	3.8.1.3
4.1.1.a	4.1.1.49
4.1.1.b	4.1.1.47
4.1.2.a	4.1.2.20
4.1.2.b	4.1.2.19
4.1.2.c	4.1.2.21
4.1.2.d	4.1.2.23
4.1.2.e	4.1.3.16
4.1.2.f	4.1.2.22
4.1.3.a	4.1.3.20
4.1.3.b	4.1.3.19
4.1.3.c	4.1.3.17
4.2.1.a	4.2.1.42
4.2.1.b	4.2.1.40
4.2.1.c	4.2.1.32
4.2.1.d	4.2.1.34
4.2.1.e	4.1.99.1
4.3.1.a	4.2.1.38
5.1.3.a	5.1.3.8
5.1.3.b	5.1.3.9
5.1.3.c	-
5.3.1.a	5.3.1.15
5.4.99.a	4.2.1.33
5.5.1.a	5.5.1.3
6.1.1.a	6.1.1.19
6.1.1.b	6.1.1.20
6.1.1.c	6.1.1.15
6.1.1.d	6.1.1.17
6.1.1.e	6.1.1.14
6.2.1.a	6.2.1.2
6.3.3.a	6.3.3.2
6.3.4.a	6.3.4.8

ENZYME DATA

ACETOIN DEHYDROGENASE

(Acetoin: NAD oxidoreductase)

Acetoin + NAD = diacetyl + reduced NAD — Ref.

Equilibrium constant

The reaction was irreversible [R] with the enzyme from beef liver (ref 1) but reversibility was demonstrated with that from rat liver (ref 2).

Molecular properties

source	value	conditions	
Beef liver	76,000 [3]	Sephadex G 100, pH 6.3	(1)

Diacetyl reductase from rat liver has been resolved into multiple forms. (2)

Specificity and Michaelis constants

source	substrate	K_m(M)	conditions	
Beef liver (purified 300 x)	diacetyl	4×10^{-5}	pH 6.1, Pi	(3)
	reduced NAD	1×10^{-4}	pH 6.1, Pi	(3)
Rat liver (purified 8 x)	diacetyl	4.8×10^{-2}	pH 6.0, Pi	(2)

The enzyme (beef liver) could utilize reduced NAD or reduced NADP equally well. Diacetyl could not be replaced by acetone; pentane-3-one; pentane-2,4-dione; hexane-2,5-dione; pyruvate; oxaloacetate; 2-oxoglutarate or acetoin. The enzyme did not require added metal ions for activity. (1)

Bacterial acetoin dehydrogenase is discussed in ref 4.

Inhibitors

The enzyme (beef liver) was inhibited by NAD, (C(reduced NAD); NC (diacetyl)) and by acetoin (C(diacetyl); NC (reduced NAD)). (3)

Abbreviations

Acetoin — 3-hydroxybutan-2-one
Diacetyl — butane-2,3-dione

References

1. Burgos, J. & Martin, R. (1972) BBA, 268, 261.
2. Gabriel, M.A., Jabara, H. & Al-Khalidi, U.A.S. (1971) BJ, 124, 793.
3. Martin, R. & Burgos, J. (1972) BBA, 289, 13.
4. Juni, E. & Heym, G.A. (1957) J. Bacteriol. 74, 757.
   Johansen, L., Larsen, S.H. & Størmer, F.C. (1973) EJB, 34, 97; 100.

MALATE DEHYDROGENASE (DECARBOXYLATING)

(L-Malate: NAD oxidoreductase (decarboxylating))

L-Malate + NAD = pyruvate + CO_2 + reduced NAD

Ref.

Equilibrium constant

$$\frac{[\text{pyruvate}]\,[CO_2]\,[\text{reduced NADP}]}{[\text{L-malate}]\,[\text{NADP}]} = 5.1 \times 10^{-2}\ \text{M}$$

(pH 7.4, $NaHCO_3$, 22-25°) (1)

Molecular weight

source	value	conditions	Ref.
Ascaris suum (roundworm)	250,000 [4]	pH 7.5; gel electrophoresis (SDS); gel filtration	(2)

Specific activity

A. suum (100 x) 15 L-malate (with NAD as the cofactor; pH 7.4, triethanolamine, 25°) (2)

Specificity and Michaelis constants

The enzyme (A. suum) could utilize NADP (0.25) in the place of NAD (1.00). It did not catalyze the decarboxylation of oxaloacetate in the presence of NAD or NADP. It required Mn^{2+} or Mg^{2+} for activity. (2)

The enzyme (Streptococcus faecalis) required Mn^{2+} (partially replaced by Mg^{2+}) and NH_4^+ for activity. (3)

source	substrate	K_m (M)	conditions	Ref.
S. faecalis	malate	1.1×10^{-4} (a)	pH 8.6, Tris	(3)
	NAD (b)	5.0×10^{-5}	pH 8.6, Tris	(3)
	pyruvate	3.6×10^{-2}	pH 8.6, Tris	(3)

(a) in the presence of Mn^{2+}. In the presence of the less effective activator Mg^{2+} a value of 6.8×10^{-4} was obtained.

(b) NAD (1.00) could be replaced by NADP (0.05).

Malate decarboxylase has also been isolated from cauliflower bud mitochondria; it has similar properties to the enzymes from S. faecalis and A. suum. It was stimulated by low concentrations of CoA. (4)

Inhibitors

The enzyme (A. suum) was inhibited by NH_4^+; high concentrations of Mg^{2+} partially abolished the inhibition. (2)

The enzyme (cauliflower) was inhibited by reduced NAD and oxaloacetate. (4)

References

1. Harary, I., Korey, S.R. & Ochoa, S. (1953) JBC, 203, 595.
2. Fodge, D.W., Gracy, R.W. & Harris, B.G. (1972) BBA, 268, 271.
3. London, J. & Meyer, E.Y. (1969) J. Bacteriol, 98, 705.
4. Macrae, A.R. (1971) BJ, 122, 495.

L-GULONATE DEHYDROGENASE

(L-Gulonate: NAD 3-oxidoreductase)

L-Gulonate + NAD = 3-Keto-L-gulonate + reduced NAD

Ref.

Equilibrium constant

The reaction is reversible, the reverse reaction being favoured. (1)

Molecular weight

source	value	conditions	
Drosophila melanogaster	63,000	sucrose gradient	(2)

Specific activity

D. melanogaster (2600 x) 12 DL-3-hydroxybutyrate (pH 8.2, Tris, 30°) (2)

Specificity and Michaelis constants

source	substrate	relative velocity	K_m(M)	conditions	
D.melanogaster	L-3-hydroxybutyrate	0.08	2.3×10^{-2}	pH 8.2, Tris, 30°	(2)
	L-gulonate	1.00	-	pH 8.2, Tris, 30°	(2)
	NAD	-	2.5×10^{-4}	pH 8.2, Tris, 30°	(2)
Pig kidney (purified 100 x)	L-gulonate	1.00	5.1×10^{-3}	pH 8.5, Tris	(1)
	L-idonate	0.59	5.3×10^{-3}	pH 8.5, Tris	(1)
	L-3-hydroxybutyrate	0.26	5.3×10^{-3}	pH 8.5, Tris	(1)
	NAD (with L-gulonate)	-	4×10^{-5}	pH 8.5, Tris	(1)
	3-keto-L-gulonate[a]	-	4.5×10^{-4}	pH 6.3, Pi	(1)
	acetoacetate[a]	-	6.4×10^{-3}	pH 6.3, Pi	(1)
	2,3-diketo-L-gulonate[a]	-	7.3×10^{-3}	pH 6.3, Pi	(1)

[a] in the reverse reaction.

The enzyme from D. melanogaster was inactive with D-gulonate, D-3-hydroxybutyrate, DL-2-hydroxybutyrate, DL-4-hydroxybutyrate, 2-butanol, DL-3-hydroxypropionate and L-threonine. (2)

The enzyme (pig kidney) is highly specific for hydroxy acids in which the 3-hydroxy group exhibits the L-configuration. Thus, L-gulonic acid (1.00) could be replaced by D-lyxonic (0.81); L-idonic (0.59); D-xylonic (0.56); D-gluconic (0.49); L-erythronic (0.33); L(+)-3-hydroxybutyric (0.26); L-threonic (0.18); D-talonic (0.17); L-altronic (0.10); D-mannonic (0.06); D-ribonic (0.05) or D-galactonic (0.01) but not by L- or D-arabonic; L-galactonic; L-talonic; D-ribonic; D-allonic; D-threonic; D(-)-3-hydroxybutyric; D-erythronic; D-idonic; L-mannonic; D-altronic; L-xylonic; D-gulonic; L-lyxonic or L-gluconic. (1)

References

1. Smiley, J.D. & Ashwell, G. (1961) JBC, 236, 357.
2. Borack, L.I. & Sofer, W. (1971) JBC, 246, 5345.

17β-HYDROXYSTEROID DEHYDROGENASE

(17β-Hydroxysteroid: NADP 17 - oxidoreductase)

Oestradiol + NADP = oestrone + reduced NADP

Ref.

Equilibrium constant

The reversibility of the reaction with the enzyme from human red blood cells has been demonstrated. (1)

Molecular weight

source	value	conditions	
Human red blood cells	75,500	Sephadex G 150, 0.1M NaCl	(1)

Specificity and Michaelis constants

source	substrate	cosubstrate	K_m (M)
Human red blood cells (purified 740-fold)[a]	oestradiol	NADP	9.1×10^{-5}
	oestradiol sulphate	NADP	3.8×10^{-4}
	NADP	oestradiol	2.7×10^{-5}
	NADP	oestradiol sulphate	2.4×10^{-5}

[a] the conditions were: pH 9.4, glycine, 37° (1)

NAD possessed one sixth to one tenth the activity of NADP. The 3-acetyl derivatives of NADP and NAD showed less activity than their parent compounds. Testosterone could replace oestradiol but oestriol was inactive. Blocking of position 17 of oestradiol by glucuronoside formation prevented enzyme activity. (1)

The enzyme is sensitive to ionic strength and is most active at sodium chloride concentrations of 0.1M and above. (1)

Inhibitors

inhibitor[a]	type	K_i (M)	
oestradiol sulphate	C(oestradiol)	2.4×10^{-4}	(1)

[a] with the enzyme from human red blood cells. The conditions were pH 9.4, glycine, 37°.

Serum albumin and organic solvents inhibited the reaction. (1)

References

1. Jacobsohn, G.M. & Hochberg, R.B. (1968) JBC, 243, 2985.

ω-HYDROXYDECANOATE DEHYDROGENASE

(10-Hydroxydecanoate: NAD oxidoreductase)

10-Hydroxydecanoate + NAD = 10-oxodecanoate + reduced NAD

Ref.

Equilibrium constant

The reaction is reversible. (1)

Specificity

The enzyme from rabbit liver (purified 5 x) is specific for ω-hydroxyacids and it oxidized 11-hydroxyundecanoate (1.00); 10-hydroxycaproate (0.90) and 9-hydroxypelargonate (0.27) but not the following: 6-hydroxycaproate, 10-hydroxystearate, 3-hydroxybutyrate, ethanol, ethanolamine, choline, serine, threonine, pantothenate, mevalonate or pyridoxine. The enzyme preparation used was contaminated with L-lactate dehydrogenase (EC 1.1.1.27); the assay conditions were pH 9.5, PPi. NADP was inactive in the place of NAD. (1)

References

1. Kamei, S., Wakabayashi, K. & Shimazono, N. (1964) J. Biochem. (Tokyo), 56, 72.

5,10-METHYLENETETRAHYDROFOLATE REDUCTASE

(5-Methyltetrahydrofolate: NAD oxidoreductase)

5-Methyltetrahydrofolate + NAD =

5,10-methylenetetrahydrofolate + reduced NAD

Ref.

Equilibrium constant

$$\frac{[\text{5,10-methylenetetrahydrofolate}]\ [\text{reduced NADP}]}{[\text{5-methyltetrahydrofolate}]\ [\text{NADP}]} = 10^{-7}$$

(pH 7.0, 30°) (1)

Molecular weight

source	value	conditions	
Pig liver	150,000-200,000	Sephadex G 200	(1)

The enzyme contained tightly bound FAD. (1)

Specificity and Michaelis constants

source	substrate	V relative	K_m (M)	conditions	
Pig liver	reduced NADP	1.00	3.3×10^{-5}	pH 6.7, Pi, 30°	(1)
(purified	reduced NAD	0.50	3.1×10^{-4}	pH 6.7, Pi, 30°	(1)
900 x)	5,10-methylene-tetrahydrofolate[a]	-	2×10^{-5}	pH 6.7, Pi, 30°	(1)
	FAD[b]	-	4×10^{-8}	-	(1)

(a) the enzyme was specific for the (+) diastereoisomer; the (-) diastereoisomer was inactive as a substrate or inhibitor. The following compounds were active in the place of 5,10-methylenetetrahydrofolate (1.00): menadione (14.3), dichlorophenolindophenol (14.3), potassium ferricyanide (7.2) and cytochrome c (1.9). FAD, FMN and O_2 were slightly or completely inactive.
(b) FMN was inactive.
The properties of the enzyme from Escherichia coli are discussed in ref. 2.

Inhibitors

The enzyme from pig liver (but not that from E. coli) was strongly inhibited by S-adenosylmethionine. The inhibition was partially reversed by S-adenosylhomocysteine. The menadione activity of the enzyme was inhibited by the folate substrates. (1)

References

1. Kutzbach, C. & Stokstad, E.L.R. (1971) BBA, 250, 459.
2. Cathou, R.E. & Buchanan, J.M. (1963) JBC, 238, 1746.

ALCOHOL DEHYDROGENASE (NAD(P))

(Alcohol: NAD(P) oxidoreductase)

Alcohol + NAD(P) = aldehyde + reduced NAD(P)

Ref.

Molecular weight

source	value	conditions	Ref.
Rat intestine	60,000-80,000	Sephadex G 200	(1)

Specificity and Michaelis constants

source	substrate	V relative	K_m (M)	conditions	Ref.
Rat intestine (purified 13 x)	retinal[a]	-	2.0×10^{-5}	pH 6.1, Pi, 37°	(1)
	reduced NAD	1	4×10^{-7}	pH 6.1, Pi, 37°	(1)
	reduced NADP	1	4×10^{-5}	pH 6.1, Pi, 37°	(1)

[a] a number of other aldehydes were also reduced. Thus, short and medium chain aliphatic aldehydes of length C-2 to C-14 were active. C-4 to C-8 aldehydes were the most active. A number of unsaturated, branched chain and aromatic aldehydes were also good substrates. Unsaturated C-18 fatty aldehydes were poor substrates and saturated aldehydes of length C-16 or greater were not reduced. Ethanol was not oxidized in the presence of NAD or NADP.

The enzyme was activated by glutathione; no other added cofactor was required. (1)

References

1. Fidge, N.H. & Goodman, D.W.S. (1968) JBC, 243, 4372.

GLYCEROL DEHYDROGENASE

(Glycerol: NADP oxidoreductase)

Glycerol + NADP = D-glyceraldehyde + reduced NADP

Ref.

Equilibrium constant

$$\frac{[\text{D-glyceraldehyde}]\ [\text{reduced NADP}]\ [\text{H}^+]}{[\text{glycerol}]\ [\text{NAD}]} = 1.67 \times 10^{-6}\ \text{M} \quad (\text{Pi},\ 37^{\circ})$$ (1)

Molecular weight

source	value	conditions	
Rabbit skeletal muscle	34,000	Sephadex G 100, pH 7.0	(1)

Specificity and Michaelis constants

source	substrate	K_m (M)	conditions	
Rabbit skeletal muscle (purified 300 x)	D-glyceraldehyde	1.5×10^{-4}	pH 7.0, Pi, 25°	(1)
	reduced NADP	1.6×10^{-5}	pH 7.0, Pi, 25°	(1)

The enzyme (rabbit skeletal muscle) could utilize the following aldehydes in the place of D-glyceraldehyde (1.00): L-glyceraldehyde (0.36); DL-glyceraldehyde-3-phosphate (0.24); dihydroxyacetone (0.04); formaldehyde (0.04); acetaldehyde (0.07); glycolaldehyde (0.44); methylglyoxal (0.80); propionaldehyde (0.45); crotonaldehyde (0.13); *n*-butyraldehyde (0.68); isobutyraldehyde (0.59); *n*-valeraldehyde (0.74); isovaleraldehyde (0.47); *n*-hexanal (0.69); *n*-heptanal (0.53); D-erythrose (0.74); D-ribose (0.05); D-lyxose (0.09); D-arabinose (0.02); L-arabinose (0.02); and glucuronolactone (0.18). The following aldehydes were reduced at very low rates: dihydroxyacetone phosphate; sodium glucuronate; D-fructose-1-phosphate; D-fructose and D-glucose. Reduced NAD possessed 10% the activity of reduced NADP with D-glyceraldehyde as substrate. The following compounds were active in the forward reaction: glycerol; 1,2-propanediol; *n*-butanol and *n*-pentanol. 1,3-Propanediol was less active and methanol; ethanol; 1,2-ethanediol, *n*-propanol; 2-butanol and *tert*-butanol were inactive. (1)

The kinetic properties of the enzyme from rat skeletal muscle are discussed in ref. 2.

References

1. Kormann, A.W., Hurst, R.O. & Flynn, T.G. (1972) BBA, 258, 40.
2. Toews, C.J. (1967) BJ, 105, 1067.

L-AMINOPROPANOL DEHYDROGENASE

(L-1-Aminopropan-2-ol: NAD oxidoreductase)

L-1-Aminopropan-2-ol + NAD = aminoacetone + reduced NAD

Ref.

Specificity and kinetic properties

source	substrate	K_m (M)	conditions	
Escherichia coli (cell free extract)	L-1-aminopropan-2-ol [(a)]	1.6×10^{-3}	pH 9.6, diethanolamine	(1)
	NAD	4×10^{-4}	pH 9.6, diethanolamine	(1)

(a) D-1-aminopropan-2-ol and L-threonine were inactive; some activity was observed with DL-3-phenylserine or DL-1-aminobutan-2-ol.

Of a range of aminopropanol analogues tested, the following were inhibitory: DL-2-hydroxy-2-phenylethylamine; DL-1-aminopropane-2,3-diol and DL-serine. The following aminopropanol analogues were inactive: DL-1,2-diaminopropane; DL-propan-1,2-diol; DL-4-amino-3-hydroxybutyrate; 3-aminopropan-1-ol; 1-aminopropane; 2-aminoethanol; DL-1-aminopropan-2,3-diol; propan-2-ol; L-2-aminopropan-1-ol; DL-3-hydroxybutyrate and DL-3-phenylserine. None of 34 common intermediary metabolites tested (sugars, amino acids and mono-di- and tri-carboxylic acids) affected the rate of aminoacetone formation from aminopropanol. Lower alcohols were also inactive but higher alcohols were slightly inhibitory. The enzyme requires K^+ for activity. (1)

A D-1-aminopropan-2-ol dehydrogenase may be present in *E. coli*. (1)

References

1. Turner, J.M. (1967) BJ, 104, 112.

GLYOXYLATE REDUCTASE (NADP)

(Glycollate: NADP oxidoreductase)

Glycollate + NADP = glyoxylate + reduced NADP

Ref.

Equilibrium constant

			Ref.
$\frac{[\text{glyoxylate}]\,[\text{reduced NADP}]\,[H^+]}{[\text{glycollate}]\,[\text{NADP}]}$	=	3.0×10^{-18} M (Tris, 25°)	(1)

Molecular weight

source	value	conditions	Ref.
Pseudomonas sp.	190,000	Sephadex G 200, pH 7.0	(1)

Specificity and kinetic properties

source	substrate	relative velocity	K_m (M)	Ref.
Pseudomonas sp. (purified 220 x)	glyoxylate[a]	1.00	1.4×10^{-2}	(1)
	hydroxypyruvate	0.71	7×10^{-3}	(1)
	reduced NADP	1.00	-	(1)
	reduced NAD	0.23	-	(1)

(a) glyoxylate (1.00) could also be replaced by glyoxal (0.18); DL-glyceraldehyde (0.08); oxalate (0.08); oxaloacetate (0.05) and 2-oxoglutarate (0.05) but not by pyruvate, glycolaldehyde, formate, oxamate, formaldehyde or acetaldehyde (conditions: pH 6.8, Pi, 25°).

Glyoxylate reductase is also present in spinach and tobacco leaves. (2)

References

1. Cartwright, L.N. & Hullin, R.P. (1966) BJ, 101, 781.
2. Zelitch, I. & Gotto, A.M. (1962) BJ, 84, 541.

MALATE DEHYDROGENASE (NADP)

(L-Malate: NADP oxidoreductase)

L-Malate + NADP = oxaloacetate + reduced NADP

Ref.

Equilibrium constant

The reaction is reversible. (1)

Molecular properties

Different molecular weight forms of the enzyme (*Zea mays*) were observed on gel filtration (Sephadex G 200). (1)

Specificity and kinetic properties

source	substrate	relative velocity	K_m (M)	conditions	
Z. mays (partially purified)	oxaloacetate	1.00	2.5×10^{-5}	pH 8.0, Tris, 22°	(1)
	reduced NADP[a]	1.00	6.0×10^{-5}	pH 8.0, Tris, 22°	(1)
	L-malate[b]	0.11	2×10^{-3}	pH 8.9, tricine, 22°	(1)
	NADP[b]	0.11	2.5×10^{-4}	pH 8.9, tricine, 22°	(1)

(a) reduced NAD was inactive.

(b) plots of activity against malate or NADP concentration were sigmoidal in shape.

The enzyme (*Z. mays*) required a thiol compound for activity. The malate dehydrogenase in green leaves was rapidly inactivated in the dark and was reactivated when the plants were illuminated. Reactivation of the enzyme extracted from darkened leaves was achieved by adding a thiol compound. (1)

References

1. Johnson, H.S. & Hatch, M.D. (1970) BJ, 119, 273.

DIMETHYLMALATE DEHYDROGENASE

(3,3-Dimethyl-D-malate : NAD oxidoreductase (decarboxylating))

3,3-Dimethyl-D-malate + NAD = 2-oxoisovalerate + CO_2 + reduced NAD

Ref.

Equilibrium constant

The reaction is reversible. (1)

Specificity and catalytic properties

source	substrate	V relative	K_m (mM)	conditions	
Pseudomonas P-2 (~100 x)	dimethyl D-malate	1.00	0.15	pH 8.5, Tris	(1)
	D-malate[(a)]	0.93	11.0	pH 8.5, Tris	(1)
	3-(n-propyl)malate	0.49	0.33	pH 8.5, Tris	(1)
	NAD	-	0.27	pH 8.5, Tris	(1)
	NH_4Cl	-	2.5	pH 8.5, Tris	(1)
	KCl	-	5.0	pH 8.5, Tris	(1)

(a) L-malate was inactive.

The dehydrogenase catalyzes an NAD independent decarboxylation of dimethyloxaloacetate. It requires Mn^{2+} (Co^{2+} was also active; Fe^{2+} and Mg^{2+} were less active and Zn^{2+} was inactive) and NH_4^+ (or K^+; Na^+ was inactive) for activity. (1)

References

1. Magee, P.T. & Snell, E.E. (1966) B, 5, 409.

DIHYDROXYISOVALERATE DEHYDROGENASE (ISOMERIZING)

(2,3-Dihydroxyisovalerate: NADP oxidoreductase (acetolactate forming))

2,3-Dihydroxyisovalerate + NADP = 2-oxo-3-hydroxyisovalerate + reduced NADP

2-oxo-3-hydroxyisovalerate = 2-acetolactate

Ref.

Equilibrium constant

Both the dehydrogenase and isomerizing reactions were reversible at high pH when catalyzed by the enzyme from Salmonella typhimurium. The enzyme from Neurospora crassa, however, was unable to catalyze the oxidation of 2,3-dihydroxyisovalerate. (1,2)

Molecular properties

source	value	conditions	
S. typhimurium	220,000	pH 7.5	(1,3)

The enzyme isolated from N. crassa contained bound lipid. (2,4)

Specific activity

S. typhimurium (450 x) 1.91 2-acetolactate (pH 7.5, Pi, 30°) (1)

Specificity and kinetic properties

source	substrate	V relative	K_m (M)	conditions	
S. typhimurium	2-acetohydroxy-butyrate[(a)]	1.00	4.5×10^{-4}	pH 7.5, Pi, 30°	(1,3)
	2-acetolactate	0.21	4.5×10^{-4}	pH 7.5, Pi, 30°	(1,3)

(a) in the reaction: 2-aceto-2-hydroxybutyrate + reduced NADP = 2,3-dihydroxy-3-methylvalerate + NADP. 2-Oxoisovalerate; 2-oxo-n-valerate; acetoin and 2-oxobutyrate were inactive.

The enzyme cannot utilize reduced NAD in the place of reduced NADP. It requires Mg^{2+} specifically. The properties of the enzyme isolated from Phaseolus radiatus seed, Escherichia coli and from N. crassa have been described. (5,6,2,4)

With the enzyme from S. typhimurium, 2,3-dihydroxyisovalerate; 2,3-dihydroxy-3-methylvalerate and NADP inhibit the reverse reaction. A number of other hydroxy- and oxo- compounds were also inhibitory. (1,3)

References

1. Arfin, S.M. & Umbarger, H.E. (1969) JBC, 244, 1118.
2. Kiritani, K., Narise, S. & Wagner, R.P. (1966) JBC, 241, 2047.
3. Arfin, S.M. (1970) Methods in Enzymology, 17A, 751.
4. Kiritani, K. & Wagner, R.P. (1970) Methods in Enzymology, 17A, 745.
5. Satyanarayana, T. & Radhakrishnan, A.N. (1965) BBA, 110, 380.
6. Radhakrishnan, A. & Snell, E.E. (1960) JBC, 235, 2316.
   Shematek, E.M., Diven, W.F. & Arfin, S.M. (1973) ABB, 158, 126; 132.

ARYLALCOHOL DEHYDROGENASE

(Arylalcohol: NAD oxidoreductase)

An aromatic alcohol + NAD = an aromatic aldehyde + reduced NAD

Ref.

Equilibrium constant

The reaction is reversible. (1)

Molecular weight

source	value	conditions	Ref.
Pseudomonas sp.	110,000	Sephadex G 200, pH 7.4	(1)

Specific activity

Pseudomonas sp. (116 x) 20.9 benzylalcohol (pH 8.5, PPi, 20°) (1)

Specificity and Michaelis constants

source	substrate	K_m (M)	conditions	Ref.
Pseudomonas sp.	benzylalcohol[a]	3.6×10^{-4}	pH 8.5, PPi, 20°	(1)
	NAD	8.0×10^{-5}	pH 8.5, PPi, 20°	(1)
	benzaldehyde[b]	2.7×10^{-5}	pH 6.5, Pi, 20°	(1)
	reduced NAD	1.8×10^{-5}	pH 6.5, Pi, 20°	(1)

(a) benzylalcohol (1.00) could be replaced by 4-hydroxymethylbenzylalcohol (1.06); 4-isopropylbenzylalcohol (0.62); 1-hydroxymethyl-4-isopropyl-1-cyclohexene (0.85); 1-hydroxymethyl-4-isopropenyl-1-cyclohexene (0.45); hexanol (0.23); heptanol (0.18) ; octanol (0.15) or butanol (0.12) but not by cyclohexanol; hydroxymethylcyclohexane; 1,4-dihydroxymethylcyclohexane; methanol; ethanol; propanol or pentanol.

(b) benzaldehyde (1.00) could be replaced by heptylaldehyde (1.10); 2-nitrobenzaldehyde (1.07); phenylacetaldehyde (1.07) or hexylaldehyde (0.81) but not by acetaldehyde or cyclohexanone.

The enzyme could not utilize NADP in the place of NAD nor reduced NADP in the place of reduced NAD. (1)

References

1. Suhara, K., Takemori, S. & Katagiri, M. (1969) ABB, 130, 422.

OXALOGLYCOLLATE REDUCTASE (DECARBOXYLATING)

(D-Glycerate: NAD(P) oxidoreductase (carboxylating))

D-Glycerate + CO_2 + NAD(P) =

oxaloglycollate + reduced NAD(P)

Ref.

Equilibrium constant

The reverse reaction is favoured. In aqueous media oxaloglycollate exists primarily in the form of dihydroxyfumarate. (1)

Molecular weight

source	value	conditions	
Pseudomonas putida	63,000	pH 7.5; amino-acid composition	(1)

Specific activity

P. putida (500 x) 2.6 dihydroxyfumarate (with reduced NAD; pH 6.0, Pi, 25°) (1)

Specificity and Michaelis constants

substrate[a]	co-substrate	V relative	K_m (M)
dihydroxyfumarate	reduced NAD	1.00	1.2×10^{-3}
hydroxypyruvate	reduced NAD	1.75	1.0×10^{-3}
glyoxylate	reduced NAD	-	3.3×10^{-3}
dihydroxyfumarate	reduced NADP	1.05	2.5×10^{-3}
hydroxypyruvate	reduced NADP	0.67	2.5×10^{-3}
glyoxylate	reduced NADP	-	8.0×10^{-3}
reduced NAD	dihydroxyfumarate	-	1.4×10^{-4}
reduced NADP	dihydroxyfumarate	-	1.5×10^{-5}

(a) with the enzyme from P. putida. The conditions were: pH 6.0, Pi, 25°.

In the presence of reduced NAD or reduced NADP, the following compounds were inactive: acetaldehyde, dihydroxyacetone, glyceraldehyde, glycolaldehyde, 2-oxobutyrate, 3-oxobutyrate, 2-oxoglutarate, oxaloacetate, propionaldehyde and pyruvate. (1)

Light absorption data

substrate	molar extinction coefficient	
dihydroxyfumarate	9200 $M^{-1}cm^{-1}$ at 290 nm in H_2O	(1)

References

1. Kohn, L.D. & Jakoby, W.B. (1968) JBC, 243, 2486.

TARTRATE DEHYDROGENASE

(Tartrate: NAD oxidoreductase)

Tartrate + NAD = oxaloglycollate + reduced NAD

Ref.

Equilibrium constant

The reaction is reversible. In the reverse reaction, both meso- and L(+)-tartrate are produced. (1)

Molecular weight

source	value	conditions	Ref.
Pseudomonas putida	145,000 [4]	pH 7.5; amino acid composition	(1)

Specific activity

P. putida (71 x) 0.71 mesotartrate (pH 8.6, Tris, 25°) (1)

Specificity and Michaelis constants

source	substrate	V relative	K_m (M)	conditions	Ref.
P. putida	mesotartrate [a]	1.0	8.3×10^{-4}	pH 8.6, Tris, 25°	(1)
	L(+)-tartrate	1.0	5.5×10^{-2}	pH 8.6, Tris, 25°	(1)
	NAD [b]	-	9×10^{-5}	pH 8.6, Tris, 25°	(1)
	dihydroxyfumarate [c]	0.01	1.1×10^{-2}	pH 7.2	(1)
	reduced NAD	-	1.4×10^{-3}	pH 7.2	(1)

[a] the following were inactive as substrates or inhibitors: D(-)-tartrate; L(-)-malate; D(+)-malate; D,L-glycerate; tartronate; 2- and 3-hydroxybutyrate; hydroxypyruvate; tartronic semialdehyde; glycollate; glycolaldehyde and citrate.

[b] NADP was inactive.

[c] dihydroxyfumarate interconverts spontaneously to oxaloglycollate. Under the assay conditions dihydroxyfumarate formation is favoured.

The enzyme requires Mn^{2+} (partially replaced by Mg^{2+} but not by Ca^{2+}) and a monovalent cation (K^+ or Rb^+) for activity. (1)

Inhibitors

Dihydroxyfumarate is a competitive inhibitor ($K_i = 8 \times 10^{-5}$ M; C (mesotartrate)). (1)

References

1. Kohn, L.D., Packman, P.M., Allen, R.H. & Jakoby, W.B. (1968) JBC, 243, 2479.

DIIODOPHENYLPYRUVATE REDUCTASE

(3,5-Diiodo-4-hydroxyphenyl-L-lactate: NAD oxidoreductase)

3,5-Diiodo-4-hydroxyphenyl-L-lactate + NAD =

3,5-diiodo-4-hydroxyphenylpyruvate + reduced NAD — Ref.

Equilibrium constant

$$\frac{[\text{3,5-diiodo-4-hydroxyphenylpyruvate}]\,[\text{reduced NAD}]\,[H^+]}{[\text{3,5-diiodo-4-hydroxyphenyl-L-lactate}]\,[\text{NAD}]}$$

$= 2.6 \times 10^{-13}$M (Pi, ambient) (1)

Molecular properties

source	value	conditions	Ref.
Rat kidney	71,000 [2][(a)]	pH 6.5	(1)

(a) there are 2 binding sites for reduced NAD per 71,000 daltons ($K_D = 2 \times 10^{-7}$M).

Specific activity

Rat kidney	206	3,5 -dibromo-4-hydroxyphenylpyruvate (pH 6.5, Pi)	(1)

Specificity and Michaelis constants

substrate	V relative	K_m (M)
3,5-dibromo-4-hydroxyphenylpyruvate(a)	1.00	8.1×10^{-5}
3,5-dichloro-4-hydroxyphenylpyruvate	0.89	1.39×10^{-4}
3,5-diiodo-4-hydroxyphenylpyruvate	0.75	1.43×10^{-4}
3-monoiodo-4-hydroxyphenylpyruvate	1.00	2.6×10^{-4}
4-hydroxyphenylpyruvate	0.01	1.2×10^{-3}
reduced NAD(b)	-	1.2×10^{-5}
3,5-diiodo-4-hydroxyphenylpyruvate(c)	-	4.3×10^{-5}
4-hydroxyphenylpyruvate	-	1.9×10^{-3}
phenylpyruvate	-	4.8×10^{-3}
reduced NAD(d)	-	5.2×10^{-5}

(a) with the enzyme from rat kidney (pH 6.5, Pi). The enzyme was also active with triiodothiopyruvate. The following compounds had very low activities: indole pyruvate, 4-hydroxyphenylpyruvate, phenylpyruvate and 4-fluorophenylpyruvate. The following were inactive: 3-amino-4-hydroxyphenylpyruvate, pyruvate, 2-oxoglutarate, glyoxylate and acetaldehyde (Ref. 1).
(b) reduced NADP was inactive.
(c) with the enzyme from dog heart (pH 6.5, Pi). The enzyme was inactive with pyruvate, 2-oxobutyrate and glyoxylate (Ref. 2 and 3).
(d) reduced NADP possessed 15% the activity of reduced NAD.

Ref.

Inhibitors

inhibitor	K_i (M)	Ref.
3,5-diiodo-2-hydroxybenzoate[a]	5.1×10^{-4}	(1)
2,5-dihydroxybenzoate[b]	7.0×10^{-5}	(2,3)

(a) with the enzyme from rat kidney (pH 6.5, Pi). The inhibition was competitive (3,5-dibromo-4-hydroxyphenylpyruvate). The following were non-inhibitory: 2,5-dihydroxybenzoate, salicylate, diiodotyrosine and dibromotyrosine.

(b) with the enzyme from dog heart (pH 6.5, Pi). The inhibition was competitive (3,5-diiodo-4-hydroxyphenylpyruvate). Pyruvate; 2,3-; 2,4-; 2,6-; and 3,5-dihydroxybenzoate did not inhibit.

References

1. Nakano, M., Tsutsumi, Y. & Danowski, T.S. (1970) JBC, 245, 4443.
2. Zannoni, V.G. & Weber, W.W. (1966) JBC, 241, 1340.
3. Zannoni, V.G. (1970) Methods in Enzymology, 17A, 665.

3-HYDROXYBENZYL-ALCOHOL DEHYDROGENASE

(3-Hydroxybenzyl-alcohol: NADP oxidoreductase)

3-Hydroxybenzyl-alcohol + NADP =

3-hydroxybenzaldehyde + reduced NADP

Ref.

Equilibrium constant

$$\frac{[\text{3-hydroxybenzaldehyde}]\ [\text{reduced NADP}]}{[\text{3-hydroxybenzyl-alcohol}][\text{NADP}]} = 0.18 \quad (\text{pH } 7.6,\ 30^{\circ})$$ (1)

Molecular properties

source	value	
Penicillium urticae	120,000	(1)

Specificity and kinetic properties

At pH 7.6 and with the enzyme from *P. urticae* the reverse rate relative to the forward rate was 6 : 1. In the presence of reduced NADP, crude extracts catalyzed the reduction of 3-hydroxybenzaldehyde (1.00); 3-methoxybenzaldehyde (1.00); acetaldehyde (0.29); benzaldehyde (0.20) and 4-hydroxybenzaldehyde (0.10) but not of 2-hydroxybenzaldehyde; acetophenone; 3-hydroxyacetophenone; 4-hydroxyacetophenone; 4-hydroxyoctaphenone or gentisaldehyde. Of the patulin pathway end products, gentisaldehyde was a feedback inhibitor. (1)

Abbreviations

Patulin 4-hydroxy-4*H*-furo[3,2-c]pyran-2(6*H*)-one

References

1. Forrester, P.I. & Gaucher, G.M. (1972) B, 11, 1108.

3-OXOACYL ACYL CARRIER PROTEIN REDUCTASE

(D-3-Hydroxyacyl-acyl carrier protein: NADP oxidoreductase)

D-3-Hydroxyacyl-ACP + NADP = 3-oxoacyl-ACP + reduced NADP

Ref.

Equilibrium constant

$$\frac{[\text{acetoacetyl-ACP}]\,[\text{reduced NADP}]\,[H^+]}{[\text{3-hydroxy-butyryl-ACP}]\,[\text{NADP}]} = 2.5 \times 10^{-8}\ M$$

(Pi, ambient) (2,3)

Specificity and Michaelis constants

source	substrate	V relative	K_m (mM)	conditions	Ref.
Escherichia coli B (purified 157 x)	acetoacetyl-ACP[(a)]	1.00	0.12	pH 7.0, Pi, ambient	(2,3)
	acetoacetyl-CoA	0.08	0.66	pH 7.0, Pi, ambient	(2,3)
	acetoacetyl-pantetheine	0.08	-	pH 7.0, Pi, ambient	(2,3)

(a) acetoacetyl-ACP could be replaced by 3-oxopentanoyl-ACP; 3-oxohexanoyl-ACP, 3-oxooctanoyl-ACP or 3-oxodecanoyl-ACP.

In the forward reaction, D-3-hydroxybutyryl-ACP could not be replaced by the L-3-isomer nor could reduced NADP be replaced by reduced NAD. The enzyme requires a high salt concentration for full activity. (2,3)

3-Oxoacyl-ACP reductase is a component of the fatty acid synthetase complex (see EC 2.3.1.38).

Abbreviations

ACP acyl carrier protein (see EC 2.3.1.39)

References

1. Prescott, D.J. & Vagelos, P.R. (1972) Advances in Enzymology, 36, 298.
2. Vagelos, P.R., Alberts, A.W. & Majerus, P.W. (1969) Methods in Enzymology, 14, 60.
3. Toomey, R.E. & Wakil, S.J. (1966) BBA, 116, 189.

4-OXOPROLINE REDUCTASE

(4-Hydroxy-L-proline: NAD oxidoreductase)

4-Hydroxy-L-proline + NAD = 4-oxoproline + reduced NAD

Ref.

Equilibrium constant

The reaction is essentially irreversible [R] with the enzyme from rabbit kidney as catalyst. (1)

Specificity and Michaelis constants

source	substrate	K_m (M)	conditions	
Rabbit kidney (purified 15 x)	4-oxoproline[a]	6.0×10^{-4}	pH 6.2, Pi, 37°	(1)
	reduced NAD[b]	8.4×10^{-4}	pH 6.2, Pi, 37°	(1)

[a] the following were inactive: Δ^1-pyrroline-2-carboxylate; Δ^1-pyrroline-5-carboxylate; 3-hydroxy-Δ^5-pyrroline-5-carboxylate; cyclohexanone; glycolaldehyde; acetaldehyde; etiocholine; androsterone; acetone; 5-aminolaevulinate; 2-oxoglutarate; pyruvate; 2-oxobutyrate; ethylacetoacetate and acetoacetate.

[b] reduced NADP was inactive.

A 4-oxoproline reductase which can utilize reduced NADP is also present in rabbit kidney. (1)

References

1. Smith, T.E. & Mitoma, C. (1962) JBC, 237, 1177.

PANTOATE DEHYDROGENASE

(D-Pantoate : NAD oxidoreductase)

D-Pantoate + NAD =

D-aldopantoate + reduced NAD

Ref.

Equilibrium constant

The reaction is reversible. The reverse direction is favoured. (1)

Specificity and catalytic properties

source	substrate	relative velocity	K_m (μM)	conditions	
Pseudomonas P-2 (purified 14 x)	D-pantoate[a]	1.0	33	pH 10, glycine 22°	(1)
	ketopantoate[b]	0.4	500	pH 10, glycine 22°	(1)

(a) D-pantoate could not be replaced by 2-hydroxy-DL-isovalerate; DL-lactate or 4-hydroxybutyrate.

(b) in the place of D-pantoate

The enzyme could not utilize NADP in the place of NAD. (1)

Abbreviations

aldopantoate	2-hydroxy-3,3-dimethyl-3-formylpropionate
ketopantoate	2-oxo-3,3-dimethyl-4-hydroxybutyrate
pantoate	2-hydroxy-3,3-dimethyl-4-hydroxybutyrate

References

1. Goodhue, C.T. & Snell, E.E. (1966) B, 5, 403.

PYRIDOXAL DEHYDROGENASE

(Pyridoxal: NAD oxidoreductase)

Pyridoxal + NAD = 4-pyridoxolactone + reduced NAD

Ref.

Equilibrium constant

The reaction is essentially irreversible [F] with the enzyme from Pseudomonas sp. MA as catalyst. (1)

Specificity and Michaelis constants

source	substrate	K_m (M)	
Pseudomonas MA (purified 4-5 x)	pyridoxal[a]	7.6×10^{-5}	(1)
	NAD	2.9×10^{-4}	(1)

[a] ω-methylpyridoxal was dehydrogenated at the same rate as pyridoxal but 5-deoxypyridoxal, pyridoxal phosphate, acetaldehyde, benzaldehyde and salicylaldehyde were inactive. NAD could not be replaced by NADP. (Conditions: pH 9.3, PPi).

The enzyme also catalyzes a dismutation reaction in the presence of NAD:

Pyridoxal + 5-deoxypyridoxal = 4-pyridoxolactone + 5-deoxypyridoxol

There is evidence that the enzyme does not require metal ions for activity. (1)

Inhibitors

4-Pyridoxolactone, 4-pyridoxate, pyridoxol and pyridoxamine are competitive (pyridoxal) inhibitors with K_i values of 0.7, 1.0, 0.8 and 1.0 mM, respectively. (pH 9.3, PPi) (1)

References

1. Burg, R.W. & Snell, E.E. (1969) JBC, 244, 2585.

CARNITINE DEHYDROGENASE

(L-Carnitine: NAD oxidoreductase)

Carnitine + NAD = 3-dehydrocarnitine + reduced NAD

Ref.

Equilibrium constant

$$\frac{[\text{3-dehydrocarnitine}]\ [\text{reduced NAD}]\ [H^+]}{[\text{carnitine}]\ [\text{NAD}]} = 1.3 \times 10^{-11}\ M\ (\text{Tris, } 30^{\circ}) \quad (1)$$

Specific activity

		Ref.
Pseudomonas aeruginosa (purified 200 x)	200 carnitine (pH 9.1, Tris, 30°)	(1,2)
	600 3-dehydrocarnitine (pH 7.2, Tris, 30°)	(1,2)

Specificity and Michaelis constants

source	substrate	K_m (M)	conditions	Ref.
P. aeruginosa	L-carnitine[a]	1.6×10^{-2}	pH 9.0, Tris, 30°	(2)
	NAD[b]	1.9×10^{-4}	pH 9.0, Tris, 30°	(2)
	3-dehydrocarnitine	1.6×10^{-3}	pH 9.0, Tris, 30°	(2)
	reduced NAD[c]	4.5×10^{-5}	pH 9.0, Tris, 30°	(2)

[a] D,L-3-hydroxybutyrate; L-lactate; ethanol; L-malate and D,L-isocitrate were inactive.

[b] NADP was inactive.

[c] reduced NADP was inactive.

The mechanism of the reaction is discussed in ref. 2.

Inhibitors

source	inhibitor[a]	K_i (M)	conditions	Ref.
P. aeruginosa	D-carnitine	8.7×10^{-2}	pH 9.1, Tris, 30°	(1)
	glycinebetaine	9.0×10^{-2}	pH 9.1, Tris, 30°	(1)
	choline	1.1×10^{-1}	pH 9.1, Tris, 30°	(1)

[a] C(L-carnitine)

The enzyme is subject to product inhibition phenomena. (2)

References

1. Aurich, H., Kleber, H.P., Sorger, H. & Tauchert, H. (1968) EJB, 6, 196.
2. Schöpp, W., Sorger, H., Kleber, H.P. & Aurich, H. (1969) EJB, 10, 56.

INDOLELACTATE DEHYDROGENASE

(Indolelactate: NAD oxidoreductase)

Indolelactate + NAD = indolepyruvate + reduced NAD

Ref.

Equilibrium constant

$$\frac{[\text{indolepyruvate}]\ [\text{reduced NAD}]\ [H^+]}{[\text{indolelactate}]\ [\text{NAD}]} = 6.3 \times 10^{-6}\ \text{M (Pi, } 30^{o})$$ (1)

Specificity and Michaelis constants

substrate	V relative	K_m (M)
indolelactate[a]	1.00	1.03×10^{-3}
4-hydroxyphenyllactate	9.95	1.77×10^{-3}
phenyllactate	9.50	2.21×10^{-3}
indoleglycollate	1.71	6.91×10^{-4}
NAD[b]	-	1.95×10^{-4}
4-hydroxyphenylpyruvate	51.65	7.5×10^{-5}
phenylpyruvate	33.06	1.06×10^{-4}
reduced NAD[c]	-	1.42×10^{-5}

[a] with the enzyme from Clostridium sporogenes (purified 13 x; from ref. 1). The following compounds were inactive as substrates: DL-lactate, pyruvate, indoleacetone and tryptophol. The kinetics with indolepyruvate were complex and plots of reaction rate versus indolepyruvate concentration were sigmoidal in shape.

[b] with indolelactate as the second substrate.

[c] with indolepyruvate as the second substrate.

Light absorption properties

Indolepyruvate has an absorption band at 340 nm (molar extinction coefficient = 4,770 $M^{-1}cm^{-1}$ at pH 7.0). (1)

References

1. Jean, R. & deMoss, R.D, (1968) Can. J. Microbiol., 14, 429.

INDANOL DEHYDROGENASE

(S(+)-1-Indanol: NAD(P) oxidoreductase)

1-Indanol + NAD(P) = 1-indanone + reduced NAD(P)

Ref.

Equilibrium constant

The reaction is irreversible [F] with the enzyme from bovine liver. (1)

Molecular weight

source	value	conditions	
Bovine liver	30,000	Sephadex G 100, pH 7.1	(1)

Specificity and Michaelis constants

substrate[(a)]	relative velocity	K_m (M)
S(+)-1-indanol	1.00	2.9×10^{-3}
(RS)-1-indanol	0.58	-
1-tetralol	0.83	-
benzosuberol	0.07	-
9-fluorenol	0.66	-
acenaphthenol	1.14	7.5×10^{-4}
trans-acenaphthene-1,2-diol	0.58	-
cyclohexen-3-ol	0.60	1.7×10^{-3}
NAD	-	8.0×10^{-4}
NADP	-	3.5×10^{-5}

(a) with the enzyme from bovine liver (purified 49 x). The conditions were: pH 9.6, glycine, 25° (ref. 1).

The following compounds were inactive: ethanol; lactate; hydroxy-hexabarbitol; crotyl alcohol; 3-buten-2-ol, cyclopentanol; 2-indanol; 2-tetralol; methylphenylcarbinol; 2-methyl methylphenylcarbinol; cyclopropylphenylcarbinol; ethylphenylcarbinol 4-chlorobenzyl alcohol; cyclohexanol; *cis*-acenaphthan-1,2-diol; *trans*-1,2-dihydroxy-1,2-dihydronaphthalene; oestradiol; pregnenolone; androsterone; corticosterone; 7α-hydroxycholesterol; 7β-hydroxycholsterol and 6β-hydroxyprogesterone.(1)

References

1. Billings, R.E., Sullivan, H.R. & McMahon, R.E. (1971) JBC, 246, 3512.

D-APIOSE REDUCTASE

(D-Apiitol: NAD 1-oxidoreductase)

D-Apiitol + NAD = D-apiose + reduced NAD

Ref.

Equilibrium constant

The reaction is reversible. (1)

Specificity and Michaelis constants

source	substrate	V relative	K_m (M)	conditions	
Aerobacter aerogenes (supernatant extract)	D-apiitol[a]	1.00	1×10^{-2}	pH 9.5, glycylglycine, 25°	(1)
	D-apiose[b]	0.17	2×10^{-2}	pH 7.5, glycylglycine, 25°	(1)

[a] D-apiitol (1.00) could be replaced by D-mannitol (0.01), ribitol (0.06) or erythritol (0.01).

[b] D-apiose (1.00) could be replaced by D-ribulose (0.08) or D-erythrose (0.01) but not by D-glucose, D-mannose, D-galactose, D-fructose, L-rhamnose, L-fucose, D-ribose, D-xylose, L-arabinose, D-arabinose, D-xylulose, D-glyceraldehyde, galactitol, sorbitol, xylitol, L-arabitol, D-arabitol or glycerol.

With the enzyme from *A. aerogenes*, NADP had 0.9% the activity of NAD and reduced NADP 1.5% the activity of reduced NAD. The enzyme does not appear to require divalent metal ions for activity. (1)

Abbreviations

D-apiose 3-*C*-hydroxymethyl-*aldehydo*-D-*glycero*-tetrose; tetrahydroxyisovaleraldehyde.

References

1. Neal, D.L. & Kindel, P.K. (1970) *J. Bacteriol* *101*, 910.
   Hanna, R., Picken, M. & Mendicino, J. (1973) *BBA*, *315*, 259.

D-ARABINOSE DEHYDROGENASE (NAD(P))

(D-Arabinose: NAD(P) 1-oxidoreductase)

D-Arabinose + NAD(P) = arabinono-γ-lactone + reduced NAD(P)

Ref.

Molecular weight

source	value	conditions	
Pseudomonad sp.	104,000 [2]	ultracentrifugation with or without guanidine-HCl; gel filtration.	(1)

Specific activity

Pseudomonad sp. (257 x) 135 D-arabinose (pH 8.5, Tris, 24°) (2)

Specificity and Michaelis constants

source	substrate	V relative	K_m (M)	conditions	
Pseudomonad sp.	D-arabinose[(a)]	-	8.2×10^{-4}	pH 8.5, Tris, 23°	(3)
	NAD	1	2.2×10^{-2}	pH 8.5, Tris, 23°	(3)
	NADP	1	2.3×10^{-3}	pH 8.5, Tris, 23°	(3)

[(a)] D-arabinose (1.00) could be replaced by L-fucose (0.62); colitose (0.47) or L-galactose (0.27) but not by D-galactose, L-arabinose, D-glucose, D- or L-xylose, D- or L-mannose, D- or L-lyxose, D-ribose, D-fucose, 2-deoxy-D-galactose, valeraldehyde, 5-hydroxyvaleraldehyde, 2,3,4-trideoxyaldopyranose, 2-deoxy-D-ribose, N-acetylglucosamine, D- or L-arabitol, cellobiose, D-erythrose, D-galactosamine, D-galacturonate, D-glucose 6-phosphate, 3-O-methyl-D-glucose, iso- or myo-inositol, lactose, melibiose, raffinose, L-rhamnose, sucrose or trehalose.

Abbreviations

Colitose 3,6-dideoxy-L-galactose

References

1. Cline, A.L. & Hu, A.S.L. (1965) JBC, 240, 4498.
2. Cline, A.L. & Hu, A.S.L. (1965) JBC, 240, 4488.
3. Cline, A.L. & Hu, A.S.L. (1965) JBC, 240, 4493.

ALDOHEXOSE DEHYDROGENASE

(D-Aldohexose: NADP 1-oxidoreductase)

D-Glucose + NADP = D-glucono-1,5-lactone + reduced NADP

Ref.

Equilibrium constant

			Ref.
[D-glucono-1,5-lactone] [reduced NADP] [H^+] / [β-D-glucose] [NADP]	=	1×10^{-7} M (Pi, 30°)	(1)

Specificity and Michaelis constants

source	substrate[a]	V relative	K_m (M)	
Gluconobacter cerinus (purified 100-fold)	D-glucose	1.00	5.3×10^{-3}	(1)
	2-deoxy-D-glucose	2.27	3.9×10^{-2}	(1)
	D-mannose	1.25	1.8×10^{-3}	(1)
	2-amino-2-deoxy-D-mannose	0.95	2.9×10^{-2}	(1)
	NADP	-	1.2×10^{-5}	(1)

[a] Conditions: pH 8.0, Tris, 30°.

The following sugars and sugar derivatives were neither substrates nor inhibitors of the enzyme: D-ribose, D-xylose, D-lyxose, D-arabinose, D-allose, D-altrose, D-gulose, D-galactose, D-talose, L-rhamnose, 3-deoxy-D-glucose, 3-deoxy-D-mannose, 6-deoxy-D-glucose, L-sorbose, D-fructose, 5-keto-D-fructose, D-manno dialdose, 3-*O*-methyl D-glucose, D-glucuronic acid, D-galacturonic acid, D-sorbitol, D-mannitol, D-glucose 1-phosphate, D-glucose 6-phosphate, D-mannose 6-phosphate, 2-amino-2-deoxy-D-galactose, 2-amino-2-deoxy-D-glucose, *N*-acetylmannosamine, *N*-acetylglucosamine, trehalose, maltose, melibiose, cellobiose, isomaltose, methyl α-D-glucoside and methyl β-D-glucoside. NAD was neither a substrate nor an inhibitor of the reaction. The enzyme did not require a divalent metal ion for activity. (1)

Inhibitors

source	inhibitor	type	K_i (M)	conditions	
G. cerinus	reduced NADP	C(NADP)	3.8×10^{-5}	pH 8.0, Tris, 37°	(1)

References

1. Avigad, G., Alroy, Y. & Englard, S. (1968) *JBC*, **243**, 1936.

D-GALACTOSE DEHYDROGENASE (NADP)

(D-Galactose: NADP 1-oxidoreductase)

D-Galactose + NADP = galactonolactone + reduced NADP

Ref.

Molecular weight

source	value	conditions	Ref.
Pseudomonad sp.	64,000 [2]	ultracentrifugation with or without guanidine-HCl; gel filtration; peptide mapping.	(1)

Specific activity

source	value	conditions	Ref.
Pseudomonad sp. (2020 x)	170	D-galactose (pH 8.5, Tris, 24°)	(2)

Specificity and Michaelis constants

source	substrate	K_m (M)	conditions	Ref.
Pseudomonad sp.	D-galactose [a]	2.2×10^{-4}	pH 8.5, Tris, 23°	(3)
	NADP [b]	1.24×10^{-5}	pH 8.5, Tris, 23°	(3)

(a) D-galactose (1.00) could be replaced by D-fucose (1.56), 2-deoxy-D-galactose (0.90) or L-arabinose (0.44) but not by L-galactose; D-arabinose; D-glucose; D- or L-xylose; D- or L-mannose; D- or L-lyxose; D-ribose; 2- or 6-deoxy-D-glucose; valeraldehyde; 5-hydroxyvaleraldehyde; 2,3,4-trideoxyaldopyranose; L-fucose; L-colitose; 2-deoxy-D-ribose; *N*-acetylglucosamine; D- or L-arabitol; cellobiose; D-erythrose; D-galactosamine; D-galacturonate; D-glucose 6-phosphate; 3-*O*-methyl-D-glucose; iso- or myoinositol; lactose; melibiose; raffinose; L-rhamnose; sucrose or trehalose.

(b) NAD was inactive.

References

1. Cline, A.L. & Hu, A.S.L. (1965) JBC, 240, 4498.
2. Cline, A.L. & Hu, A.S.L. (1965) JBC, 240, 4488.
3. Cline, A.L. & Hu, A.S.L. (1965) JBC, 240, 4493.

ALDOSE DEHYDROGENASE

(D-Aldose: NAD 1-oxidoreductase)

D-Aldose + NAD = D-aldonolactone + reduced NAD

Ref.

Equilibrium constant

The reaction is reversible. (1)

Molecular weight

source	value	conditions	Ref.
Pseudomonad sp1	140,000 [4]	ultracentrifugation with or without guanidine-HCl; gel filtration; peptide mapping.	(2)

Specific activity

Pseudomonad sp1 (86 x)	41	D-galactose (pH 8.5, Tris, 24°)	(3)
Pseudomonad sp2 (335 x)	66	D-glucose (pH 8.1, Tris, 25°)	(1)

Specificity and Michaelis constants

source	substrate	V relative	K_m (mM)
Pseudomonad sp2 (purified 335 x)	D-glucose[(a)]	1.88	0.86
	D-galactose	1.42	1.6
	D-mannose	1.24	4.5
	2-deoxy-D-glucose	1.07	1.6
	D-fucose	1.00	5.8
	2-deoxy-D-galactose	0.94	6.3
	D-altrose	0.79	2.4
	6-deoxy-D-glucose	0.61	-
	D-allose	0.47	13.0
	3,6-dideoxy-D-galactose	0.23	-
	NAD[(b)]	-	0.08

[(a)] the conditions were: pH 8.1, Tris, 25°. The following were inactive as substrates or inhibitors; L-fucose, L-rhamnose, L-glucose, L-galactose, L-mannose, D- and L-fructose, L-sorbose, D-fructose 6-phosphate, D- and L-xylose, D-lyxose, D-ribose, 2-deoxy-D-ribose, D- and L-arabinose, D-xylulose, DL-glyceraldehyde, D-mannitol, D-glucitol, *myo*-inositol, D- and L-arabitol, xylitol, ribitol, maltose, cellobiose, sucrose, lactose, melezitose, turanose, melibiose, trehalose, raffinose, D-glucuronate, D-galacturonate, D-galactose 6-phosphate, D-glucose 6-phosphate, D-glucosamine, *N*-acetyl-D-glucosamine, α-methylglucoside, 6-deoxy-D-allose, 6-iodo-6-deoxy-D-galactose, 2-acetamido-6-deoxy-D-allose or 2-acetamido-6-deoxy-D-altrose (ref. 1).

Ref.

Specificity and Michaelis constants continued

(b) NADP was inactive as a substrate or inhibitor.

The kinetic properties of the aldose dehydrogenase from *Pseudomonad* sp 1 will be found in Ref. 4. This enzyme was also inactive with NADP as cofactor. The enzyme from *Pseudomonad* sp 2 did not require divalent metal ions for activity. (1)

References

1. Dahms, A.S. & Anderson, R.L. (1972) JBC, 247, 2222.
2. Cline, A.L. & Hu, A.S.L. (1965) JBC, 240, 4498.
3. Cline, A.L. & Hu, A.S.L. (1965) JBC, 240, 4488.
4. Cline, A.L. & Hu, A.S.L. (1965) JBC, 240, 4493.

L-FUCOSE DEHYDROGENASE

(6-Deoxy-L-galactose: NAD 1-oxidoreductase)

L-Fucose + NAD = L-fucono-1,5-lactone + reduced NAD

Ref.

Equilibrium constant

The reaction is essentially irreversible [F] as L-fucono-1,5-lactone hydrolyses spontaneously to L-fuconate. (1)

Molecular weight

source	value	conditions	Ref.
Sheep liver	96,000	Sephadex G 200	(2)

Specificity and Michaelis constants

source	substrate	V relative	K_m (mM)
Pig liver (purified 300 x)	L-fucose [(a)]	1.00	0.32
	D-arabinose	1.08	2.1
	D-lyxose	0.77	48
	L-xylose	0.63	44
	3-amino-D-arabinose	0.57	8
	L-galactose	0.48	8
	NAD [(b)]	-	0.02
Sheep liver (purified 45 x)	L-fucose [(c)]	1.00	0.074
	D-arabinose	1.40	0.40
	NAD [(b)]	-	0.19

(a) the following were inactive: D-arabitol, L-arabinose; D-fucose; D-mannose; D-galactose; L-rhamnose; D-fructose; lactose; melibiose; D-galactono-1,4-lactone; D-gulono-1,4-lactone; D-glucosamine; D-galactosamine; *N*-acetyl-D-glucosamine; *N*-acetyl-D-galactosamine and D-glucose. The conditions were: pH 8.0, Tris, 37° (1).

(b) NADP was inactive.

(c) L-fucose (1.00) could also be replaced by 2-deoxy-D-ribose (0.42) and D-ribose (0.10) but not by D-glucose, D-galactose, D-xylose or L-arabinose (conditions: pH 8.5, Tris, 25°, ref. 2)

References

1. Schachter, H., Sarney, J., McGuire, E.J. & Roseman, S. (1969) JBC, 244, 4785.
2. Mobley, P.W., Metzger, R.P. & Wick, A.N. (1970) ABB, 139, 83. Maijub, A.G., Pecht, M.A., Miller, G.R. & Carper, W.R. (1973) BBA, 315, 37.

5-KETO-D-FRUCTOSE REDUCTASE

(L-Sorbose: NADP 5-oxidoreductase)

L-Sorbose + NADP = 5-keto-D-fructose + reduced NADP

Ref.

Equilibrium constant

The reaction was essentially irreversible [R] with the enzyme from bakers' yeast. (1)

Specificity and Michaelis constants

substrate[a]	V relative	K_m (M)	
reduced NADP	-	4.6×10^{-5}	(1)
5-keto-D-fructose	1.00	1×10^{-3}	(1)
5-keto-D-fructose phosphate	0.02	2.9×10^{-4}	(1)

(a) with the enzyme from bakers' yeast (purified 300 x). The conditions were: pH 7.5, Tris, 30°.

The enzyme from bakers' yeast was inactive with the following compounds: D-fructose; D-tagatose; D-glucose; acetone; acetaldehyde; 2-oxoglutarate; pyruvate; D-mannose; L-arabinose; D-fucose; 2-deoxy-D-ribose; D-xylose and L-rhamnose. Reduced NADP could not be replaced by reduced NAD. (1)

Inhibitors

source	inhibitor	type	K_i (M)	conditions	
Bakers' yeast	NADP	C (reduced NADP)	1.3×10^{-4}	pH 7.5, Tris, 30°	(1)

The enzyme (bakers' yeast) was not inhibited by L-sorbose. (1)

References

1. Englard, S., Kaysen, G. & Avigad, G. (1970) JBC, 245, 1311.

5-KETO-D-FRUCTOSE REDUCTASE (NADP)

(D-Fructose: NADP 5-oxidoreductase)

D-Fructose + NADP = 5-Keto-D-fructose + reduced NADP

Ref.

Equilibrium constant

$$\frac{[\text{5-keto-D-fructose}]\ [\text{reduced NADP}]\ [H^+]}{[\text{D-fructose}]\ [\text{NADP}]} = 5.08 \times 10^{-11}\ M$$

(Tris, 30°) (1)

Specific activity

Gluconobacter cerinus (1200 x) 627 5-keto-D-fructose (pH 7.4, Tris, 30°) (1)

Specificity and Michaelis constants

source	substrate	V relative	K_m (M)	conditions	
G. cerinus	D-fructose[a]	1	7.0×10^{-2}	pH 7.3, Tris, 30°	(1)
	NADP	1	1.3×10^{-3}	pH 7.4, Tris, 30°	(1)
	5-keto-D-fructose[b]	140	4.5×10^{-3}	pH 7.4, Tris, 30°	(1)
	reduced NADP[c]	140	1.8×10^{-6}	pH 7.4, Tris, 30°	(1)

[a] D-fructose could not be replaced by glucose, mannose, sorbose, xylose, xylitol, mannitol or sorbitol.

[b] 5-keto-D-fructose could not be replaced by dihydroxyacetone, tagatose, sorbose, glucose, mannose or DL-glyceraldehyde.

[c] reduced NAD was inactive.

Inhibitors

source	inhibitor	type	K_i (M)	conditions	
G. cerinus	D-fructose	C (5-keto-D-fructose)	8.2×10^{-2}	pH 7.4, Tris, 30°	(1)

The enzyme was inhibited by NADP in the reverse reaction. (1)

References

1. Avigad, G., Englard, S. & Pifko, S. (1966) JBC, 241, 373.

UDP-N-ACETYLGLUCOSAMINE DEHYDROGENASE

(UDP-2-acetamido-2-deoxy-D-glucose: NAD 6-oxidoreductase)

UDP-2-acetamido-2-deoxy-D-glucose + 2NAD + H_2O =

UDP-2-acetamido-2-deoxy-D-glucuronate + 2 reduced NAD

Ref.

Specificity and Michaelis constants

source	substrate	K_m (M)	
Achromobacter georgiopolitanum	UDP-N-acetyl-glucosamine[a]	5×10^{-4}	(1)
	NAD[b]	1.5×10^{-3}	(1)

[a] UDP-glucose was oxidized at 10% the rate with UDP-N-acetylglucosamine. (pH 9.0, glycine, 30°).

[b] NADP was reduced at 10% the rate with NAD.

Inhibitors

The enzyme (A. georgiopolitanum) was inhibited by reduced NAD (C (NAD); $K_i = 5 \times 10^{-5}$M at pH 9.0, glycine, 30°). (1)

References

1. Fan, D-F., John, C.E., Zalitis, J. & Feingold, D.S. (1969) ABB, 135, 45.

POLYOL DEHYDROGENASE (NADP)

(Xylitol: NADP 1-oxidoreductase (D-xylose-forming))

Xylitol + NADP = D-xylose + reduced NADP

Ref.

Equilibrium constant

$$\frac{[\text{D-xylose}]\,[\text{reduced NADP}]\,[H^{+}]}{[\text{xylitol}]\,[\text{NADP}]} = 3.4 \times 10^{-10}\ \text{M}\ (25°)$$ (1)

Molecular properties

source	value	conditions	
Candida utilis	69,700	pH 6.8, in the presence of mercaptoethanol; amino acid composition.	(1)

Specific activity

C. utilis (30 x)	37	D-xylose (pH 7.4, glygly, 25°)	(1)

Specificity and Michaelis constants

source	substrate	relative velocity	K_m (M)	conditions	
C. utilis	D-xylose[a]	1.00	2.8×10^{-2}	pH 7.4, glygly, 25°	(1)
	D-glyceraldehyde	2.00	2.1×10^{-3}	pH 7.4, glygly, 25°	(1)
	D-erythrose	1.80	9.2×10^{-4}	pH 7.4, glygly, 25°	(1)
	L-arabinose	1.36	4.2×10^{-2}	pH 7.4, glygly, 25°	(1)
	reduced NADP	-	1.6×10^{-5}	pH 7.4, glygly, 25°	(1)
	xylitol[b]	0.014	5.3×10^{-1}	pH 8.0, TEA, 25°	(1)
	NADP	-	1.4×10^{-4}	pH 8.0, TEA, 25°	(1)

(a) D-xylose (1.00) could also be replaced by D-ribose (0.43); D-galactose (0.45); D-glucose (0.13); D-glycero-D-galactoheptose (0.12); 2-deoxy-D-glucose (0.05); D-mannose (0.04) or 2-deoxy-D-ribose (0.01). The following were inactive: L-xylose; D-arabinose; D-lyxose; D-fructose; L-glyceraldehyde; D-glucosamine; D-galactosamine and D-ribose 5-phosphate.

(b) xylitol (1.00) could be replaced by galactitol (1.87), L-arabitol (1.67), sorbitol (1.46), ribitol (0.58), erythritol (0.40) or glycerol (0.36) but not by ethanol.

The enzyme (C. utilis) showed no activity with NAD or reduced NAD. (1)

Polyol dehydrogenase is also present in calf lens. (2)

References

1. Scher, B.M. & Horecker, B.L. (1966) ABB, 116, 117.
2. Hayman, S. & Kinoshita, J.H. (1965) JBC, 240, 877.

15-HYDROXYPROSTAGLANDIN DEHYDROGENASE

(11α,15-Dihydroxy-9-oxoprost-13-enoate:
NAD 15-oxidoreductase)

11α,15-Dihydroxy-9-oxoprost-13-enoate + NAD =
11α-hydroxy-9,15-dioxoprost-13-enoate + reduced NAD

Ref.

Specificity and Michaelis constants

source	substrate	K_m(M)	conditions	
Pig lung	PGE	5.6×10^{-6}	pH 7.3, Pi, 44°	(1)
(purified 14 x)	NAD	8×10^{-4}	pH 7.3, Pi, 44°	(1)

The enzyme was active with all prostaglandins except those containing the dienone chromophore i.e. PGB compounds and their 19-hydroxylated derivatives. The oxidation was specific for the secondary alcohol group at carbon atom 15. The following compounds were inactive: steroid alcohols, hydroxyacids and several substrates of other dehydrogenases.(1,2)

A recent review of prostaglandins will be found in ref. 3.

Abbreviations

PGE Prostaglandin E
PGB Prostaglandin B

References

1. Anggård, E. & Samuelsson, B. (1969) Methods in Enzymology, 14, 215.
2. Anggård, E. & Samuelsson, B. (1966) Arkiv Kemi, 25, 293.
3. Hinman, J.W. (1972) Ann. Rev. Biochem, 41, 161.

SEQUOYITOL DEHYDROGENASE

(5D-5-O-Methyl-*myo*-inositol: NAD oxidoreductase)

5D-5-O-Methyl-*myo*-inositol + NAD =
5D-5-O-methyl-2,3,5/4,6-pentahydroxycyclohexanone + reduced NAD

Ref.

Molecular weight

source	value	conditions	
Trifolium incarnatum[a]	34,000	Sephadex G 100 pH 7.2	(1)

(a) sequoyitol and D-pinitol dehydrogenase activities.

Specificity and kinetic properties

T. incarnatum possesses an enzyme system which catalyzes the epimerization of sequoyitol to D-pinitol. The reaction proceeds *via* a keto intermediate, 5D-5-O-methyl-2,3,5/4,6-PHC and is catalyzed by 2 dehydrogenases (or one protein with 2 dehydrogenase activities):

Sequoyitol dehydrogenase (EC 1.1.1.143):

sequoyitol + NAD = 5D-5-O-methyl-2,3,5/4,6-PHC + reduced NAD.

D-Pinitol dehydrogenase (EC 1.1.1.142):

D-pinitol + NADP = 5D-5-O-methyl-2,3,5/4,6-PHC + reduced NADP.

Both the dehydrogenase reactions and the overall epimerization reaction are reversible. The two activities have not been separated. They differ in their stereospecificity with respect to their coenzyme substrates - the sequoyitol dehydrogenase being B-specific and the D-pinitol dehydrogenase A-specific. The sequoyitol dehydrogenase requires an SH-compound for full activity but the D-pinitol dehydrogenase was slightly inhibited by SH-compounds. (1)

substrate[a]	Activity with reduced NADP	Activity with reduced NAD
D-5-O-methyl-2,3,5/4,6-PHC	100	100
D-2,3,5/4,6-PHC	57.5	0
L-3-O-methyl-2,3,5/4,6-PHC	6.5	0
L-2,3,5/4,6-PHC	8.0	0
DL-2,3,4,6/5-PHC	14.5	1.5
2,4,6/3,5-PHC	47.0	3.0
dihydroxyacetone	8.0	0
D-fructose	0	0

(a) the conditions were: pH 6.1, Pi. The enzyme used had been purified 270 x.

Abbreviations

sequoyitol	5D-5-O-methyl-myo-inositol
D-pinitol	5D-5-O-methyl-chiro-inositol
PHC	pentahydroxycyclohexanone

References

1. Ruis, H. & Hoffmann-Ostenhof, O. (1969) EJB, 7, 442.

16α-HYDROXYSTEROID DEHYDROGENASE

(16α-Hydroxysteroid: NAD(P) 16-oxidoreductase)

16α-Hydroxysteroid + NAD(P) = 16-oxosteroid + reduced NAD(P)

Ref.

Equilibrium constant

The reaction is reversible. (1,2)

Specificity and Michaelis constants

substrate[a]	V relative	K_m (μM)
16-oxo-17β-oestradiol	1.00	89
16-oxo-17α-oestradiol	1.16	70
16-oxotestosterone	0.81	280
reduced NAD	1.00	8.3
reduced NADP	1.40	41

(a) with the enzyme (purified 18 x) from rat kidney (conditions: pH 7.0, Pi, 37°. The assay mixture contained 20% glycerol). 16-Oxo-17-β-oestradiol (1.00) could also be replaced by 16-oxooestrone (1.32); oestr-16-one (0.09) and 16-oxoandrost-5-enediol (0.06) but not by oestrone or androst-4-ene-3,17-dione. (See ref. 1 & 2).

In the forward reaction the following were oxidized to the corresponding 16-oxosteroids: oestriol; 3-methoxyoestra-1,3,5(10)-triene-16α, 17β-diol; 17-epioestriol; 16α-oestradiol; 16α, 17β-dihydroxyandrost-4-en-3-one and androst-5-ene-3β, 16α, 17β-triol. The following were inactive: 16α-hydroxyandrost-4-ene-3, 17-dione and 3β, 16α-dihydroxyandrost-5-en-17-one. (2)

The enzyme (rat kidney) did not require divalent metal ions for activity. (2)

References

1. Meigs, R.A. & Ryan, K.J. (1969) Methods in Enzymology, 15, 753.
2. Meigs, R.A. & Ryan, K.J. (1966) JBC, 241, 4011.

20α-HYDROXYSTEROID DEHYDROGENASE

(20α-Hydroxysteroid: NAD(P) 20-oxidoreductase)

17α, 20α-Dihydroxypregn-4-en-3-one + NAD(P) = 17α-hydroxyprogesterone + reduced NAD(P)

Ref.

Equilibrium constant

With the enzyme from rat testes, the forward reaction could not be demonstrated. (2)

Molecular weight

source	value	conditions	Ref.
Porcine testes	35,000 [1]	pH 7.0; Sephadex G 150, pH 8.0; gel electrophoresis (SDS); sucrose density gradient.	(1)

Specific activity

source	value	conditions	Ref.
Porcine testes (1300 x)	0.08	17α-hydroxyprogesterone (pH 7.0, Pi, 37°)	(1)

Specificity and Michaelis constants

source	substrate	K_m(M)	conditions	Ref.
Porcine	reduced NAD	7.5×10^{-4}	pH 7.0, Pi, 37°	(1)
testes	reduced NADP	5.2×10^{-5}	pH 7.0, Pi, 37°	(1)
Rat	17α-hydroxyprogesterone	6.3×10^{-5}	pH 7.4, Tris, 37°	(2)
testes	reduced NADP	2.5×10^{-5}	pH 7.4, Tris, 37°	(2)

The enzyme (porcine testes) was active with 17α-hydroxyprogesterone (1.00), corticosterone (0.03), 17α-hydroxycorticosterone (0.03) and progesterone (0.01) but not with oestrone or lactate. It did not require divalent metal ions for activity. (1)

References

1. Sato, F., Takagi, Y. & Shikita, M. (1972) JBC, 247, 815.
2. Shikita, M., Inano, H. & Tamaoki, B. (1967) B, 6, 1760.

21-HYDROXYSTEROID DEHYDROGENASE (NADP)

(21-Hydroxysteroid: NADP 21-oxidoreductase)

Cortisol + NADP = 21-dehydrocortisol + reduced NADP

Ref.

Equilibrium constant

The reaction is reversible. The reverse reaction is favoured. (1)

Specificity and Michaelis constants

substrate	$\bar{V}$ relative	K_m (μM)
21-dehydrocortisone[a]	1.00	461
21-dehydrocortisol	0.53	177
21-dehydro-11-deoxycortisol	0.65	137
21-dehydro-Δ^1-cortisol	0.47	137
21-dehydro-11-deoxycorticosterone	1.33	118
21-dehydrocorticosterone	1.42	233
reduced NADP[b]	-	21

[a] with the enzyme from sheep liver (purified 32 x). Conditions: pH 5.9, Pi, 30° (ref. 1). The following compounds were also reduced: benzaldehyde, salicylaldehyde, methylglyoxal and glyoxal. The following were inactive: cortisol, cortisone, deoxycorticosterone, aldosterone, epiandrosterone, oestrone, testosterone, acetaldehyde and pyruvate.

[b] with 21-dehydrocortisone as the substrate. Conditions: pH 6.3, Pi, 30° (ref. 1).

Inhibitors

The enzyme (sheep liver) was inhibited, in the reverse reaction, by NADP and to a less extent by NAD (C(reduced NADP)). (1)

References

1. Monder, C. & White, A. (1965) JBC, 240, 71.

SEPIAPTERIN REDUCTASE

(7,8-Dihydrobiopterin: NADP oxidoreductase)

7,8-Dihydrobiopterin + NADP = sepiapterin + reduced NADP

Ref.

Equilibrium constant

$$\frac{[\text{sepiapterin}]\ [\text{reduced NADP}]\ [H^+]}{[\text{dihydrobiopterin}]\ [\text{NADP}]} = \text{about } 1 \times 10^{-9}\ (25^{\circ})$$ (1)

Molecular weight

source	value	conditions	
Horse liver	47,000	Sephadex G100	(1)

Specificity and Michaelis constants

source	substrate	K_m(M)	conditions	
Horse liver	sepiapterin	2.1×10^{-5}	pH 6.4, Pi, 25°	(1)
(purified	reduced NADP	1.4×10^{-5}	pH 6.4, Pi, 25°	(1)
5000 x)	dihydrobiopterin	5.7×10^{-5}	pH 10.4, glycine, 25°	(1)
	NADP	7.7×10^{-5}	pH 10.4, glycine, 25°	(1)

With the enzyme from horse liver, reduced NAD possessed 25% the activity of reduced NADP and NAD 20% the activity of NADP. The enzyme was highly specific for sepiapterin (1.00) which could be replaced only to a limited extent by isosepiapterin (0.02) and xanthopterin B_2 (0.03). The enzyme was also highly specific for dihydrobiopterin. (1)

Inhibitors

The enzyme (horse liver) was inhibited by acetic acid and by other aliphatic monocarboxylic acids. The inhibition was NC(sepiapterin). The enzyme was also inhibited by a number of unconjugated pteridines. (1)

Light absorption data

Sepiapterin has an absorption band at 420nm (molar extinction coefficient = 10,400 $M^{-1}\ cm^{-1}$ at pH 6.8). (1)

Abbreviations

Dihydrobiopterin (7,8-dihydrobiopterin)	2-amino-4-hydroxy-6-(L-*erythro*-1',2'-dihydroxypropyl)-7,8-dihydropteridine
Sepiapterin	2-amino-4-hydroxy-6-lactyl-7,8-dihydropteridine
Isosepiapterin	2-amino-4-hydroxy-6-propionyl-7,8-dihydropteridine
Xanthopterin B_2	2,4-dihydroxy-6-lactyl-7,8-dihydropteridine

References

1. Katoh, S. (1971) ABB, 146, 202.

GDP-4-OXO-D-RHAMNOSE REDUCTASE

(GDP- D -rhamnose: NAD(P) 4-oxidoreductase)

GDP-D-rhamnose (or GDP-D-talomethylose) + NAD(P) =

GDP-4-oxo-D-rhamnose + reduced NAD(P)

Ref.

Equilibrium constant

The forward reaction could not be demonstrated with the enzyme from "soil bacterium GS". (1)

Molecular weight

source	value	conditions	
Soil bacterium GS (ATCC 19241)	60,000	Sephadex G150, pH 7.5	(1)

Specificity and Michaelis constants

source	substrate[(a)]	K_m (M)
Soil bacterium GS (purified about 300 x)	GDP-4-oxo-D-rhamnose[(b)]	5.0×10^{-5}
	GDP-4-oxo-D-rhamnose[(c)]	2.5×10^{-5}
	reduced NAD	2.7×10^{-5}
	reduced NADP	2.8×10^{-5}

(a) the conditions were: pH 7.0, Pi (1)
(b) with reduced NAD as cofactor
(c) with reduced NADP as cofactor

A single enzyme protein is responsible for the formation of both GDP-D-rhamnose and GDP-D-talomethylose from GDP-4-oxorhamnose. The enzyme does not catalyze the interconversion of GDP-D-talomethylose and GDP-D-rhamnose. (1)

Inhibitors

source	inhibitor	type	K_i (M)	conditions	
Soil bacterium GS	GDP-mannose	C (GDP-4-oxorhamnose)	7.2×10^{-4}	pH 7.0, Pi	(1)

Abbreviations

D-rhamnose	6-deoxy-D-mannose
D-talomethylose	6-deoxy-D-talose

GDP-4-oxo-D-rhamnose reductase may be identical with GDP-D-talomethylose dehydrogenase (EC 1.1.1.135).

References

1. Winkler, N.W. & Markovitz, A. (1971) JBC, 246, 5868.

UREIDOGLYCOLLATE DEHYDROGENASE

(S-Ureidoglycollate: NAD(P) oxidoreductase)

Ureidoglycollate + NAD(P) = oxalureate + reduced NAD(P)

Ref.

Equilibrium constant

The reverse reaction could not be demonstrated with the enzyme from *Arthrobacter allantoicus* as catalyst. (1)

Specificity and Michaelis constants

substrate[a]	second substrate	V relative	K_m (M)	
RS-ureidoglycollate[b]	NAD	-	9.1×10^{-3}	(1)
NAD	RS-ureidoglycollate	1.00	2.2×10^{-3}	(1)
NADP	RS-ureidoglycollate	1.02	1.0×10^{-4}	(1)

[a] with the enzyme from *A. allantoicus* (purified 50 x). Conditions: pH 8.1, Tris, 30°.

[b] the enzyme was specific for S-ureidoglycollate: R-ureido-glycollate was inactive.

The following compounds were inactive in the place of ureidoglycollate: L-malate, D-malate, DL-isocitrate, D-lactate, L-lactate, 3-hydroxy-butyrate, tartronate, glycerate, 2-hydroxypyridine-5-carboxylate, glycollate, DL-ureidolactate, glyoxylate and parabanate. The enzyme did not require divalent metal ions for activity. (1)

Inhibitors

inhibitor[a]	type	K_i (M)
glyoxylate	NC(ureidoglycollate)	8.3×10^{-2}

[a] with the enzyme from *A. allantoicus*. The conditions were: Tris, pH 8.1, 30°.

The following compounds were also inhibitory: L-lactate and glycollate. D-Lactate, DL-malate and oxalate were noninhibitory. (1)

References

1. van der Drift, C., van Helvoort, P.E.M. & Vogels, G.D. (1971) ABB, 145, 465.

UDP-N-ACETYLENOLPYRUVYL GLUCOSAMINE REDUCTASE

(UDP-N-acetyl-2-amino-2-deoxy-3-O-lactylglucose:
NADP oxidoreductase)

UDP-N-acetylmuramate + NADP =
UDP-N-acetylenolpyruvyl glucosamine + reduced NADP

Ref.

Specificity and Michaelis constants

source	substrate	K_m (M)	conditions	Ref.
Enterobacter cloacae (purified 151 x)	UDP-N-acetyl-enolpyruvyl-glucosamine	4.0×10^{-5}	pH 8.0, Tris, 37°	(1)
	reduced NADP[a]	4.9×10^{-5}	pH 8.0, Tris, 37°	(1)

[a] reduced NADP (1.00) could be replaced by sodium dithionite (1.07), sodium borohydride (0.15) or reduced NAD (0.05).

The enzyme contained tightly bound FAD which was required for activity. The FAD could not be replaced by FMN. (1)

Abbreviations

UDP-N-acetylmuramate — UDP-N-acetyl-2-amino-2-deoxy-3-O-lactylglucose

References

1. Taku, A., Gunetileke, K.G. & Anwar, R.A. (1970) JBC, 245, 5012.
   Taku, A. & Anwar, R.A. (1973) JBC, 248, 4971.

HOMOISOCITRATE DEHYDROGENASE

(Homoisocitrate: NAD oxidoreductase (decarboxylating))

Homoisocitrate + NAD = 2-oxoadipate + CO_2 + reduced NAD

Ref.

Equilibrium constant

The ratio of the rate of the forward reaction to the rate of the reverse reaction is 8. (1)

Molecular weight

source	value	conditions	Ref.
Bakers' yeast	48,000	Biogel P-100, pH 7.6	(1)

Specificity and Michaelis constants

source	substrate	K_m (M)	conditions	Ref.
Bakers' yeast (purified 500 x)	NAD	3.3×10^{-4}	pH 7.0, HEPES, 30°	(1)
	homoisocitrate[a]	less than 10^{-5}	pH 7.0, HEPES, 30°	(1)
	2-oxoadipate	1.5×10^{-3}	pH 7.0, HEPES, 30°	(1)
	reduced NAD	6.5×10^{-5}	pH 7.0, HEPES, 30°	(1)

(a) the following were inactive: ethanol, isocitrate, malate and glutamate. This enzyme did not decarboxylate oxaloglutarate.

The enzyme (bakers' yeast) required Mg^{2+} (or Mn^{2+}) and K^+ for activity. AMP did not stimulate the reaction. (1)

Abbreviations

homoisocitrate 1-hydroxy-1,2,4-butanetricarboxylate

References

1. Rowley, B. & Tucci, A.F. (1970) ABB, 141, 499.

GALACTOSE 6-PHOSPHATE DEHYDROGENASE

(Galactose-6-phosphate: NAD oxidoreductase)

Galactose 6-phosphate + NAD =
oxidized galactose 6-phosphate + reduced NAD

Ref.

Specificity and Michaelis constants

source	substrate	K_m (M)	conditions	
Goat liver (purified 16 x)	galactose 6-phosphate[a]	6×10^{-4}	pH 8.1, Tris.	(1)
	NAD[b]	1.2×10^{-4}	pH 8.1, Tris.	(1)

[a] glucose 6-phosphate; galactose 1-phosphate; fructose 6-phosphate; glucose; galactose; mannose and mannose 6-phosphate were inactive.

[b] NAD (1.00) could be replaced by NADP (0.12), 3-acetyl pyridine DPN (0.16) or 3-acetyl pyridine deamino DPN (0.15).

Light absorption data

NAD(P) has an absorption band at 340 nm (extinction coefficient = 6,200 $M^{-1}cm^{-1}$). (2)

References

1. Ray, M., Pal, D. & Bhaduri, A. (1972) FEBS lett, 25, 239.
2. Publication No. 719 of the National Academy of Sciences - National Research Council. Washington 1960, Criteria CoE-4 and -5.

3-HYDROXYBUTYRYL-CoA DEHYDROGENASE (NADP)

(L(+)-3-Hydroxybutyryl-CoA: NADP oxidoreductase)

3-Hydroxybutyryl-CoA + NADP = acetoacetyl-CoA + reduced NADP

Ref.

Equilibrium constant

The reaction is reversible. (1)

Molecular weight

source	value	conditions	Ref.
Clostridium kluyveri	217,500 [8]	gel electrophoresis (SDS); Sephadex G 200, pH 7.0	(1)

Specific activity

			Ref.
C. kluyveri (19 x)	445	acetoacetyl-CoA (pH 6.5, Pi, 25°)	(1)
	36	3-hydroxybutyryl-CoA (pH 9.0, glycine, 25°)	(1)

Specificity and Michaelis constants

source	substrate	K_m (μM)	conditions	Ref.
C. kluyveri	acetoacetyl-CoA (a)	50	pH 6.5, Pi, 25°	(1)
	reduced NADP (b)	70	pH 6.5, Pi, 25°	(1)

(a) acetoacetyl-CoA (1.00) could be replaced by acetoacetyl-dephospho CoA (0.47); N-acetyl-S-acetoacetylcysteamine (0.15) or acetoacetyl-pantotheine (0.10).

(b) with reduced NAD in the place of reduced NADP, less than 0.25% the activity was obtained. Neither reduced NAD nor NAD inhibited reduced NADP utilization.

In the forward reaction (pH 9.0, glycine), L(+)-3-hydroxybutyryl-CoA (1.00) could be replaced by 3-hydroxyvaleryl-CoA (0.02) or 3-hydroxy-caproyl-CoA (0.02) but not by D-3-hydroxybutyryl-CoA.

References

1. Madan, V.K., Hillmer, P. & Gottschalk, G. (1973) EJB, 32, 51.

GLYCEROL DEHYDROGENASE (NADP, DIHYDROXYACETONE FORMING)

(Glycerol: NADP oxidoreductase)

Glycerol + NADP = dihydroxyacetone + reduced NADP

Ref.

Equilibrium constant

The reaction is reversible. (1)

Specificity and Michaelis constants

source	substrate	K_m (M)	conditions	
Dunaliella parva (cell free extract)	glycerol[a]	1.4	pH 9.0, ambient	(1)
	NADP[b]	1.5×10^{-5}	pH 9.0, ambient	(1)
	dihydroxyacetone[c]	8×10^{-4}	pH 7.5, ambient	(1)
	reduced NADP[d]	4×10^{-6}	pH 7.5, ambient	(1)

(a) glycerol (1.00) could be replaced by 2,3-butanediol (0.12); 1,2-propanediol (0.10); *iso*-erythritol (0.09); *meso*-erythritol (0.04) or 1,2-ethanediol (0.03) but not by D-sorbitol or D-mannitol.

(b) NAD was inactive.

(c) less than 5% the activity of dihydroxyacetone were obtained with the following: D,L-glyceraldehyde; dihydroxyacetonephosphate; D,L-ribulose; sedoheptulose and methylglyoxal.

(d) reduced NADP (1.00) could be replaced by reduced NAD (0.09).

References

1. Ben-Amotz, A. & Avron, M. (1973) FEBS lett, 29, 153.

HEXOSE OXIDASE

(D-Hexose: oxygen 1-oxidoreductase)

β-D-Glucose + O_2 = D-glucono-δ-lactone + H_2O_2

Ref.

Molecular properties

source	value	conditions	Ref.
Chondrus crispus (red alga)	130,000	Sephadex G 200, pH 6.8; amino acid composition	(1)

The enzyme contains 12 gm atoms of copper per 130,000 daltons. It contains no FAD. It is a glycoprotein, containing about 70% carbohydrate (mainly galactose and xylose). (1)

Specific activity

			Ref.
C. crispus (117 x)	4.1	glucose (pH 6.3, Pi, 25°; with o-dianisidine and peroxidase (EC 1.11.1.7))	(1)

Specificity and Michaelis constants

source	substrate	relative velocity	K_m (mM)	Ref.
C. crispus	D-glucose[a]	1.00	4	(1)
	D-galactose	0.82	8	(1)

(a) for the conditions used see "specific activity". D-Glucose (1.00) could also be replaced by maltose (0.40); cellobiose (0.32); lactose (0.22); D-glucose 6-phosphate (0.10); D-mannose (0.08); 2-deoxy-D-glucose (0.08); 2-deoxy-D-galactose (0.06); D-fucose (0.02); D-glucuronate (0.02) or D-xylose. L-Glucose was not oxidized.

Hexose oxidase from Iridophycus flaccidum and Euthora cristata have similar properties to the enzyme from C. crispus. (1,2)

References

1. Sullivan, J.D. & Ikawa, M. (1973) BBA, 309, 11.
2. Bean, R.C. & Hassid, W.Z. (1956) JBC, 218, 425.

L-SORBOSE OXIDASE

(L-Sorbose: oxygen 5-oxidoreductase)

L-Sorbose + O_2 = 5-Keto-D-fructose + H_2O_2

Ref.

Specificity and Michaelis constants

source	substrate	K_m (M)	electron acceptor
Trametes sanguinea (purified 25 x)	L-sorbose[a]	2.2×10^{-2}	oxygen
	L-sorbose	1.0×10^{-1}	2,6-dichlorophenol-indophenol
	D-glucose	1.2×10^{-3}	oxygen
	D-glucose	1.8×10^{-3}	2,6-dichlorophenol-indophenol

[a] with 2,6-dichlorophenolindophenol as the electron acceptor L-sorbose (1.00) could be replaced by D-glucose (1.55); D-galactose (0.71); D-xylose (0.42) or D-maltose (0.15) but not by D-fructose; D-mannose; D-arabinose; L-arabinose; sucrose; D-sorbitol; D-mannitol; *i*-erythritol; D-gluconate or D-glucurono-γ-lactone (conditions: pH 6.0, Pi, 30°). (1)

Light absorption data

See EC 1.1.99.10

References

1. Yamada, Y., Iizuka, K., Aida, K. & Uemura, T. (1967) J. Biochem (Tokyo), 62, 223.

PYRIDOXOL 4-OXIDASE

(Pyridoxol: oxygen 4-oxidoreductase)

Pyridoxol + O_2 = pyridoxal + H_2O_2

Ref.

Specificity and Michaelis constants

source	substrate	K_m (M)	conditions	
Pseudomonas MA-1 (purified 12 x)	pyridoxol [a]	4.3×10^{-5}	pH 8.0, Pi, 26°	(1,2)
	FAD	2.7×10^{-6}	pH 8.0, Pi, 26°	(1,2)
	2,6-dichloro-phenolindophenol	8.3×10^{-6}	pH 8.0, Pi, 26°	(1)

[a] pyridoxol (1.00) could be replaced by 5-deoxypyridoxol (0.65); 2-demethyl-2 ethylpyridoxol (0.52); pyridoxol phosphate (0.12); isopyridoxol (0.11); 2-methyl-4,5-bis(hydroxymethyl)pyridine (0.06); 2-methyl-3-amino-4,5-bis(hydroxymethyl)pyridine (0.04), pyridoxamine (0.03) or pyridoxamine phosphate (0.03) but not by 4-deoxypyridoxol or 2-methyl-3-hydroxy-5,6-bis(hydroxymethyl)pyridine.

The enzyme (Pseudomonas MA-1) required FAD (FMN was inactive) for activity. Either O_2 or 2,6-dichlorophenolindophenol could serve as electron acceptor. (1,2)

Pyridoxol 4-oxidase is thought to be absent in mammalian tissues where pyridoxol is converted to pyridoxal by a slow non-enzymic process which involves Fe^{2+} and substrates of aldehyde oxidase (which produce H_2O_2). (3)

Light absorption data

See EC 1.1.99.10

Abbreviation

Pyridoxol (≡ pyridoxin) — 5-hydroxy-6-methyl-3,4-pyridine dimethanol

References

1. Sundaram, T.K. & Snell, E.E. (1969) JBC, 244, 2577.
2. Burg, R.W. (1970) Methods in Enzymology, 18A, 646.
3. Shane, B. & Snell, E.E. (1972) ABB, 153, 333.

ALCOHOL OXIDASE

(Alcohol: oxygen oxidoreductase)

A primary alcohol + O_2 = an aldehyde + H_2O_2

Ref.

Molecular weight

source	value	conditions	
Basidiomycetes sp.	greater than 300,000[a]	Sephadex G 200, pH 7.5	(1)
Kloeckera sp. No. 2201 (yeast)	570,000[b]	pH 7.5	(2,3)

[a] the enzyme contains 1 mole FAD per 85,000 daltons. It is not a glycoprotein.
[b] the enzyme contains 6 moles FAD per 570,000 daltons.

Specific activity

Basidiomycetes sp.	157 methanol	(pH 7.5, Pi, 25°)	(1)
Yeast (12.5 x)	11 methanol	(pH 7.5, Pi, 37°)	(2)

Specificity and Michaelis constants

source	substrate	relative velocity	K_m (M)
Basidiomycetes sp.[a]	methanol	1.00	1.52×10^{-3}
	ethanol	0.28	1.0×10^{-2}
	n-propanol	0.05	5.46×10^{-2}
	n-butanol	0.02	1.33×10^{-1}
Yeast[b]	methanol	1.00	1.25×10^{-3}
	ethanol	1.06	2.5×10^{-3}

[a] the conditions were: pH 7.5, Pi, 25°. The following were also oxidized (methanol = 1.00): 2-propyn-1-ol (0.45); allylalcohol (0.17); 2-chloroethanol (0.06); 2-bromoethanol (0.04) and n-amylalcohol (0.01). iso-Butanol; anisylalcohol; sec-butanol; benzylalcohol; phenethyl alcohol; hydracrylonitrile; ethylene glycol; methyl cellusolve; glycollate; ethanolamine; 2-mercaptoethanol; choline; 1,3-propanediol; 1,2-propanediol; 1,4-butanediol; glycerine and DL-serine were not oxidized (Ref. 1).
[b] the conditions were: pH 7.5, Pi, 37°. The following were also oxidized (methanol = 1.00): allylalcohol (0.67); n-propanol (0.48); n-butanol (0.31); 2-chloroethanol (0.28); 2-bromoethanol (0.20); 3,4-cis-hexen-1-ol (0.10); ethyleneglycol (0.03), iso-propanol (0.03); iso-butanol (0.02); 2-phenyl-ethanol (0.01); n-hexanol (0.01); n-octanol (0.01) and iso-amylalcohol (0.01) but not 2-aminoethanol propyleneglycol; glycerol; tert-butanol; tert-amylalcohol or benzylalcohol (Ref. 3).

Alcohol oxidase does not require metal ions for activity. (1,2)

References

1. Janssen, F.W. & Ruelius, H.W. (1968) BBA, 151, 330.
2. Tani, Y., Miya, T., Nishikawa, H. & Ogata, K. (1972) Agr. Biol. Chem, (Tokyo), 36, 68.
3. Tani, Y., Miya, T. & Ogata, K. (1972) Agr. Biol. Chem. (Tokyo), 36, 76.
Sahm, H. & Wagner, F. (1973) EJB, 36, 250.

LACTATE-MALATE TRANSHYDROGENASE

(L-Lactate: oxaloacetate oxidoreductase)

L-Lactate + oxaloacetate = L-malate + pyruvate

Ref.

Equilibrium constant

$$\frac{[\text{L-malate}]\,[\text{pyruvate}]}{[\text{L-lactate}]\,[\text{oxaloacetate}]} = $$

1.8 (pH 7.5, 30°) (1)
about 5 (pH 7.8, 24°) (2)

Molecular properties

source	value	conditions	
Micrococcus lactilyticus	99,000 [3 or 4]	sucrose gradient	(3)

The enzyme contains 3 moles NAD per 99,000 daltons; it also contains tightly bound pyruvate. Its amino acid composition is known. (1,2,3)

Specific activity

M. lactilyticus (5 x)	368	L-malate (pH 7.8, Tris, 25°)	(3)

Specificity and kinetic constants

substrate[(a)]	second substrate	V relative	K_m (M)	
L-lactate	oxaloacetate	1.0	2.5×10^{-3}	(3)
oxaloacetate	L-lactate	1.0	4.8×10^{-5}	(3)
L-malate	pyruvate	2.8	6.0×10^{-4}	(3)
pyruvate	L-malate	2.8	1.0×10^{-3}	(3)

(a) with the enzyme from *M. lactilyticus* (conditions: pH 7.8, Tris, 25°).

A large number of 2-oxo- and 2-hydroxy acids are substrates. Thus, malate (1.00) could be replaced by *dl*-2-hydroxybutyrate (1.00) or *dl*-3-hydroxybutyrate (0.014). D-Lactate; 2-methyllactate and isocitrate were inactive. Increase in the chain length of the 2-hydroxyacid resulted in a decrease in activity; thus 2-hydroxycaproate had only 0.2% the activity of L-malate. The enzyme requires no coenzyme or metal ion for activity. Added reduced NAD or NADP (or artificial hydrogen carrier) could not be coupled to the transhydrogenation. High concentrations of oxaloacetate were inhibitory as were the two products of the forward and reverse reactions. The kinetics of the enzyme have been investigated. (1,2,3,4)

Light absorption data

Oxaloacetate has an absorption band at 258 nm (molar extinction coefficient = 840 M^{-1} cm^{-1}). (1)

References

1. Allen, S.H.G. (1966) *JBC*, *241*, 5266.
2. Dolin, M.I. (1969) *JBC*, *244*, 5273.
3. Allen, S.H.G. & Patil, J.R. (1972) *JBC*, *247*, 909.
4. Allen, S.H.G. (1969) *Methods in Enzymology*, *13*, 262.
   Allen, S.H.G. (1973) *EJB*, *35*, 338.

ALCOHOL DEHYDROGENASE (ACCEPTOR)

(Primary alcohol: (acceptor) oxidoreductase)

Primary alcohol + acceptor = aldehyde + reduced acceptor

Ref.

Molecular properties

source	value	conditions	Ref.
Pseudomonas sp. M 27	146,000	pH 7.2	(1)

The amino acid composition of the dehydrogenase, which is a glycoprotein, is known. It contains a prosthetic group, which is a pteridine derivative, but no metal ions. (1,2)

Specific activity

Pseudomonas M 27 (10 x)	0.74	methanol (pH 9.0, Tris; electron acceptor = PMS + 2,6-dichlorophenol indophenol)	(1)

Specificity

The enzyme (Pseudomonas M 27) was active with methanol (1.00); ethanol (1.00); propan-1-ol (0.80); butan-1-ol (0.83); pentan-1-ol (0.50); hexan-1-ol (0.50); heptan-1-ol (0.54); octan-1-ol (0.48); nonan-1-ol (0.60); decan-1-ol (0.65); undecan-1-ol (0.60); 2-chloroethanol (0.80); 2-bromoethanol (0.65); 3-chloropropan-1-ol (0.75); ethane-1,2-diol (0.83); propane-1,3-diol (0.80); 2-methoxyethanol (0.85); 2-ethoxyethanol (0.83); 2-(2'-hydroxyethoxy) ethanol (0.60); 2-phenylethanol (0.90); 3-phenylpropan-1-ol (0.68); allyl alcohol (0.81); crotyl alcohol (0.52); geraniol (0.33); and cinnamyl alcohol (0.92). The following were inactive: amino alcohols, secondary and tertiary alcohols and alcohols of the type R.C ⋮ C.CH_2OH. (3)

PMS could not be replaced as primary hydrogen acceptor by NAD; NADP; cytochrome c; ferricyanide; pyocyanine; methylene blue; 2,6-dichlorophenol indophenol; benzyl viologen; methyl viologen; Janus green; phenosafranine; neutral red; resazurine; menaphthone or menapthone bisulphite. (3)

The dehydrogenase requires ammonia for full activity. (1)

Light absorption data

See EC 1.1.99.10.

Abbreviations

PMS N-methylphenazonium methosulphate

References

1. Anthony, C. & Zatman, L.J. (1967) BJ, 104, 953.
2. Anthony, C. & Zatman, L.J. (1967) BJ, 104, 960.
3. Anthony, C. & Zatman, L.J. (1965) BJ, 96, 808.

PYRIDOXOL 5'-DEHYDROGENASE

(Pyridoxol: (acceptor) 5'-oxidoreductase)

Pyridoxol + acceptor = isopyridoxal + reduced acceptor

Ref.

Specificity and Michaelis constants

source	substrate	K_m (M)	conditions	Ref.
Pseudomonas IA (purified 3 x)	pyridoxol[a]	4.6×10^{-4}	pH 6.0, Pi, 26°; electron acceptor = 2,6-dichlorophenol indophenol.	(1)
	FAD[b]	5×10^{-7}		

[a] pyridoxol (1.00) could be replaced by pyridoxamine (0.11); isopyridoxal and 4-deoxypyridoxol showed slight activities but pyridoxol phosphate and 4-pyridoxoate were inactive.
[b] FMN was inactive.

The enzyme did not react directly with O_2, with or without methylene blue. (1)

Light absorption data

See 1.1.99.10

Abbreviations

Pyridoxol (or pyridoxine) — 5-hydroxy-6-methyl-3,4-pyridine dimethanol

References

1. Sundaram, T.K. & Snell, E.E. (1969) JBC, 244, 2577.

GLUCOSE DEHYDROGENASE (Aspergillus)

(D-Glucose: (acceptor) 1-oxidoreductase)

D-Glucose + acceptor = D-glucono-1,5-lactone + reduced acceptor

Ref.

Molecular properties

source	value	conditions	
Aspergillus oryzae	118,000	Sephadex G 100, pH 6.5	(1)

The enzyme contains 1 mole of FAD per 118,000 daltons. It also contains 24% carbohydrate. Its amino acid composition is known. (1)

Specific activity

A. oryzae (550 x)	891	glucose (pH 6.5, Pi; electron acceptor = 2,6 dichlorophenol indophenol)	(1)

Kinetic properties

The enzyme (A. oryzae) could utilize certain redox dyes (e.g. 2,6-dichlorophenol indophenol) and quinones as electron acceptors but O_2, NAD and NADP were inactive. Xylose, mannose and galactose were less effective substrates than glucose. (1,2)

Light absorption data

Oxidized 2,6-dichlorophenol indophenol has an absorption band at 600 nm with a molar extinction coefficient of 21,000 M^{-1} cm^{-1} at pH 7.6 and 18,500 M^{-1} cm^{-1} at pH 6.5 (Ref. 2) and 14,000 M^{-1} cm^{-1} (Ref. 3) or 11,850 M^{-1} cm^{-1} (Ref. 4) at pH 6.0. For further information on the extinction coefficient of 2,6-dichlorophenolindophenol see Ref. 5.

References

1. Bak, T.G. (1967) BBA, 139, 277.
2. Bak, T.G. & Sato, R. (1967) BBA, 139, 265.
3. Armstrong, J.Mc.D. (1964) BBA, 86, 194.
4. van Beeumen, J. & deLey, J. (1968) EJB, 6, 331.
5. Bergmeyer, H-U. (1965) Methods of Enzymatic Analysis, Academic Press: New York & London. p 435.

D-FRUCTOSE 5-DEHYDROGENASE

(D-Fructose: (acceptor) 5-oxidoreductase)

D-Fructose + acceptor =

5-keto-D-fructose + reduced acceptor

Ref.

Specificity and Michaelis constants

source	substrate	K_m (M)	conditions	
Gluconobacter cerinus (purified 45 x)	D-fructose[a]	1.0×10^{-2}	pH 5.5, Pi, 15°, with 2,6-dichlorophenolindophenol as electron acceptor	(1)
	D-glucose	1.0×10^{-3}		

[a] D-fructose (1.00) could be replaced by D-glucose (2.00), D-mannose (0.22), D-galactose (0.54), D-xylose (0.70), L-arabinose (0.43), D-gluconate (0.37), maltose (0.08), cellobiose (0.06) or melibiose (0.11) but not by L-sorbose, D-arabinose, D-sorbitol, D-mannitol, dulcitol, myo-inositol, adonitol, i-erythritol, glycerol, 2-keto-D-gluconate, 5-keto-D-gluconate, D-glucurono-γ-lactone, α,α'-trehalose or sucrose.

The natural electron acceptor for D-fructose 5-dehydrogenase may be cytochrome c. 2,6-Dichlorophenolindophenol could not be replaced by methylviologen; safranin T; Janus green; phenosafranin; cresyl violet; nile blue; triphenyltetrazolium chloride; methylene blue; toluidine blue; cresyl blue; toluylene blue, or ferricyanide. The activity of the enzyme was not affected by divalent metal ions. (1)

Light absorption data

See EC 1.1.99.10

References

1. Yamada, Y., Aida, K. & Uemura, T. (1967) J. Biochem (Tokyo), 61, 636.

SORBOSE DEHYDROGENASE

(L-Sorbose: (acceptor) 5-oxidoreductase)

L-Sorbose + acceptor =

5-keto-D-fructose + reduced acceptor

Ref.

Specificity and Michaelis constants

source	substrate	K_m(M)	conditions	
Gluconobacter suboxydans (a particulate fraction)	L-sorbose[a]	4.3×10^{-3}	pH 6.5, Pi, 20° (electron acceptor = 2,6-dichlorophenol-indophenol)	(1)

[a] L-sorbose (1.00) could be replaced by D-glucose (3.82), D-mannose (1.90), D-galactose (1.80), D-xylose (3.55), D-arabinose (0.08), L-arabinose (3.18), D-gluconate (1.35), 2-keto-D-gluconate (0.64), 5-keto-D-gluconate (0.09), D-sorbitol (7.28), D-mannitol (3.28), *i*-erythritol (2.38), glycerol (3.27) or maltose (2.82) but not by D-fructose, dulcitol or sucrose. (1)

The enzyme could not utilize oxygen as electron acceptor. (1)

Light absorption data

See EC 1.1.99.10

References

1. Sato, K., Yamada, Y., Aida, K. & Uemura, T. (1969) *J. Biochem (Tokyo)*, **66**, 521.

D-GLUCOSIDE 3-DEHYDROGENASE

(D-Aldohexoside: (acceptor) 3-oxidoreductase)

Sucrose + acceptor = 3-keto-α-D-glucosyl-β-D-fructofuranoside + reduced acceptor

Ref.

Equilibrium constant

$$\frac{[\text{3-Ketolactose}]\,[\text{ferrocytochrome } c]^2\,[H^+]^2}{[\text{lactose}]\,[\text{ferricytochrome } c]^2} = 1 \times 10^{-11} M^2 \ (25^{\circ})$$ (1)

Molecular properties

source	value	conditions	Ref.
Agrobacterium tumefaciens	85,000	pH 7.2	(1)

A sedimentation constant of 3.6 S is recorded in Ref. 2. The enzyme contains bound FAD. (1,2)

Specific activity

A. tumefaciens (584 x) — 7.6 lactose (pH 6.04, Pi, 30°; with 2,6-dichlorophenolindophenol as electron acceptor) (1)

Specificity and kinetic properties

source	substrate[a]	relative velocity	K_m (M)
A. tumefaciens	cellobiose	1.00	2.0×10^{-4}
	lactobionate	1.00	2.1×10^{-4}
	lactose	0.84	1.7×10^{-3}
	D-glucose	0.84	2.9×10^{-3}
	maltose	0.70	2.8×10^{-3}
	sucrose	0.60	4.1×10^{-3}
	D-galactose	0.56	2.5×10^{-2}
	maltobionate	0.46	3.6×10^{-4}

[a] the conditions were: pH 6.04, Pi, 30°; electron acceptor = 2,6-dichlorophenol indophenol. The following could replace cellobiose (1.00): methyl-β-D-glucose (0.85); lactulose (0.85); methyl-α-D-glucose (0.84); p-arbutine (0.84); glucose 1-phosphate (0.62); trehalose (0.30); β-melibiose (0.25); 2-deoxy-D-glucose (0.22); leucrose (0.12); melezitose (0.10); methyl-β-D-thiogalactose (0.10); D-mannose (0.10); raffinose (0.07); D-glucosamine (0.06); anhydro-1,6-D-glucose (0.04); 2-deoxy-D-galactose (0.02). The following compounds were active as electron acceptors: cytochrome c; 2,6-dichlorophenolindophenol; 5-methylphenazinium methyl sulphate and

Ref.

ferricyanide. Neutral red, benzylviologen; nile blue; brilliant alizarin blue; cresylviolet; phenosaphranine and riboflavin were inactive. NAD, NADP, FMN or FAD, either alone or in the presence of Mg^{2+} or Fe^{2+}, were inactive. (1)

The enzyme was inhibited by Ca^{2+}. (2)

Light absorption data

See EC 1.1.99.10.

References

1. van Beeumen, J. & deLey, J. (1968) EJB, 6, 331.
2. Hayano, K. & Fukui, S. (1967) JBC, 242, 3665.

GLUTARATE SEMIALDEHYDE DEHYDROGENASE

(Glutarate-semialdehyde: NAD(P) oxidoreductase)

Glutarate semialdehyde + NAD(P) + H_2O =

glutarate + reduced NAD(P)

Ref.

Molecular properties

source	value	conditions	Ref.
Pseudomonas putida	170,000	sucrose gradient	(1,2,3)

Glutarate semialdehyde dehydrogenase is a separate protein from 2-oxoglutarate dehydrogenase (EC 1.2.1.26): the former enzyme is constitutive, the latter inductive. (3)

Specific activity

P. putida (33 x) 6.5 glutarate semialdehyde (with NADP; pH 8.5, PPi) (3)

Specificity and Michaelis constants

substrate[a]	second substrate	V relative	K_m (M)
glutarate semialdehyde	NADP	1.00	2.7×10^{-5}
2-oxoglutarate semialdehyde	NADP	0.95	6.9×10^{-5}
NADP	glutarate semialdehyde	0.93	6.3×10^{-5}
NAD	glutarate semialdehyde	0.85	2.6×10^{-4}
NADP	2-oxoglutarate semialdehyde	0.72	4.9×10^{-5}
NAD	2-oxoglutarate semialdehyde	0.68	2.9×10^{-4}

(a) with the enzyme from P. putida; conditions: pH 8.5, PPi (Ref. 3).

The enzyme (P. putida) acts on a number of aliphatic aldehydes; of those tested only DL-glyceraldehyde, butyraldehyde, 2-hydroxybutyraldehyde, valeraldehyde, 5-hydroxyvaleraldehyde, glutaraldehyde and heptaldehyde had V values 5-18% that of glutarate semialdehyde: the others were inactive. (3)

References

1. Adams, E. & Rosso, G. (1967) JBC, 242, 1802.
2. Adams, E. & Rosso, G. (1966) BBRC, 23, 633.
3. Chang, Y.F. & Adams, E. (1971) Methods in Enzymology, 17B, 166.

LACTALDEHYDE DEHYDROGENASE

(L-Lactaldehyde: NAD oxidoreductase)

L-Lactaldehyde + NAD + H_2O = lactate + reduced NAD

Ref.

Equilibrium constant

With the enzyme from Escherichia coli K12 as catalyst, the reaction was essentially irreversible [F]. (1)

Molecular weight

source	value	conditions	
E. coli	100,000	sucrose gradient	(1)

Specific activity

E. coli (200 x)	5.1	L-lactaldehyde (pH 10.5, sodium carbonate)	(1)

Specificity and Michaelis constants

source	substrate	K_m (M)	conditions	
E. coli	L-lactaldehyde[a]	1×10^{-2}	pH 10.5, sodium carbonate	(1)
	NAD[b]	1×10^{-4}	pH 10.5, sodium carbonate	(1)

[a] L-lactaldehyde (1.00) could be replaced by propionaldehyde (0.12); D-lactaldehyde (0.05); pyruvaldehyde (0.03) and DL-glyceraldehyde (0.03) but not by acetaldehyde.

[b] NAD (1.00) could be replaced by the 3-acetylpyridine-(0.25) or the thionicotinamide-(0.06) analogue of NAD but not by NADP.

The properties of lactaldehyde dehydrogenase are compared with those of other less specific aldehyde dehydrogenases acting on L-lactaldehyde in Ref. 1. Lactaldehyde dehydrogenase requires a thiol compound (but no metal ions - e.g. K^+, Mg^{2+}) for full activity. (1)

References

1. Sridhara, S. & Wu, T.T. (1969) JBC, 244, 5233.

2-OXOALDEHYDE DEHYDROGENASE

(2-Oxoaldehyde: NAD(P) oxidoreductase)

2-Oxoaldehyde + NAD(P) + H_2O =

2-oxoacid + reduced NAD(P)

Ref.

Specificity and catalytic properties

substrate[a]	co-substrate	V relative	K_m (mM)
methylglyoxal[b]	NAD	1.00	4.50
	NADP	0.30	0.40
glyoxal	NAD	0.16	1.70
	NADP	0.24	2.22
phenylglyoxal	NAD	0.90	1.50
	NADP	0.11	0.66
hydroxypyruvaldehyde	NAD	0.50	1.30
	NADP	0.23	0.90
NAD[c]	methylglyoxal	1.00	0.19
NADP	methylglyoxal	0.70	0.02
3-acetylpyridine-adenine dinucleotide	methylglyoxal	1.17	1.00
thionicotinamide-adenine dinucleotide	methylglyoxal	0.70	0.67

(a) with the enzyme from sheep liver (purified 70 x). Assay conditions: pH 8.5, Tris, 30°, Ref (1).

(b) acetaldehyde; glutaric dialdehyde; glyoxylic acid hydrate; D-glyceraldehyde; glycolaldehyde; 1-phenyl-1,2-propanedione and malonic dialdehyde were inactive.

(c) nicotinamide hypoxanthine dinucleotide; 3-pyridinealdehyde adenine dinucleotide; 3-acetylpyridine hypoxanthine dinucleotide and 3-pyridinealdehyde hypoxanthine dinucleotide were inactive.

Sheep liver may contain 2 enzymes that oxidize 2-oxoaldehydes: one specific for NAD and the other specific for NADP. (1)

References

1. Monder, C. (1967) JBC, 242, 4603.

2-OXOGLUTARATE-SEMIALDEHYDE DEHYDROGENASE

(2,5-Dioxovalerate: NAD(P) oxidoreductase)

2-Oxoglutarate semialdehyde + NAD(P) + H_2O =
2-oxoglutarate + reduced NAD(P)

Ref.

Molecular properties

source	value	conditions	
Pseudomonas putida	115,000	sucrose gradient	(1,2)

2-Oxoglutarate semialdehyde dehydrogenase is a separate protein from glutarate semialdehyde dehydrogenase (EC 1.2.1.20): the former enzyme is inductive, the latter constitutive. (1,2)

Specific activity

P. putida (40 x)	53	2-oxoglutarate semialdehyde (with NADP; pH 8.5, PPi, 25°)	(1,2)

Specificity and Michaelis constants

source	substrate	V relative	K_m(M)	conditions	
P. putida	2-oxoglutarate semialdehyde[a]	-	1×10^{-5}	pH 8.5, PPi, 25°	(1)
	glutaraldehyde[b]	-	5×10^{-2}	pH 8.5, PPi, 25°	(1)
	NADP	2.5	1×10^{-4}	pH 8.5, PPi, 25°	(1)
	NAD	1.0	1×10^{-3}	pH 8.5, PPi, 25°	(1)

(a) 2-oxoglutarate semialdehyde (1.00) could be replaced by glutarate semialdehyde (0.15); glutaraldehyde (0.15); heptaldehyde (0.18); butyraldehyde (0.14); propionaldehyde (0.11); octylaldehyde (0.11); decylaldehyde (0.10); hexylaldehyde (0.08); valeraldehyde (0.08); 4-hydroxy-2-oxoglutarate semialdehyde (0.008); 5-hydroxyvaleraldehyde (0.03); isobutyraldehyde (0.01); or succinate semialdehyde (0.004) but not by formaldehyde; acetaldehyde; malonate dialdehyde; DL-glyceraldehyde; pyruvate aldehyde; 2-hydroxybutyraldehyde or 2,3-dimethylvaleraldehyde.

(b) both aldehyde groups were oxidized.

Inhibitors

The enzyme (P. putida) was inhibited by 4-hydroxy-2-oxoglutarate semialdehyde (Ki = 3.5×10^{-5}M, C(2-oxoglutarate semialdehyde)).

References

1. Adams, E. & Rosso, G. (1967) JBC, 242, 1802.
2. Adams, E. (1971) Methods in Enzymology, 17B, 303.

METHYLMALONATE SEMIALDEHYDE DEHYDROGENASE (ACYLATING)

(Methylmalonate-semialdehyde: NAD

oxidoreductase (CoA-propionylating))

Methylmalonate semialdehyde + CoA + NAD =

propionyl-CoA + CO_2 + reduced NAD

Molecular weight

source	value	conditions	Ref.
Pseudomonas aeruginosa	132,000 [2]	Sephadex G 200, pH 7.0; amino acid composition	(1)

Specific activity

P. aeruginosa (380 x)	5	methylmalonate semialdehyde (pH 9.2, Tris, 37°)	(1)

Specificity and Michaelis constants

substrate[(a)]	V relative	K_m (M)
with CoA as acyl acceptor		
methylmalonate semialdehyde	1.00	1.9×10^{-5}
propionaldehyde	0.70	6.1×10^{-3}
NAD	-	8.2×10^{-5}
CoA	-	2.2×10^{-5}
with mercaptoethanol as acyl acceptor		
methylmalonate semialdehyde	1.03	3.5×10^{-5}
propionaldehyde	1.86	5.1×10^{-2}
NAD	-	8.7×10^{-5}
mercaptoethanol	-	5.4×10^{-4}

(a) with the enzyme from P. aeruginosa. The conditions were: pH 9.2, Tris, 37°. (Ref. 1).

Methylmalonate semialdehyde could be replaced by propionaldehyde but not by acetaldehyde. CoA could be replaced by mercaptoethanol.

For further information on the enzyme see Ref. 2.

References

1. Bannerjee, D., Sanders, L.E. & Sokatch, J.R. (1970) JBC, 245, 1828.
2. Sokatch, J.R., Sanders, L.E. & Marshall, V.P. (1968) JBC, 243, 2500.

ARYLALDEHYDE DEHYDROGENASE (NADP)

(Arylaldehyde: NADP oxidoreductase (ATP forming))

An aromatic aldehyde + NADP + ADP + Pi + H_2O

= an aromatic acid + reduced NADP + ATP

Ref.

Equilibrium constant

The reaction is essentially irreversible [R] with the enzyme from *Neurospora crassa* as catalyst. (1)

Molecular weight

source	value	conditions	
N. crassa	120,000	Sephadex G 200, pH 7.5	(1)

Specificity and Michaelis constants

source	substrate	K_m (M)	conditions	
N. crassa	salicylate[a]	1.67×10^{-4}	pH 8.0, Tris, 30°	(1)
(purified 300 x)	ATP[b]	1.52×10^{-4}	pH 8.0, Tris, 30°	(1)

[a] salicylate (1.00) could be replaced by benzoate (2.86). A number of other aromatic acids were also active but the enzyme did not catalyze the reduction of the aliphatic carboxyl group.

[b] ATP (1.00) could be replaced by ITP (0.05) but not by CTP, GTP or UTP.

The enzyme requires Mg^{2+} and a thiol compound for full activity. The kinetics observed with reduced NADP as substrate were complex and a Km value was not obtained. With respect to the cofactor NADP, the enzyme is of the "B" type. (1)

References

1. Gross, G.G. & Zenk, M.H. (1969) EJB, 8, 413.

AMINOADIPATE SEMIALDEHYDE DEHYDROGENASE

(L-2-Aminoadipate-6-semialdehyde: NAD(P) oxidoreductase)

L-2-Aminoadipate 6-semialdehyde + NAD(P) + H_2O =

L-2-aminoadipate + reduced NAD(P)

Ref.

Equilibrium constant

With the enzyme from *Pseudomonas putida*, the reaction was essentially irreversible [F]. In solution, L-2-aminoadipate 6-semialdehyde is in equilibrium with Δ'-piperideine 6-carboxylate. (1)

Specificity and Michaelis constants

source	substrate	K_m (M)	conditions	
P. putida (purified 20 x)	Δ'-piperideine 6-carboxylate[(a)]	2.0×10^{-5}	pH 7.0, Pi, 30°	(1)
	NAD[(b)]	2.0×10^{-4}	pH 7.0, Pi, 30°	(1)

(a) *n*-butyraldehyde was a poor substrate and the following were inactive: Δ'-pyrroline 5-carboxylate; acetaldehyde; D-glyceraldehyde; glycolaldehyde; DL-pipecolate; L-proline; 5-hydroxy-L-pipecolate and hydroxy-L-proline.

(b) a number of analogues tested were active.

The enzyme did not have a metal requirement. (1)

Inhibitors

The enzyme (*P. putida*) was inhibited by ATP; by *n*-butyraldehyde; acetaldehyde and by negatively charged α-amino acids (neutral or basic amino acids did not inhibit). (1)

References

1. Calvert, A.F. & Rodwell, V.W. (1966) JBC, 241, 409.

RETINAL DEHYDROGENASE

(Retinal: NAD oxidoreductase)

Retinal + NAD + H_2O = retinoate + reduced NAD

Ref.

Equilibrium constant

The enzyme from rat intestinal mucosa did not catalyze the reverse reaction. (1)

Molecular weight

source	value	conditions	
Rat intestinal mucosa	80,000	Sephadex G 200, pH 7.7	(1)

The enzyme contained about 2 moles of iron per 80,000 daltons. It had a small absorption band at 410 nm.

Specific activity

Rat intestinal mucosa (160 x) 8×10^{-2} retinal (pH 7.7, Pi, 37°) (1)

Specificity and Michaelis constants

source	substrate	K_m (M)	conditions	
Rat intestinal mucosa	retinal	3.0×10^{-4}	pH 7.7, Pi, 37°	(1)

The enzyme oxidized all-*trans*-retinal (1.00) and 13-*cis*-retinal (0.71) but it was inactive with all-*trans*-retinol, all-*trans*-retinoate, β-carotene and xanthine.

The enzyme did not utilize atmospheric oxygen and the reaction can proceed under aerobic or anaerobic conditions. The rate of the reaction was maximal in the presence of glutathione, NAD, FAD and Fe^{2+}. (1)

Inhibitors

Reduced NAD inhibited the reaction (NC (retinal)). The inhibition was not reversed by NAD. (1)

References

1. Moffa, D.J., Lotspeich, F.J. & Krause, R.F. (1970) JBC, 245, 439.

N-ACETYLGLUTAMATE SEMIALDEHYDE DEHYDROGENASE

(N-Acetyl-L-glutamate 5-semialdehyde:
NADP oxidoreductase (phosphorylating))

N-Acetyl-L-glutamate 5-semialdehyde + Pi + NADP =
N-acetyl-5-L-glutamylphosphate + reduced NADP

Ref.

Molecular weight

source	value	conditions	Ref.
Escherichia coli	$S_{20,w}$ = 7.9S	pH 8.0. The buffer contained 1 x 10^{-3}M mercaptoethanol and 1 x 10^{-4}M EDTA.	(1)

Highly purified preparations of the enzyme are unstable. Thus, the active dehydrogenase of $S_{20,w}$ = 7.9S slowly breaks down to an inactive material of $S_{20,w}$ = 2.9S. (1)

Specificity and Michaelis constants

source	substrate	K_m (M)	conditions	Ref.
E. coli (purified 300 x)	N-acetyl-L-glutamate-5-semialdehyde	6 x 10^{-4}	pH 9.3, glycine, 25°	(1)
	NADP	1.7 x 10^{-4}	pH 9.3, glycine, 25°	(1)
	Pi [(a)]	3 x 10^{-3}	pH 9.3, glycine, 25°	(1)

(a) arsenate could replace Pi

Inhibitors

The enzyme (E. coli) was inhibited by Fe^{2+}, Mn^{2+} or Ni^{2+}. (1)

References

1. Vogel, H.J. & McLellan, W.L. (1970) Methods in Enzymology, 17A, 255.

PHENYLACETALDEHYDE DEHYDROGENASE

(Phenylacetaldehyde: NAD oxidoreductase)

Phenylacetaldehyde + NAD + H_2O = phenylacetate + reduced NAD

Ref.

Equilibrium constant

The reaction is essentially irreversible [F] with the enzyme from *Achromobacter eurydice*. (1,2)

Molecular weight

source	value	Ref.
A. eurydice	$s_{20,w}$ = 5.6 S	(1,2)

Specific activity

				Ref.
A. eurydice	(41 x)	19	phenylacetaldehyde (pH 8.9, PPi, 25°)	(1,2)

Specificity and Michaelis constants

source	substrate	K_m (M)	conditions	Ref.
A. eurydice	NAD (a)	7×10^{-5}	pH 8.9, PPi, 25°	(1,2)
	K^+ (b)	4×10^{-4}	pH 8.9, PPi, 25°	(1,2)

(a) NADP possessed 10% the activity of NAD.

(b) K^+, Rb^+, Na^+, Li^+, NH_4^+ and Cs^+ activated the enzyme in this order of effectiveness but Mg^{2+} and Ca^{2+} did not.

The enzyme is highly specific for phenylacetaldehyde. Low activities occurred in the presence of indoleacetaldehyde, propionaldehyde or *n*-butyraldehyde but acetaldehyde, benzaldehyde and crotonaldehyde were inactive. High concentrations of phenylacetaldehyde were inhibitory; a maximum reaction rate was observed at 1×10^{-5}M. (1,2)

References

1. Fujioka, M., Asakawa, H. Wada, H. & Yamano, T. (1966) *Koso Kagaku Shimpojiumu*, *18*, 106.
2. Fujioka, M., Morino, Y. & Wada, H. (1970) *Methods in Enzymology*, *17A*, 593.

PYRUVATE SYNTHASE

(Pyruvate: ferredoxin oxidoreductase (CoA-acetylating))

Pyruvate + CoA + oxidized ferredoxin =

acetyl-CoA + CO_2 + reduced ferredoxin

Ref.

Equilibrium constant

The forward reaction is favoured. (1)

Molecular properties

source	value	conditions	Ref.
Clostridium acidi-urici [a]	240,000	pH 7.4	(1)
Chlorobium thiosulfatophilum	300,000	sucrose gradient	(5)

[a] multiple forms were observed on electrophoresis. The highly purified enzyme contains 6 atoms non-haem iron, 3 moles sulphide and 0.5-0.8 moles of thiaminepyrophosphate per 240,000 daltons. The enzyme has a chromophore which absorbs at 400 nm.

Specificity and kinetic constants

It is thought that the primary electron acceptor in the reaction is the chromophore absorbing at 400 nm and not ferredoxin. The enzyme requires Mg^{2+} for activity.

The enzyme catalyzes 3 reactions:

1. The exchange of CO_2 with the carboxyl group of pyruvate. This reaction requires catalytic amounts of CoA and is inhibited by various electron acceptors.
2. The oxidation of pyruvate in the presence of various electron acceptors.
3. The reductive synthesis of pyruvate from acetyl-CoA and CO_2 in the presence of reduced ferredoxin (reaction, above). (1,2,3)

source	substrate	reaction [a]	relative velocity	K_m (M)
C. acidi-urici (purified 30 x)	oxidized ferredoxin [b]	3	1.00	-
	FAD	3	1.57	1×10^{-3}
	FMN	3	1.37	-
	rubredoxin	3	0.42	-
	pyruvate (with FAD)	3	-	1×10^{-3}
	CoA (with FAD)	3	-	$\sim 5 \times 10^{-5}$
	pyruvate	1 [c]	-	2.2×10^{-3}
	bicarbonate	1 [c]	-	3.5×10^{-3}

Specificity and kinetic constants continued

(a) pH 6.8, Pi, 25° (Ref. 2 and 3).

(b) NAD and NADP were inactive.

(c) CoA = 2×10^{-5}M. At CoA = 2×10^{-6}M, Km (pyruvate) = 2.2×10^{-3}M and Km (bicarbonate) = 1.54×10^{-2}M.

The mechanism of the reaction is discussed in Ref. 3. Pyruvate synthase has also been isolated from *Micrococcus lactilyticus* (Ref. 4) and a variety of anaerobes (Ref. 3).

Light absorption data

Reduced ferredoxin (from *C. acidi-urici*) has an absorption band at 390 nm (molar extinction coefficient = 19,600 at pH 7.1). (1)

References

1. Uyeda, K. & Rabinowitz, J.C. (1971) JBC, 246, 3111.
2. Uyeda, K. & Rabinowitz, J.C. (1971) JBC, 246, 3120.
3. Raeburn, S. & Rabinowitz, J.C. (1971) ABB, 146, 21.
4. Whiteley, H.R. & McCormick, N.G. (1963) J. Bacteriol, 85, 382.
5. Gehring, U. & Arnon, D.I. (1972) JBC, 247, 6963.

2-OXOGLUTARATE SYNTHASE

(2-Oxoglutarate: ferredoxin oxidoreductase (CoA-succinylating))

2-Oxoglutarate + CoA + oxidized ferredoxin = succinyl-CoA + CO_2 + reduced ferredoxin

Ref.

Equilibrium constant

The reaction is reversible. (1)

Molecular weight

source	value	conditions	Ref.
Chlorobium thiosulfatophilum	220,000	sucrose gradient and Sephadex G 200	(1)

Specificity and kinetic properties

source	substrate	K_m (M)	conditions	Ref.
C. thiosulfatophilum (purified 114 x)	succinyl-CoA[a]	2.5×10^{-4}	pH 7.7, HEPES, 20° anaerobic	(1)
	2-oxoglutarate[b]	2.6×10^{-3}	pH 8.7, HEPES, 20° anaerobic	(1)

(a) with a ferredoxin as electron donor. Ferredoxins isolated from different sources differed in their effectiveness as electron donor; those from the photosynthetic bacteria Chromatium, C. thiosulfatophilum and Rhodospirillum rubrum were the most effective, ferredoxins from Clostridium pasteurianum and Bacillus polymyxa were much less active and ferredoxins from spinach leaves and the blue-green algae (Anabaena cylindrica and Anacystics nidulans) were inactive. Ferredoxin could not be replaced by methyl viologen, benzyl viologen or azotoflavin.

(b) with 2,3,5-triphenyltetrazolium (1.00) as the electron acceptor. Other electron acceptors were FAD (1.14), FMN (1.00), ferredoxin from C. pasteurianum (0.25), ferredoxin from Chromatium (0.14) and azotoflavin (0.04). NAD and NADP were inactive.

In the forward reaction, CoA could not be replaced by pantetheine; glutathione; cysteine; mercaptoethanol or dithiothreitol. In the presence of either CoA or reduced ferredoxin, 2-oxoglutarate synthase catalyses an active ^{14}C exchange between [^{14}C] bicarbonate and 2-oxoglutarate. Neither the synthetic nor the exchange reaction was effected by TPP or Mg^{2+}. The enzyme is inactivated by molecular oxygen. (1)

References

1. Gehring, U. & Arnon, D.I. (1972) JBC, 247, 6963.

2-FUROYL-CoA HYDROXYLASE

(2-Furoyl-CoA: acceptor 5-oxidoreductase (hydroxylating))

2-Furoyl-CoA + acceptor + H_2O =

5-Hydroxy-2-furoyl-CoA + reduced acceptor

Ref.

Equilibrium constant

5-Hydroxy-2-furoyl-CoA tautomerizes nonenzymically to 5-oxo-Δ^2-dihydro-2-furoyl-CoA and the overall reaction is essentially irreversible [F]. (1)

Molecular properties

source	value	conditions	Ref.
Pseudomonas putida F2	3.27×10^6	in Pi buffer, pH 7.1, $S_{20,w}$ = 42.9S; in SDS, $S_{20,w}$ = 3.26S	(1)

The enzyme contains copper, but no flavin compound or iron. It has no light absorption above 273 nm. (1)

Specificity and kinetic properties

source	substrate	K_m (M)	conditions	Ref.
P. putida (purified 60 x)	2-furoyl-CoA	2.02×10^{-5}	anaerobic reduction of methylene blue; pH 7.5, Pi, 30°	(1)

The enzyme (P. putida) can use methylene blue, nitro blue tetrazolium or a membrane fraction from P. putida as electron acceptor. It has no requirement for reduced NAD(P). (1)

Light absorption data

Methylene blue has an absorption band at 665 nm (molar extinction coefficient = 41,800 $M^{-1}cm^{-1}$ at pH 7.5). (1)

References

1. Kitcher, J.P., Trudgill, P.W. & Rees, J.S. (1972) BJ, 130, 121.

CUCURBITACIN B Δ^{23}-REDUCTASE

(23,24-Dihydrocucurbitacin B: NAD(P) Δ^{23}-oxidoreductase)

23,24-Dihydrocucurbitacin B + NAD(P) = cucurbitacin B + reduced NAD(P)

Ref.

Equilibrium constant

The reverse reaction is strongly favoured. The ratio of the rate of the reverse to the rate of the forward reaction = 1 : 0.02. (1)

Molecular weight

source	value	conditions	
Cucurbita maxima (Green hubbard fruit)	$S_{20,w}$ = 2.96S	pH 5.6	(2)

Specific activity

C. maxima (142 x)	5812	cucurbitacin B (pH 6.6, maleate, 25°)	(2)

Specificity and Michaelis constants

source	substrate	relative velocity	K_m (M)	conditions	
C. maxima	cucurbitacin B[a]	-	8.01×10^{-4}	pH 6.6, maleate, 25°	(1)
	reduced NADP	1.00	6.05×10^{-5}	pH 6.6, maleate, 25°	(1)
	reduced NAD	0.39	-	pH 6.6, maleate, 25°	(1)

[a] cucurbitacin B (1.00) could be replaced by the β-D-glucoside of cucurbitacin E (1.31), cucurbitacin E (0.2), cucurbitacin D (0.12), cucurbitacin F (0.03), cucurbitacin A (0.03) or cucurbitacin C (0.02). Desmosterol and lanosterol were not reduced.

The enzyme requires Mn^{2+} for activity; Fe^{2+} and Zn^{2+} were less effective. (2)

Abbreviations

Cucurbitacin B — 2,16,20,25-tetrahydroxy-9-methyl-19-nor-lanosta-5,23-diene-3,11,22-trione 25-acetate.

References

1. Schabort, J.C. & Potgieter, D.J.J. (1968) BBA, 151, 47.
2. Schabort, J.C., Potgieter, D.J.J. & de Villiers, V. (1968) BBA, 151, 33.

ENOYL ACYL CARRIER PROTEIN REDUCTASE

(Acyl-acyl carrier protein: NADP oxidoreductase)

Acyl-ACP + NADP = 2,3-dehydroacyl-ACP + reduced NADP

Ref.

Equilibrium constant

The reaction (with either reduced NADP or reduced NAD) is essentially irreversible [R] with the enzyme from Escherichia coli as the catalyst. (1,2)

Specificity and kinetic properties

Enoyl-ACP reductase is a component of the fatty acid synthetase complex of E. coli (see EC 2.3.1.38). There are two reductases, one specific for reduced NADP and the other specific for reduced NAD (EC 1.3.1.9). The two enzymes have not been separated. The ratio reduced NADP activity to reduced NAD activity was 2.3 in a purified preparation (250 x). The preparation had no flavin. It did not have any transhydrogenase activity (EC 1.6.1.1). The reduced NADP specific enzyme is specific for ACP derivatives, thus crotonoyl-ACP could not be replaced by crotonoyl-CoA. This enzyme could utilize ACP derivatives of chain lengths C_4 to C_{16}; maximum rates were obtained with crotonoyl-ACP and 2-hexenoyl-ACP. The reduced NAD specific enzyme is active on both ACP and CoA derivatives (crotonoyl-ACP, Km = 40 µM; crotonoyl-CoA, Km = 2.5 mM) and it could utilize a number of 2,3-dehydroacyl derivatives. Maximum rates were obtained with 2-hexenoyl-ACP and 2-octenoyl-ACP. (1,2)

Abbreviations

ACP acyl carrier protein (see EC 2.3.1.39)

References

1. Weeks, G. & Wakil, S.J. (1969) Methods in Enzymology, 14, 66.
2. Weeks, G. & Wakil, S.J. (1968) JBC, 243, 1180.

_o_-COUMARATE REDUCTASE

(2-Hydroxyphenylpropionate: NAD oxidoreductase)

2-Hydroxyphenylpropionate + NAD =

_o_-coumarate + reduced NAD

Ref.

Equilibrium constant

The reaction was essentially irreversible [R] with the enzyme from _Arthrobacter_ sp. as catalyst. (1)

Specificity and kinetic properties

The enzyme from _Arthrobacter_ sp. (purified 20 x) is highly specific for _o_-coumarate (2-hydroxycinnamate). Thus, 4-hydroxycinnamate; 2,4-dihydroxycinnamate; 4-hydroxy-3-methoxycinnamate; 3,4-dihydroxycinnamate and cinnamate had less than 2% the activity observed with 2-hydroxycinnamate and the following were inactive: 2-nitrocinnamate; 3-nitrocinnamate; 4-nitrocinnamate; 3,5-dimethoxy-4-hydroxycinnamate and _o_-_cis_-coumarinic acid lactone. Reduced NADP had 30% of the activity observed with reduced NAD. The Km for 2-hydroxycinnamate was 17.3 μM (pH 7.3, Pi, 27°). (1)

The enzyme did not require added divalent metal ions (Mg^{2+}, Mn^{2+} or Ca^{2+}) for activity. (1)

References

1. Levy, C.C. & Weinstein, G.D. (1964) _B_, _3_, 1944.

BILIVERDIN REDUCTASE

(Bilirubin: NAD(P) oxidoreductase)

Bilirubin + NAD(P) = biliverdin + reduced NAD(P)

Ref.

Equilibrium constant

Only the reverse reaction has been observed with the enzyme from guinea-pig liver. Biliverdin can be reduced chemically (i.e. in the absence of the reductase) by cysteine, glutathione or ascorbate but not by reduced NAD. (1)

Specificity and Michaelis constants

source	substrate	K_m (M)	conditions	
Guinea-pig liver (10 x)	biliverdin	1×10^{-6}	pH 7.4, Pi	(1)
	reduced NAD[(a)]	2.4×10^{-4}	pH 7.4, Pi	(1)

(a) reduced NAD (1.0) could be replaced by reduced NADP (0.2).

Biliverdin reductase has also been purified from human liver. (1)

Inhibitors

The enzyme (guinea-pig liver) was inhibited by NAD or NADP. (1)

Light absorption data

Biliverdin has an absorption band at 670 nm (molar extinction coefficient = 12,100 $M^{-1}cm^{-1}$ at pH 7.4). Bilirubin does not absorb at 670 nm; instead it has an absorption band at 440 nm (molar extinction coefficient = 32,000 $M^{-1}cm^{-1}$ at pH 7.4). (1)

References

1. Singleton, J.W. & Laster, L. (1965) JBC, 240, 4780.

DIHYDRODIPICOLINATE REDUCTASE

(Δ'-Piperideine-2,6-dicarboxylate: NAD(P) oxidoreductase)

Tetrahydrodipicolinate + NAD(P) =

2,3-dihydrodipicolinate + reduced NAD(P)

Ref.

Equilibrium constant

The equilibrium position at neutral pH lies very far in the direction of tetrahydrodipicolinate formation. (1)

Molecular properties

source	value	conditions	
Escherichia coli	110,000	Sephadex G 200, pH 7.9	(1)

The enzyme is free of flavin compounds.

Specific activity

E. coli (1870 x)	300	2,3-dihydrodipicolinate (pH 7.5, Tris.)	(1)

Specificity and Michaelis constants

source	substrate	K_m (M)	conditions	
E. coli	2,3-dihydrodipicolinate	9×10^{-6}	pH 7.5, Tris.	(1)

With an enzyme preparation from E. coli (purified about 150 x) reduced NAD had 47% the activity of reduced NADP. (2)

The enzyme (E. coli) required neither flavin compounds nor metal ions for activity. (1)

Inhibitors

Dipicolinate and isophthalate are competitive inhibitors (with respect to 2,3-dihydrodipicolinate) of the enzyme from E. coli. (1)

References

1. Tamir, H. (1971) Methods in Enzymology, 17B, 134.
2. Farkas, W. & Gilvarg, C. (1965) JBC, 240, 4717.

3,5-CYCLOHEXADIENE-1,2-DIOL-1-CARBOXYLATE DEHYDROGENASE

(3,5-Cyclohexadiene-1,2-diol-1-carboxylate:
NAD oxidoreductase (decarboxylating))

3,5-Cyclohexadiene-1,2-diol-1-carboxylate + NAD =
catechol + CO_2 + reduced NAD

Ref.

Equilibrium constant

The reaction is essentially irreversible [F] with the enzyme from *Alcaligenes eutrophus* as catalyst. (1)

Molecular weight

source	value	conditions	Ref.
A. eutrophus	94,600 [4]	pH 8.0. In the presence or absence of NAD, reduced NAD or DHB	(1)

Specific activity

A. eutrophus	(190 x)	150	DHB (pH 8.0, Tris, 25°)	(1)

Specificity and Michaelis constants

source	substrate	K_m (M)	conditions	Ref.
A. eutrophus	DHB [a]	2×10^{-4}	pH 8.0, Tris, 25°	(1)
	NAD [b]	1.5×10^{-4}	pH 8.0, Tris, 25°	(1)

[a] the following could not replace DHB nor did they inhibit the reaction: benzoate; *cis*-3,5-cyclohexadiene-1,2-diol and malate.

[b] NADP was inactive although high concentrations were inhibitory.

The enzyme was not stimulated by TPP, biotin, CoA, pyridoxal phosphate, Mg^{2+}, Mn^{2+} or Zn^{2+}. (1)

Inhibitors

High concentrations of certain inorganic salts were inhibitory (e.g. Na_2SO_4, $(NH_4)_2SO_4$). (1)

Abbreviations

DHB 3,5-cyclohexadiene-1,2-diol-1-carboxylate. Also called dihydrodihydroxybenzoate.

References

1. Reiner, A.M. (1972) JBC, 247, 4960.

COPROPORPHYRINOGEN OXIDASE

(Coproporphyrinogen: oxygen oxidoreductase (decarboxylating))

Coproporphyrinogen-III + O_2 = protoporphyrinogen-III + $4CO_2$

Ref.

Molecular weight

source	value	conditions	Ref.
Rat liver mitochondria	80,000	Sephadex G 100, pH 7.4	(1)

Specificity and Michaelis constants

source	substrate	K_m (M)	conditions	Ref.
Rat liver (purified 18 x)	coproporphyrinogen-III[a]	3×10^{-5}	pH 7.4, Tris, 38°	(1)

[a] the following were inactive: coproporphyrin I, II and III; coproporphyrinogen I and II; uroporphyrinogen III; mesoporphyrin IX; mesoporphyrinogen IX and haematoporphyrin; 2,4-diacetyldeuteroporphyrin IX and their corresponding porphyrinogens.

The enzyme (rat liver) did not contain any identifiable prosthetic group (lipoate, flavin, haem group). It did not require pyridoxal phosphate or any metal ions as cofactor. Oxygen could not be replaced by any other electron acceptor tested. (1)

References

1. del C. Battle, A.M., Benson, A. & Rimington, C. (1965) BJ, 97, 731.

L-LEUCINE DEHYDROGENASE

(L-Leucine: NAD oxidoreductase (deaminating))

L-Leucine + H_2O + NAD = 2-oxo-4-methyl-pentanoate + NH_3 + reduced NAD

Ref.

Equilibrium constant

$$\frac{[\text{2-oxo-4-methyl-pentanoate}]\,[NH_4^+]\,[\text{reduced NAD}]\,[H^+]}{[\text{L-leucine}]\,[\text{NAD}]} = 1.11 \times 10^{-13}\ M^2 \quad (\text{glycine, } 25^o)$$ (1)

Molecular weight

source	value	Ref.
Bacillus sphaericus	280,000	(2)

Specific activity

			Ref.
B. sphaericus (64 x)	20	L-leucine (pH 11.3, glycine, 25°)	(2)

Specificity and Michaelis constants

source	substrate	second substrate	V relative	K_m (M)
B. sphaericus	L-leucine [a]	NAD	-	1.2×10^{-3}
	NAD	L-leucine	-	4.0×10^{-4}
B. subtilis	L-leucine [b]	NAD	1.00	6.2×10^{-3}
	L-valine	NAD	0.80	2.0×10^{-2}
	L-isoleucine	NAD	0.64	5.2×10^{-3}
	DL-norleucine	NAD	0.24	-
	DL-norvaline	NAD	0.21	-
	2-oxoisovalerate	reduced NAD	-	2.2×10^{-3}
	2-oxoisocaproate	reduced NAD	-	3.3×10^{-3}
	NAD	L-leucine	-	1.6×10^{-4}
	reduced NAD	2-oxoisocaproate	-	1.2×10^{-4}
	NH_4^+	2-oxoisocaproate	-	1.3×10^{-2}

---

[a] from Ref. 2. (pH 11.3, glycine, 25°). The following were also active: L-valine; L-isoleucine; L-norleucine and L-norvaline but not D-leucine, D-valine, L-alanine or L-glutamate.

[b] from Ref. 1 & 3. (forward reaction: pH 11.3, glycine, 25°; reverse reaction: pH 9.3, Tris, 25°). L-Alanine; DL-serine; DL-threonine; L-glutamate, and DL-aspartate were inactive in the forward reaction and 2-oxobutyrate; 2-aminobutyrate; 2-oxoglutarate and pyruvate in the reverse reaction.

Abbreviations

2-oxoisocaproate 2-oxo-4-methyl-pentanoate

References

1. Sanwal, B.D. & Zink, M.W. (1961) ABB, 94, 430.
2. Soda, K., Misono, H., Mori, K. & Sakato, H. (1971) BBRC, 44, 931.
3. Zink, M.W. & Sanwal, B.D. (1962) ABB, 99, 72.

L-erythro-3,5-DIAMINOHEXANOATE DEHYDROGENASE

(L-erythro-3,5-Diaminohexanoate: NAD oxidoreductase (deaminating))

L-erythro-3,5-Diaminohexanoate + NAD + H_2O =

3-oxo-5-aminohexanoate + NH_3 + reduced NAD

Ref.

Equilibrium constant

$$\frac{[KAH]\ [NH_4^+]\ [\text{reduced NAD}]\ [H^+]}{[\text{L-erythro-3,5-DAH}]\ [NAD]} = 4 \times 10^{-10}\ M^2 \quad (Pi,\ 26^o)$$ (1)

Molecular weight

source	value	conditions	
Clostridium SB4	140,000 [4] (a)	Sephadex G 150, pH 6.5	(1)

(a) at high pH or in solutions of low ionic strength a dimer of molecular weight 68,000 was formed; this possessed 30% the activity of the tetramer.

Specific activity

Clostridium SB4 (30 x) 34 L-erythro-3,5-DAH (pH 8.9, ethanolamine-HCl, 25°) (1)

Specificity and kinetic properties

source	substrate	K_m (M)	conditions	
Clostridium SB4 (a)	KAH (b)	2.6×10^{-4}	pH 7.0, Pi, 23°	(1)
	reduced NAD	7.4×10^{-5}	pH 7.0, Pi, 23°	(1)
	NH_4Cl (c)	1.4×10^{-1}	pH 7.0, Pi, 23°	(1)
	L-erythro-3,5-DAH (d)	1.8×10^{-4}	pH 6.8, Na-EDTA	(1)
	NAD (e)	2.8×10^{-4}	pH 6.8, Na-EDTA	(1)

(a) The data are for the tetrameric form of the enzyme.
(b) Pyruvate, oxaloacetate and acetoacetate were inactive.
(c) Methylamine and hydroxylamine were inactive.
(d) D-erythro-3,5-DAH. DL-threo-3,5-DAH; L-lysine; L-β-lysine; β-alanine and L-3-aminobutyrate were inactive.
(e) NAD (1.00) could be replaced by acetylpyridine NAD (0.70) or NADP (0.013).

Inhibitors

The enzyme was inhibited by reduced NAD (Ki = 4×10^{-6}M, C(NAD)) and D-erythro-3,5-DAH (Ki = 5×10^{-4}; NC(L-erythro-3,5-DAH)). (1)

Abbreviations

DAH	diaminohexanoate
KAH	3-oxo-5-aminohexanoate

References

1. Baker, J.J., Jeng, I. & Barker, H.A. (1972) JBC, 247, 7724.

2,4-DIAMINOPENTANOATE DEHYDROGENASE

(2,4-Diaminopentanoate: NAD(P) oxidoreductase (deaminating))

2,4-Diaminopentanoate + H_2O + NAD(P) =

2-amino-4-oxopentanoate + reduced NAD(P) + NH_3

Ref.

Equilibrium constant

$$\frac{[\text{2-amino-4-oxopentanoate}]\ [\text{reduced NAD}]\ [NH_4^+]\ [H^+]}{[\text{2,4-diaminopentanoate}]\ [\text{NAD}]\ [H_2O]} = 1.04 \times 10^{-14}\ \text{M}\ \ (\text{PPi},\ 28°)$$ (3)

Molecular weight

source	value	conditions	Ref.
Clostridium sticklandii	72,000 [2]	pH 7.0; gel electrophoresis (SDS); amino acid composition	(2)

Specific activity

C. sticklandii	(51 x)	272.5	2,4-diaminopentanoate (with NADP; pH 8.5, Tris, 25°)	(2)

Specificity and catalytic properties

source	substrate[(a)]	V relative	K_m (mM)	Ref.
C. sticklandii (100 x)	2,4-diaminopentanoate[(b)]	1.000	1.8	(1,3)
	NAD	-	1.8	(1)
	NADP	-	0.15	(1)
	2,5-diaminohexanoate	0.001	2.5	(1)
	NAD	-	3.3	(1)
	NADP	-	0.28	(1)

(a) pH 9.6, PPi.

(b) the following compounds were inactive: L-arginine; L-alanine; β-alanine; L-citrulline; L-2,4-diaminobutyrate; 4-aminopentanoate; L-lysine; L-ornithine; L-β-ornithine, L-norvaline and 3,5-diamino-hexanoate.

The dehydrogenase does not require Mg^{2+} or ATP or any other cofactor tested for activity. It was not inhibited by EDTA. With 2,5-diamino-hexanoate as the substrate, the product of the reaction, 2-amino-5-oxohexanoate, cyclizes nonenzymically to Δ'-pyrroline-2-methyl-5-carboxylate. (1,3)

References

1. Stadtman, T.C. (1973) Advances in Enzymology, 38, 441.
2. Somack, R. & Costilow, R.N. (1973) JBC, 248, 385.
3. Tsuda, Y. & Friedmann, H.C. (1970) JBC, 245, 5914.

TYRAMINE OXIDASE

(Tyramine: oxygen oxidoreductase (deaminating))

Tyramine + O_2 + H_2O =

4-hydroxyphenylacetaldehyde + NH_3 + H_2O_2

Molecular weight

source	value	conditions	Ref.
Sarcina lutea	129,000	pH 7.0; in the presence of dithiothreitol	(1)

The enzyme contains 2 moles of FAD per 129,000 daltons. No metal could be detected. (1)

Specific activity

S. lutea	(1430 x)	41.2	tyramine (pH 7.0, Pi, 30°)	(1)

Specificity

Tyramine oxidase from S. lutea oxidized tyramine and dopamine about equally well but a number of other amines and amino acids were not attacked (2). Further details of the kinetic properties of the enzyme will be found in Ref. 3.

References

1. Kumagai, H., Matsui, H., Ogata, K. & Yamada, H. (1969) BBA, 171, 1.
2. Yamada, H., Uwajima, T., Kumagai, H., Watanabe, M. & Ogata, K. (1967) BBRC, 27, 350.
3. Yamada, H., Uwajima, T., Kumagai, H., Watanabe, M. & Ogata, K. (1967) Agr. Biol. Chem. (Tokyo), 31, 890.

PUTRESCINE OXIDASE

(1,4-Diaminobutane: oxygen oxidoreductase (deaminating))

Putrescine + O_2 + H_2O = 4-aminobutyraldehyde + NH_3 + H_2O_2

Ref.

Equilibrium constant

The reaction is irreversible.[F]. 4-Aminobutyraldehyde condenses non-enzymically to Δ'-pyrroline. (1)

Molecular weight

source	value	conditions	Ref.
Micrococcus rubens	88,000 [2]	pH 7.0; gel electrophoresis (SDS); amino acid composition	(2)

The enzyme contains 1 mole FAD per 82,000 daltons.

Specific activity

				Ref.
M. rubens	(250 x)	59	putrescine (pH 8.6, Tris, 20°)	(2)
M. rubens	(230 x)	9	putrescine (pH 8.0, Tris, 30°)	(1)

Specificity and Michaelis constants

source	substrate	K_m (M)	conditions	Ref.
M. rubens	putrescine (a)	3.8×10^{-5}	pH 8.6, Tris, 20°	(2)
	oxygen	1.24×10^{-4}	pH 8.6, Tris, 20°	(2)
	putrescine (b)	2.3×10^{-4}	pH 8.0, Tris, 30°	(1)
	spermidine (c)	2.3×10^{-4}	pH 8.0, Tris, 30°	(1)

(a) putrescine (1.00) could be replaced by cadaverine (0.07) and spermidine. The enzyme was inert towards sulphite but it had a low oxidase activity towards mercaptoethanol.

(b) cadaverine was also oxidized but histamine, agmatine and spermine were inactive.

(c) spermidine was oxidized to 4-aminobutyraldehyde and 1,3-diaminopropane without the formation of NH_3.

Inhibitors

source	inhibitor	type	K_i (M)	conditions	Ref.
M. rubens	cadaverine	NC(putrescine)	2.8×10^{-4}	pH 8.6, Tris, 20°	(2)

References

1. Yamada, H. (1971) Methods in Enzymology, 17B, 726.
2. DeSa, R.J. (1972) JBC, 247, 5527.

D-AMINO ACID DEHYDROGENASE

(D-Amino-acid: (acceptor) oxidoreductase (deaminating))

A D-amino acid + H_2O + acceptor =

a 2-oxo acid + NH_3 + reduced acceptor

Ref.

Specificity and Michaelis constants

Pseudomonas fluorescens contains two distinct D-amino acid dehydrogenases. Both contain tightly bound FAD. One (enzyme MB, purified 39 x) utilized methylene blue as electron acceptor and the other (enzyme DCIP, purified 66 x) 2,6-dichlorophenolindophenol. Neither enzyme could utilize oxygen, NAD, NADP, FMN or FAD. They did not require added metal ions for activity. The enzymes oxidized a number of D-amino acids, including D-kynurenine, but excluding D-aspartate and D-glutamate. For both enzymes the Km values for D-amino acids were in the region of $3\text{-}5 \times 10^{-4}$M. (1)

source	enzyme	substrate	K_m (M)	
P. fluorescens	MB	methylene blue[a]	8×10^{-6}	(1)
	DCIP	2,6-dichlorophenol indophenol[b]	7×10^{-5}	(1)

(a) methylene blue (1.00) could be replaced by pyocyanine (0.95); phenazine methosulphate (0.71) or ferricyanide (0.43) but not by 2,6-dichlorophenolindophenol. (Conditions: pH 7.5, Pi, 37°).

(b) 2,6-dichlorophenolindophenol (1.00) could be replaced by ferricyanide (0.49); pyocyanine (0.97) or phenazine methosulphate (1.31) but not by methylene blue. (Conditions: pH 7.5, Pi, 37°)

Light absorption data

See EC 1.1.99.10 and 1.2.99.aa

References

1. Tsukada, K. (1966) *JBC*, *241*, 4522.

GLYCINE-CYTOCHROME c REDUCTASE

(Glycine: cytochrome c oxidoreductase (deaminating))

Glycine + H_2O + 2 ferricytochrome c =

glyoxylate + NH_3 + 2 ferrocytochrome c

Ref.

Equilibrium constant

With the enzyme from Nitrobacter agilis as the catalyst the reaction is essentially irreversible [F]. (1)

Molecular weight

source	value	conditions	
N. agilis	69,000	Sephadex G 200 (pH 7.8) and sucrose gradient	(1)

Specificity and Michaelis constants

source	substrate	K_m (M)	conditions	
N. agilis (purified 250 x)	glycine	8×10^{-4}	pH 6.5, Pi.	(1)
	cytochrome c	1.6×10^{-5}	pH 6.5, Pi.	(1)

The enzyme (N. agilis) is highly specific for glycine which could not be replaced by any of the D- or L-forms of the protein amino acids nor by glyoxylate; glycylglycine; glycollate; 2-oxoglutarate; pyruvate; acetate; formate; betaine or sarcosine. It is also highly specific for oxidized cytochrome c which could not be replaced by ferricyanide; 2,6-dichloroindophenol; 2,3'6-trichloroindophenol; methylene blue; benzyl viologen; phenazine, methosulphate, NAD, FAD, FMN or molecular oxygen. (1)

The enzyme does not require any added cofactor for activity. (1)

Light absorption data

Cytochrome c has an absorption band at 550 nm (molar extinction coefficient, reduced cytochrome c minus oxidized cytochrome c = 21,000 $M^{-1}cm^{-1}$ at pH 6.5). (1)

References

1. Sanders, H.K., Becker, G.E. & Nason, A. (1972) JBC, 247, 2015.

FORMYLTETRAHYDROFOLATE DEHYDROGENASE

((-)-10-Formyltetrahydrofolate: NADP oxidoreductase)

(-)-10-Formyltetrahydrofolate + NADP + H_2O =

(-)-tetrahydrofolate + CO_2 + reduced NADP

Ref.

Equilibrium constant

The reaction is irreversible [F] with the enzyme from pig liver as catalyst. From thermodynamic data an equilibrium constant of 1.6×10^8 has been calculated. (1)

Molecular weight

source	value	conditions	
Pig liver	320,000	gel filtration	(1)

Specificity and Michaelis constants

source	substrate	K_m (M)	conditions	
Pig liver (purified 100 x)	NADP[a]	3.5×10^{-6}	pH 7.7, Tris, 30°	(1)
	(-)-10-formyltetra-hydrofolate[b]	8.2×10^{-6}	pH 7.7, Tris, 30°	(1)

(a) NAD was inactive.
(b) (-)-10-formyltetrahydrofolate could not be replaced by (+)-10-formyltetrahydrofolate or 5-formyltetrahydrofolate.

In the absence of NADP the enzyme (pig liver) catalyzes a hydrolytic cleavage of 10-formyltetrahydrofolate to formate and tetrahydrofolate at 15-30% the rate of the oxidative reaction. (1)

Inhibitors

source	inhibitor[a]	K_i (M)	conditions	
Pig liver	(-)-tetrahydrofolate	1×10^{-6}	pH 7.7, Tris, 30°	(1)
	(+)-tetrahydrofolate	1×10^{-5}	pH 7.7, Tris, 30°	(1)

(a) C((-)-10-formyltetrahydrofolate)

The following compounds did not inhibit: (+)-10-formyltetrahydrofolate; 5-methyltetrahydrofolate; 5-formyltetrahydrofolate; folate; aminopterin and tetrahydroaminopterin. (1)

References

1. Kutzbach, C. & Stokstad, E.L.R. (1971) Methods in Enzymology, 18B, 793.

SACCHAROPINE DEHYDROGENASE

(N^6-(1,3-Dicarboxypropyl)-L-lysine: NAD oxidoreductase (L-lysine forming))

Saccharopine + NAD + H_2O =

2-oxoglutarate + L-lysine + reduced NAD

Ref.

Equilibrium constant

The reaction is reversible. (1)

Molecular weight

source	value	conditions	Ref.
Bakers' yeast	49,000	sucrose density gradient	(1)

The enzyme has a light absorption band at about 420 nm. (1)

Specific activity

Bakers' yeast (708 x) 23 L-lysine (pH 7.0, Pi, 25°) (1)

Specificity and Michaelis constants

source	substrate	K_m (M)	conditions	Ref.
Bakers' yeast	L-lysine (a)	1.2×10^{-2}	pH 7.0, Pi, 25°	(1)
	2-oxoglutarate (b)	4.4×10^{-4}	pH 7.0, Pi, 25°	(1)
	reduced NAD (c)	4.6×10^{-5}	pH 7.0, Pi, 25°	(1)

(a) L-ornithine; DL-N^2-acetyllysine; DL-N^6-acetyllysine; N^6-methyllysine and NH_4^+ were inactive.
(b) pyruvate; oxobutyrate; oxaloacetate and 2-oxoadipate were inactive.
(c) reduced NAD (1.00) could be replaced by reduced NADP (0.05) but in the forward reaction NADP was inactive in the place of NAD.

Inhibitors

A number of α-amino acids [C(L-lysine); UC(reduced NAD or 2-oxoglutarate)] and α-oxoacids [NC(L-lysine, reduced NAD or 2-oxoglutarate)] were inhibitors. Of the amino acids tested those having 5 or 6 carbon atoms were the most effective inhibitors. Dicarboxylic acids such as L-glutamate, L-aspartate or 2-aminoadipate did not inhibit. (2)

References

1. Saunders, P.P. & Broquist, H.P. (1966) JBC, 241, 3435.
2. Fujioka, M. & Nakatani, Y. (1972) EJB, 25, 301.

AMINOADIPATE SEMIALDEHYDE-GLUTAMATE REDUCTASE

(N^6-(1,3-Dicarboxypropyl)-L-lysine: NAD(P) oxidoreductase (L-glutamate-forming))

Saccharopine + NAD(P) + H_2O =

L-2-aminoadipate 6-semialdehyde + L-glutamate + reduced NAD(P)

Ref.

Equilibrium constant

The reaction is reversible. (1)

Molecular weight

source	value	conditions	Ref.
Bakers' yeast	73,000	sucrose density gradient	(1)

Michaelis constants

source	substrate	K_m (M)	conditions	Ref.
Bakers' yeast	saccharopine	9.2×10^{-4}	pH 9.5, glycine, 25°	(1)
(purified 129 x)	NADP	2.2×10^{-5}	pH 9.5, glycine, 25°	(1)
Human liver	saccharopine[a]	5×10^{-4}	pH 8.8, Tris, 23°	(2)
(purified 122 x)	NAD[b]	4×10^{-4}	pH 8.8, Tris, 23°	(2)

(a) saccharopine could not be replaced by 2-oxoglutarate; DL-2-aminoadipate; L-aspartate; L-lysine; N^6-acetyl-L-lysine; DL-2-amino-6-hydroxycaproate; 5-aminovalerate; DL-pipecolate or 6-aminocaproate.

(b) NAD (1.00) could be replaced by NADP (0.05).

With the yeast enzyme, NAD and NADP were equally effective in the forward reaction but in the reverse reaction reduced NAD had only 10% the activity of reduced NADP. (1)

References

1. Jones, E.E. & Broquist, H.P. (1966) JBC, 241, 3430.
2. Hutzler, J. & Dancis, J. (1970) BBA, 206, 205.

NICOTINATE DEHYDROGENASE

(Nicotinate: NADP 6-oxidoreductase (hydroxylating))

Nicotinate + H_2O + NADP = 6-hydroxynicotinate + reduced NADP

Ref.

Equilibrium constant

The forward reaction is strongly favoured. (1)

Molecular weight

source	value	conditions	Ref.
Clostridium sp.	300,000	pH 7.5	(1)

The enzyme contains at least 11 moles of non-haem iron, 6 moles of labile sulphide and 1.5 moles of FAD per 300,000 daltons. (1)

Specific activity

Clostridium sp. (24 x)	28.8	nicotinate (pH 7.5, Pi, 23°. Anaerobic conditions were maintained by the presence of 3 mM $FeSO_4$, 10 mM glutathione and 20 mM PPi).	(1)

Specificity and Michaelis constants

source	substrate	K_m (M)	conditions	
Clostridium sp.	nicotinate	1.1×10^{-4}	pH 7.5, Pi, 23° anaerobic	(1)
	NADP	2.8×10^{-5}	pH 7.5, Pi, 23° anaerobic	(1)

In addition to its hydroxylase activity (1.0), the enzyme also catalyzes the reduction of artificial electron acceptors by reduced NADP (diaphorase activity; 4.0). It also possesses reduced NADP oxidase activity (1.0). (1)

References

1. Holcenberg, J.S. & Stadtman, E.R. (1969) JBC, 244, 1194.

6-HYDROXY-L-NICOTINE OXIDASE

(6-Hydroxy-L-nicotine: oxygen oxidoreductase)

6-Hydroxy-L-nicotine + H_2O + O_2 =

[6-hydroxypyridyl(3)]-(4-N-methylaminopropyl)-ketone + H_2O_2

Ref.

Molecular weight

source	value	Ref.
Arthrobacter oxidans	93,000 [2]	(1)

The enzyme contains 2 moles FAD per 93,000 daltons. (1)

Specific activity

			Ref.
A. oxidans (62 x)	4.5	L-6-hydroxynicotine (pH 7.5, Tris, 22°)	(1)

Specificity and kinetic properties

source	substrate	K_m (M)	conditions	Ref.
A. oxidans	L-6-hydroxynicotine [a]	2×10^{-5}	pH 7.5, Tris, 22°	(2)

(a) D-6-hydroxynicotine was a competitive inhibitor (L-6-hydroxynicotine) but not a substrate.

Light absorption data

See EC 1.5.3.6

References

1. Dai, V.D., Decker, K. & Sund, H. (1968) EJB, 4, 95.
2. Decker, K. & Bleeg, H. (1965) BBA, 105, 313.

6-HYDROXY-D-NICOTINE OXIDASE

(6-Hydroxy-D-nicotine: oxygen oxidoreductase)

6-Hydroxy-D-nicotine + H_2O + O_2 =

[6-hydroxypyridyl(3)]-(4-N-methylaminopropyl)-ketone + H_2O_2

Molecular weight

source	value	conditions	Ref.
Arthrobacter oxidans	53,000 [1]	pH 7.5, with or without guanidine-HCl	(1)

The enzyme contains 1 mole of FAD per 53,000 daltons.

Specific activity

A. oxidans (270 x)	22.5	D-6-hydroxynicotine (pH 9.2, glycine, 30°)	(1)

Specificity and Michaelis constants

source	substrate	V relative	K_m (M)	Ref.
A. oxidans	D-6-hydroxynicotine (a)	1.00	5×10^{-5}	(1)
	D-6-aminonicotine	1.43	2×10^{-4}	(1)

(a) the following were inactive: L-6-hydroxynicotine; DL-2-hydroxynicotine; D-nicotine; D,L-2-aminonicotine; D,L-anabasine; D-proline; L-nicotine; L-nornicotine and N-methylpyrrolidine. Conditions: pH 9.2, glycine, 30°.

The enzyme (A. oxidans) could utilize methylene blue or 2,6-dichlorophenolindophenol but not ferricyanide, cytochrome c or benzylviologen as electron acceptor. (1)

Inhibitors

source	inhibitor (a)	K_i (M)
A. oxidans	L-6-hydroxynicotine	1.5×10^{-3}
	D,L-2-hydroxynicotine	1.7×10^{-3}
	[6-hydroxypyridyl(3)-]-(4-N-methylamino propyl)-ketone	5.0×10^{-5}
	3-(4-aminobutyl)-pyridine	7×10^{-4}

(a) C(D-6-hydroxynicotine). Conditions: pH 9.2, glycine, 30°. (1)

Light absorption data

[6-Hydroxypyridyl(3)-]-(4-N-methylaminopropyl)-ketone has an absorption band at 334 nm (molar extinction coefficient = 20,700 $M^{-1}cm^{-1}$ at pH 9.2). (1)

References

1. Brühmüller, M., Möhler, H. & Decker, K. (1972) EJB, 29, 143.
   Brühmüller, M. & Decker, K. (1973) EJB, 37, 256.

SPERMIDINE DEHYDROGENASE

(1-(3-Aminopropylamino)-4-aminobutane:
acceptor (donor-cleaving))

Spermidine + oxidized acceptor =
1,3-diaminopropane + 4-aminobutyraldehyde
+ reduced acceptor

Ref.

Equilibrium constant

The reaction is irreversible [F]. 4-Aminobutyraldehyde condenses nonenzymically to Δ'-pyrroline. (1)

Molecular weight

source	value	conditions	
Sarratia marcescens	76,000 [1]	pH 7.2	(2)

The enzyme contains 1 mole of iron-protoporphyrin IX and 1 mole of FAD per 76,000 daltons. The amino acid composition of the enzyme has been determined. (2)

Specific activity

S. marcescens	(4670 x)	420	spermidine (with ferricyanide as the acceptor; pH 7.2, Pi, 37°)	(2)

Specificity and Michaelis constants

source	substrate	relative velocity	K_m (M)	conditions	
S. marcescens	spermidine[(a)]	1.0	$<5 \times 10^{-7}$	pH 7.2, Pi, 37°	(2)
	spermine	0.17	2×10^{-5}	pH 7.2, Pi, 37°	(2)

(a) spermidine (1.00) could be replaced by monoacetylspermidine B (0.60); N,N-bis(3-aminopropyl)-1,3-propanediamine (0.48); N-(3-aminopropyl)-1,3-propanediamine (0.36) or N-(3-hydroxypropyl)-1,4-diaminobutane (0.07) but not by monoacetylspermidine.

Spermidine dehydrogenase (S. marcescens) can not utilize oxygen directly; instead it requires an added electron acceptor such as ferricyanide, phenazine methosulphate, dichloroindophenol or cytochrome c. NAD; ferredoxin and coenzyme Q were inactive. (1,2)

Light absorption data

The molar extinction coefficient of potassium ferricyanide at 400 nm is 960 $M^{-1}cm^{-1}$ (pH 7.0) and that of 2,6-dichloroindophenol at 600 nm is 18,400 $M^{-1}cm^{-1}$ (pH 6.5). (2)

References

1. Tabor, H. & Tabor, C.W. (1972) Advances in Enzymology, 36, 225.
2. Tabor, C.W. & Kellogg, P.D. (1970) JBC, 245, 5424.

N-METHYLGLUTAMATE DEHYDROGENASE

(N-Methyl-L-glutamate:
(acceptor) oxidoreductase (demethylating))

N-Methyl-L-glutamate + H_2O + acceptor =
L-glutamate + formaldehyde + reduced acceptor

Ref.

Specificity and Michaelis constants

source	substrate	V relative	K_m (M)
Pseudomonas MA (purified 5 x)	N-methyl-L-glutamate[(a)]	1.00	4.3×10^{-5}
	N-methyl-DL-valine	1.08	2.0×10^{-4}
	N-methyl-L-isoleucine	1.07	2.0×10^{-4}
	N-methyl-L-phenylalanine	1.08	2.5×10^{-3}
	N-methyl-L-aspartate	1.04	1.3×10^{-3}
	N-methyl-L-alanine	1.03	6.0×10^{-3}
	N-methyl-L-serine	1.03	3.6×10^{-2}
	N-methyl-L-glycine	1.00	1.0×10^{-1}
	potassium ferricyanide[(b)]	1.00	1.4×10^{-3}
	2,6-dichlorophenolindo-phenol	5.46	5.4×10^{-5}
	phenazine methosulphate	16.40	1.1×10^{-5}

(a) with 2,6-dichlorophenolindophenol as electron acceptor. N,N-Dimethylglutamate and N-methyl-D-alanine were inactive. Conditions: pH 7.4, Pi, 30° (Ref. 1).

(b) with N-methylglutamate as electron donor.

The reaction mechanism of the enzyme is discussed in Ref. 1.

Inhibitors

A number of carboxylic acids were competitive inhibitors (N-methyl-glutamate). L-Glutamate and formaldehyde, products of the reaction, did not inhibit. (1)

Light absorption data

See EC 1.2.99.a and EC 1.1.99.10

References

1. Hersh, L.B., Stark, M.J., Worthen, S. & Fiero, M.K. (1972) ABB, 150, 219.

REDUCED NADP-CYTOCHROME c_2 REDUCTASE

(Reduced NADP: ferricytochrome c_2 oxidoreductase)

Reduced NADP + 2 ferricytochrome c_2 =

NADP + 2 ferrocytochrome c_2

Ref.

Molecular properties

source	value	conditions	
Rhodopseudomonas spheroides	43,000	see below	(1)

The enzyme contains 1 mole of FAD per 43,000 daltons. It contains no metal component. When the enzyme was subjected to gel filtration (Sephadex G 100, pH 8.0), or ultracentrifugation, 3 molecular forms were obtained. (Molecular weights 67,000; 43,000 and 20,000 daltons) (1)

Specificity and Michaelis constants

source	substrate	relative velocity	K_m (M)	
R. spheroides (purified 115 x)	cytochrome c_2 [a]	1.00	3.7×10^{-5}	(1)
	2,6-dichloroindophenol	2.17	1.25×10^{-5}	(1)
	ferricyanide	0.25	1.25×10^{-4}	(1)

(a) with reduced NADP as the electron donor. Conditions: pH 7.5, Tris, 25°. Horse heart cytochrome c had 5% the activity of cytochrome c_2 and R. spheroides cytochrome c-553 was inactive.

The enzyme (R. spheroides) could not utilized reduced NAD in the place of reduced NADP nor did it reduce NADP with reduced benzyl viologen as electron donor. (1)

Inhibitors

The enzyme (R. spheroides) was inhibited by reduced NAD (Ki = 5.5×10^{-5}M; C(reduced NADP)) but not by NAD or NADP. Thyroxine was a very effective inhibitor. No protection towards thyroxine was afforded by preincubating the enzyme with reduced NADP. (1)

Light absorption data

substance	wavelength	molar extinction coefficient [a]
cytochrome c_2	550 nm	24,600 $M^{-1}cm^{-1}$
horse heart cytochrome c	550 nm	27,700 $M^{-1}cm^{-1}$
2,6-dichloroindophenol	600 nm	20,000 $M^{-1}cm^{-1}$
potassium ferricyanide	420 nm	100 $M^{-1}cm^{-1}$

(a) at pH 7.5 (Ref. 1).

References

1. Sabo, D.J. & Orlando, J.A. (1968) JBC, 243, 3742.

CoAS-SGLUTATHIONE REDUCTASE (REDUCED NADP)

(Reduced NADP: CoAS-Sglutathione oxidoreductase)

Reduced NADP + CoAS-Sglutathione = NADP + CoA + glutathione

Ref.

Equilibrium constant

The reaction is irreversible [F] with the enzyme from *Saccharomyces cerevisiae* as catalyst. (1)

Molecular weight

source	value	conditions	
S. cerevisiae	108,000	sucrose gradient	(1)

Specificity and Michaelis constants

source	substrate	K_m (M)	conditions	
S. cerevisiae (purified 143 x)	CoAS-Sgluta-thione (a)	2×10^{-4}	pH 5.5, Pi, 24°	(1)

(a) the following were inactive: CoAS-Scysteine; glutathione S-S-cysteine and cysteine S-S-cysteine.

Inhibitors

The enzyme (*S. cerevisiae*) was inhibited by Pi. (1)

Abbreviations

L-glutathione (reduced) L-γ-glutamyl-L-cysteinylglycine

References

1. Ondarza, R.N., Abney, R. & López-Colomé, A.M. (1969) BBA, 191, 239.

MONODEHYDROASCORBATE REDUCTASE (reduced NAD)

(Reduced NAD : monodehydroascorbate oxidoreductase)

Reduced NAD + 2 monodehydroascorbate = NAD + 2 ascorbate

Molecular weight

source	value	conditions	Ref.
Neurospora crassa	66,000	Sephadex G 100, pH 8.0	(1)

Specific activity

				Ref.
N. crassa	(760 x)	40	monodehydroascorbate (pH 7.0, Tris, 30°)	(1)

Specificity and catalytic properties

source	substrate	K_m (μM)	conditions	Ref.
N. crassa	reduced NAD[a]	12	pH 7.0, Tris, 30°	(1)
	monodehydroascorbate[b]	1.2	pH 7.0, Tris, 30°	(1)
Rat liver	reduced NAD[a]	25	pH 6.6, Pi, 30°	(2)
microsomes	monodehydroascorbate	2.2	pH 6.6, Pi, 30°	(2)

(a) reduced NADP was inactive.

(b) the enzyme had low activities towards 2,6-dichlorophenolindophenol, cytochrome b_5 and cytochrome c.

References

1. Schulze, H-U., Schott, H-H. & Staudinger, H. (1972) Hoppe-Seylers' Z. Physiol. Chem., 353, 1931.
2. Schulze, H-U. & Staudinger, H. (1971) Hoppe-Seylers' Z. Physiol. Chem., 352, 309.

FERREDOXIN-NADP REDUCTASE

(Reduced NADP : ferredoxin oxidoreductase)

Reduced NADP + oxidized ferredoxin = NADP + reduced ferredoxin

Ref.

Equilibrium constant

The reaction is reversible. (1)

Molecular properties

source	value	conditions	
Spinach chloroplasts	40,000	ultracentrifugation	(2)

The enzyme contains one mole of FAD per 40,000 daltons. It forms 1 : 1 complexes with ferredoxin and with NADP. With spinach ferredoxin and at a low ionic strength, a K_D of 7.6 μM was obtained; at high ionic strengths the reductase-ferredoxin complex dissociated. There are conflicting views as to the physiological importance of this complex (see Ref. 3 for a discussion). The binding of NADP to the enzyme was also dependent on ionic strength; at 0.23, K_D = 260 μM and at 0.01, K_D = 51 μM. The presence of ferredoxin or NADP (but not NAD, 2'-AMP or NMN) caused a perturbation in the visible spectrum of the reductase. (2,3)

Specificity and catalytic properties

Ferredoxin-NADP reductase has been highly purified from spinach chloroplasts. The enzyme has also reduced NADP-diaphorase and reduced NADP-NAD transhydrogenase activities. Thus, it catalyzes the oxidation of reduced NADP by ferredoxin, NAD, menadione, ferricyanide, indophenol dyes, FMN, FAD, pyocyanine, plastocyanine or cytochrome f (but not cytochrome c). In the presence of molecular oxygen, the superoxide anion radical is produced. (1,2,3,4)

The catalytic and molecular properties of the enzyme are effected by ionic strength (also see molecular properties, above). Thus, at low ionic strengths, ferredoxin inhibits the diaphorase activity of the enzyme (partially competitive (reduced NADP); Ki = 2×10^{-9}M) whereas at high ionic strengths activation occurs. (2,3)

The mechanism of the reaction catalyzed by ferredoxin-NADP reductase has been investigated. (2,3,5,4)

Ferredoxin

Ferredoxins are non-haem iron containing proteins, with a very low redox potential, which occur in higher plants, algae and certain bacteria (photosynthetic and anaerobic). The sequences of a number of ferredoxins have been determined. Their molecular weights vary considerably; thus spinach ferredoxin has a molecular weight of 10,482 whereas ferredoxin from *Clostridium pasteurianum* has a molecular

Ref.

Ferredoxin continued

weight of 5,499 daltons. The ferredoxins contain 4 to 7 iron atoms per mole of protein and they absorb in the visible region of the light spectrum: spinach ferredoxin has an absorption band at 420 nm (molar extinction coefficient = 9,400 $M^{-1}cm^{-1}$) and ferredoxin from *C. pasteurianum* has an absorption band at 390 nm (molar extinction coefficient = 24,500 $M^{-1}cm^{-1}$). (2,6)

References

1. Shin, M., Tagawa, K. & Arnon, D.I. (1963) BZ, 338, 84.
2. Foust, G.P., Mayhew, S.G. & Massey, V. (1969) JBC, 244, 964.
3. Nakamura, S. & Kimura, T. (1971) JBC, 246, 6235.
4. Nelson, N. & Neumann, J. (1969) JBC, 244, 1926.
5. Nakamura, S. & Kimura, T. (1972) JBC, 247, 6462.
6. Dayhoff, M.O. (1972) Atlas of Protein Sequence and Structure, Vol. 5, National Biomedical Research Foundation: Washington, p D35.

RUBREDOXIN REDUCTASE

(Reduced NAD : rubredoxin oxidoreductase)

Reduced NAD + oxidized rubredoxin = NAD + reduced rubredoxin

Ref.

Molecular weight

source	value	conditions	Ref.
Pseudomonas oleovorans	55,000 [1]	pH 7.3; gel electrophoresis (SDS); amino acid composition.	(1)

The enzyme contains 1 mole FAD (but no metal or other cofactor) per 55,000 daltons. (1)

Specific activity

P. oleovorans (270 x) 82 reduced NAD (pH 7.8, Tris, 30°) (1)

Specificity and catalytic properties

Fatty acid and alkane hydroxylation in P. oleovorans requires three protein components: rubredoxin reductase (EC 1.6.7.2); rubredoxin and alkane 1-monooxygenase (EC 1.14.15.3). Rubredoxin reductase can utilize the low molecular weight rubredoxins (∿6000 daltons) of anaerobic bacteria (e.g. Peptostreptococcus elsdenii, Clostridium pasteurianum and Desulfovibrio gigas) as well as the larger (molecular weight ~19,000 daltons) rubredoxin of P. oleovorans. The dissociation constant for the reductase - oxidized rubredoxin complex is 0.21 μM. Oxidized rubredoxin (1.00) could be replaced by ferricyanide (1.35) or dichlorophenolindophenol (0.03) but not by any of the following: cytochrome c; thionicotinamide-NAD; coenzyme Q_6 and Q_{10} or by nonhaem proteins such as spinach ferredoxin, putidoredoxin or adreno-doxin. Reduced NADP had low activity. (1)

The properties of rubredoxin have been described; see EC 1.14.15.3.

References

1. Ueda, T., Lode, E.T. & Coon, M.J. (1972) JBC, 247, 2109; 5010.

DIHYDROPTERIDINE REDUCTASE

(Reduced NADP: 6,7-dihydropteridine oxidoreductase)

Reduced NADP + 6,7-dihydropteridine =

NADP + 5,6,7,8-tetrahydropteridine

Ref.

Specificity and kinetic properties

substrate (a)	V relative	K_m (μM)
2-amino-4-hydroxy-6-methyl dihydropteridine	1.00	30
2-amino-4-hydroxy-7-methyl dihydropteridine	1.00	18
2-amino-4-hydroxy-6,7-dimethyl dihydropteridine	1.00	59
dihydropteroate	0.67	30
dihydropteroylglutamate	0.52	30

(a) with the enzyme from rat liver (purified 800 x). Conditions: pH 6.8, Tris, 30° (Ref. 1).

Inhibitors

inhibitor (a)	K_i (μM)
2,4-diaminopteroate	29.5
2,4-diaminopteroylglutamate	24
2,4-diamino-7,8-dihydropteroylglutamate	23
quinoid 2,4-diamino-dihydropteroylglutamate	22

(a) with the enzyme from rat liver (Ref. 1). The inhibitors were competitive (6,7-dihydropteridine substrate). The following were non-inhibitory: 2,4-diamino-6-methyl-pteridine; 4-(2',4'-diamino-pteridin-6'-yl)-butane-1,2,3,4-tetra-ol; 4-(2',4'-diamino-pteridin 6'-yl)-2,3,4-trihydroxy-butanoate and 4-(2',4'-diaminopteridine-6'-yl)-2,3,4-trihydroxy-butane-1-phosphate.

References

1. Lind, K.E. (1972) EJB, 25, 560.

NITRITE REDUCTASE (CYTOCHROME)

(Nitric oxide: ferricytochrome c oxidoreductase)

Nitric oxide + H_2O + 2 ferricytochrome c =

nitrite + 2 ferrocytochrome c

Ref.

Molecular properties

source	value	conditions	
Pseudomonas denitrificans	149,000	pH 7.0	(1)
The enzyme is a copper protein			(1)

Specific activity

P. denitrificans (45 x)	4 nitrite (with reduced cytochrome c-553; pH 7.0, Pi)	(1)

Specificity

The enzyme (P. denitrificans) catalyzes nitrite reduction (1.00) and oxygen consumption (0.15) in the presence of cytochrome c-553 and hydroxylamine oxidation in the presence of nitrite (0.27). In the nitrite reduction reaction, reduced cytochrome c-553 could be replaced by 2,6-dichlorophenolindophenol, thionine, brilliant cresyl blue or methylene blue. (2)

References

1. Iwasaki, H., Shidara, S., Suzuki, H. & Mori, T. (1963) J. Biochem (Tokyo), 53, 299.
2. Miyata, M. & Mori, T. (1969) J. Biochem (Tokyo), 66, 463.

FERREDOXIN-NITRITE REDUCTASE

(Ammonia : ferredoxin oxidoreductase)

Ammonia + 3 oxidized ferredoxin = nitrite + 3 reduced ferredoxin

Molecular properties

source	value	conditions	Ref.
Chlorella fusca[a]	63,000 [1]	gel filtration; gel electrophoresis (SDS); amino acid composition	(1)
Spinach leaf[b]	60,000	Sephadex G 200	(2)
Cucurbita pepo (vegetable marrow; leaf)	62,000	Sephadex G 100 and G 200	(4)

(a) two forms have been obtained. These have identical molecular weights but they differ in charge. Each form contains 2 gm atoms of iron per 63,000 daltons. Manganese and copper were absent. The enzyme probably does not have a flavin prosthetic group.
(b) the enzyme contains iron but probably not a flavin prosthetic group.

Specific activity

		Ref.
C. fusca (500 x)	51.7 nitrite (with reduced methyl viologen as electron donor; pH 8.2, Tris, 30°)	(1)
Spinach leaf (500 x)	3.2 nitrite (with reduced methyl viologen as electron donor; pH 8.0, Tris, 30°)	(2,3)
C. pepo (∿1000 x)	47 nitrite (with reduced ferredoxin as electron donor; pH 7.5, Pi, 27°)	(4)

Specificity and catalytic properties

source	substrate	relative velocity	K_m (μM)	conditions	Ref.
Spinach leaf	nitrite	-	100	pH 8.2, Tris, 30°	(2,3)
	reduced ferredoxin[a]	1.0	10	pH 8.2, Tris, 30°	(2,3)
	methyl viologen	1.5	100	pH 8.2, Tris, 30°	(2,3)

(a) reduced ferredoxin (1.0) could also be replaced by menadione (0.1) but not by FMN; FAD; phenazine methosulphate or reduced NAD(P).

The nitrite reductase from C. fusca was also inactive with reduced NAD(P). This enzyme reduced both nitrite (1.00) and hydroxylamine (0.17) to ammonia (it was inactive with sulphite) and it could utilize ferredoxin, flavodoxin or methyl viologen as electron donor. (1)

The nitrite reductase from C. pepo had very low activity with hydroxylamine. (4)

The properties of ferredoxin have been described (see EC 1.6.7.1).

References

1. Zumft, W.G. (1972) BBA, 276, 363.
2. Losada, M. & Paneque, A. (1971) Methods in Enzymology, 23A, 487.
3. Ramírez, J.M., del Campo, F.F., Paneque, A. & Losada, M. (1966) BBA, 118, 58.
4. Hucklesby, D.P. & Hewitt, E.J. (1970) BJ, 119, 615.

NITRITE REDUCTASE

(Nitric-oxide: (acceptor) oxidoreductase)

Nitric oxide + H_2O + acceptor = nitrite + reduced acceptor

Ref.

Molecular properties

source	value	conditions	Ref.
Achromobacter cycloclastes[a]	69,000	Sephadex G 150, pH 7.0	(1)

(a) the enzyme contains 2 copper atoms per 69,000 daltons. It contains no cytochrome or flavin compound.

Nitrite reductase has also been purified (600 x) from Pseudomonas aeruginosa. This enzyme contains copper, cytochrome c and FAD. (2)

Specific activity

A. cycloclastes	150	nitrite (reduced acceptor = ascorbate-phenazine methosulphate, pH 6.2, Pi)	(1)

Specificity and Michaelis constants

source	substrate	K_m (M)	conditions	Ref.
A. cycloclastes	nitrite[a]	5×10^{-4}	reduced acceptor = ascorbate-phenazine methosulphate; pH 6.2, Pi	(1)

(a) nitrite (1.00) could be replaced by hydroxylamine (0.023) but not by nitric oxide.

The enzyme (A. cycloclastes) also converts nitrite or hydroxylamine to nitrous oxide. (1)

The enzyme (P. aeruginosa) could utilize the following as electron donors in the reverse reaction: reduced-FAD; FMN; riboflavin; pyocyanine; methylene blue and 1,4-naphthoquinone. Reduced naphthoquinone reduced nitrite non-enzymically. In the presence of reduced pyocyanine (pH 7.1, Pi, 30°) the Km for nitrite was 31 μM. This enzyme has an absolute requirement for Pi. (2)

References

1. Iwasaki, H. & Matsubara, T. (1972) J. Biochem (Tokyo), 71, 645.
2. Walker, G.C. & Nicholas, D.J.D. (1961) BBA, 49, 350.

TRIMETHYLAMINE DEHYDROGENASE

(Trimethylamine: (acceptor) oxidoreductase (demethylating))

Trimethylamine + acceptor + H_2O =

dimethylamine + formaldehyde + reduced acceptor

Ref.

Molecular weight

source	value	conditions	
Bacterium 4B 6	160,000 [2]	gel filtration	(1)

The enzyme has an absorption maximum at 445 nm (1)

Specific activity

Trimethylamine dehydrogenase (from Bacterium 4B 6) has been purified (30 x) to homogeneity. (1)

Specificity and Michaelis constants

The kinetics of the reaction have been studied under anaerobic conditions (pH 8.5) using phenazine methosulphate as the primary electron acceptor and 2,6-dichlorophenolindophenol as the final electron acceptor. Of 42 compounds tested, the following were active in the place of trimethylamine: ethyldimethylamine, diethylmethylamine, 2-aminoethyldimethylamine, 2-hydroxyethyldimethylamine, 2-chloroethyldimethylamine and diethylamine. The Km for trimethylamine is 2 μM. (1)

Light absorption data

See EC 1.1.99.10

References

1. Colby, J. & Zatman, L.J. (1971) BJ, 121, 9P.

GLUTATHIONE CoAS-SG TRANSHYDROGENASE

(Coenzyme A: oxidized-glutathione oxidoreductase)

CoA + GS-SG = CoAS-SG + G·SH

Ref.

Equilibrium constant

$$\frac{[\text{CoAS-SG}]\ [\text{GSH}]}{[\text{CoA}]\ [\text{GSSG}]} = 1.25 \quad (\text{pH } 6.9,\ 25^{\circ})$$ (1)

Molecular weight

source	value	conditions	Ref.
Bovine kidney	12,300	Sephadex G 100, pH 7.6	(1)

Specificity and Michaelis constants

source	substrate	K_m (M)	conditions	Ref.
Bovine kidney	GSH	2.5×10^{-4}	pH 7.6, Pi, 25°	(1)
(purified 184 x)	CoAS-SG[a]	6.1×10^{-5}	pH 7.6, Pi, 25°	(1)

(a) CoAS-SG (1.00) could be replaced by pantetheine-glutathione (1.15); thioethanolamine-glutathione (1.05) or cysteine-glutathione (0.98). Little or no activity was observed with homocysteine-glutathione, ribonuclease, insulin or lipoate.

The transhydrogenase also catalyzes a GSH-GSSG exchange reaction. (1)

Abbreviations

GSH (reduced glutathione)	L-γ-glutamyl-L-cysteinylglycine
GS-SG	oxidized glutathione

References

1. Chang, S.H. & Wilken, D.R. (1966) JBC, 241, 4251.

GLUTATHIONE-CYSTINE TRANSHYDROGENASE

(Glutathione: cystine oxidoreductase)

2 Glutathione + cystine = oxidized glutathione + 2 cysteine

Ref.

Molecular weight

source	value	conditions	Ref.
Bakers' yeast	15,000	Sephadex G 75, pH 5.8 (in the presence of 1 mM glutathione)	(1)

Specific activity

Bakers' yeast (1340 x)	61.6	β-hydroxyethyl disulphide (pH 7.8, Pi, 23°)	(1)

Specificity

At low substrate concentrations, the reaction rate was greater with L-cystine than with D-cystine; diacetyl-L-cystine; L-cystine diamide; L-cystinylglycine; L-homocystine or β-hydroxyethyl disulphide. Kinetic constants were not obtained because of experimental difficulties. Insulin was inactive. (1)

Abbreviations

glutathione L-γ-glutamyl-L-cysteinyl-glycine

References

1. Nagai, S. & Black, S. (1968) JBC, 243, 1942.

SULPHITE REDUCTASE

(Hydrogen-sulphide: (acceptor) oxidoreductase)

Hydrogen sulphide + acceptor + $3H_2O$ =

sulphite + reduced acceptor

Ref.

Molecular weight

source	value	conditions
Spinach leaf[a]	84,000	Sephadex G 200 (1,2,3)

(a) the enzyme is composed of two protein fractions. One is the enzyme proper of molecular weight 84,000 (containing 0.76 gm atoms of iron per 84,000 gm of enzyme protein. Magnesium, molybdenum, manganese and flavins (as FMN) were absent). The other fraction could be replaced by any of several proteins including bovine serum albumin, or thiol and disulphide compounds or RNA. This second fraction appears to stabilize the enzyme in the presence of substrate.

Specific activity

Spinach leaf (500 x)	2.3	reduced methyl viologen (acceptor = sulphite; pH 7.7, Pi, 25°)	(1,3)

Specificity and Michaelis constants

source	substrate[a]	K_m (M)	conditions	
Spinach leaf	sulphite[b]	6.7×10^{-4}	pH 7.75, Pi, 25°	(1,3)
	sulphite	2.1×10^{-5}	pH 7.2, tricine, 25°	(1,3)
	hydroxylamine	9.1×10^{-3}	pH 7.75, Pi, 25°	(1,3)

(a) with reduced methyl viologen as the donor. 6 Molecules of reduced methyl viologen are oxidized per molecule of hydrogen sulphide produced.
(b) sulphite (1.0) could be replaced by hydroxylamine (1.1) but not by nitrite; cytochrome-c; thiosulphate; cystine; cysteate; cysteine sulphinate; lipoamide; oxidized glutathione or S-sulphocysteine. Reduced methyl viologen could not be replaced by reduced NAD(P) or by reduced ferredoxin. The enzyme did not require any added cofactor.

Spinach leaf also contains a ferredoxin-linked sulphite reductase. This enzyme can also utilize reduced methyl viologen and it has been purified (30 x). (1)

Inhibitors

The enzyme (spinach leaf) was inhibited by carbon monoxide and by high concentrations of sulphite. (1,3)

Light absorption data

The molar extinction coefficient of reduced methyl viologen at 604 nm = 14,000 $M^{-1}cm^{-1}$ (pH 7.7). (1)

References

1. Asada, K., Tamura, G. & Bandurski, R.S. (1971) Methods in Enzymology, 17B, 528.
2. Asada, K., Tamura, G. & Bandurski, R.S. (1968) BBRC, 30, 554.
3. Asada, K., Tamura, G. & Bandurski, R.S. (1969) JBC, 244, 4904.

IRON-CYTOCHROME c REDUCTASE

(Ferrocytochrome c: iron oxidoreductase)

Ferrocytochrome c + ferric ions =

ferricytochrome c + ferrous ions

Ref.

Molecular properties

source	value	conditions	
Ferrobacillus ferrooxidans	about 100,000	Sephadex G 100	(1)

The enzyme contains bound cytochrome c and cytochrome b. No enzymic function could be found for the cytochrome b. (1)

Specificity and Michaelis constants

source	substrate	K_m (M)	conditions	
F. ferrooxidans (purified 20 x)	ferricytochrome c[a]	2.75×10^{-5}	pH 5.7, acetate	(1)
	ferrous iron	2.6×10^{-3}	pH 5.7, acetate	(1)

(a) ferricytochrome c (1.0) could be replaced by dichlorophenol-indophenol (1.0) but not by potassium ferricyanide.

Inhibitors

The enzyme was inhibited by ferrocytochrome c and ferric iron in the reverse reaction. (1)

Light absorption data

The transformation of ferricytochrome c to ferrocytochrome c is accompanied by an increase in absorption at 650 nm (molar extinction coefficient = 21,000 $M^{-1}cm^{-1}$ at pH 5.7). (1)

References

1. Yates, M.G. & Nason, A. (1966) JBC, 241, 4872.

2-AMINOPHENOL OXIDASE

(2-Aminophenol: oxygen oxidoreductase)

2-Aminophenol + O_2 = 2-quinoneimine + H_2O_2

Ref.

Equilibrium constant

The reaction is essentially irreversible [F]. The dehydrogenation of 2-aminophenol leads to the formation of 2-quinoneimine which condenses with 2-aminophenol to give isophenoxazine. (1)

Molecular properties

The enzyme (*Pycnoporus coccineus*) contains FMN and Mn^{2+} both of which are easily removed. Both are required for enzymic activity. (2)

Specific activity

P. coccineus	(1253 x)	4	2-aminophenol (pH 5.0, Pi-citrate 30°)	(2)

Specificity and Michaelis constants

source	substrate	K_m (M)	conditions	
P. coccineus	2-aminophenol [(a)]	4.35×10^{-4}	pH 5.0, Pi-citrate, 30°	(3)
	FMN [(b)]	2.64×10^{-4}	pH 5.0, Pi-citrate, 30°	(2)
Bauhenia monandra (purified 395 x)	2-aminophenol [(c)]	7.5×10^{-4}	pH 6.2, Pi, 30°	(4)

(a) 3-hydroxykynurenine, 3-hydroxyanthranilate and pyrocatechol were inactive.

(b) FAD was inactive (but see Ref. 3).

(c) 3-hydroxyanthranilate, 3-hydroxykynurenine, 4-aminophenol and catechol were inactive as substrates but active as inhibitors.

The enzyme from *B. monandra* showed no requirement for added cofactor or metal ions. (4)

Inhibitors

The enzyme (*P. coccineus* and *B. monandra*) was inhibited by ascorbate and by a number of reduced agents. (2,4)

Abbreviations

isophenoxazine 2-amino-3*H*-isophenoxazin-3-one

References

1. Nair, P.M. & Vaidyanathan, C.S. (1964) BBA, 81, 507.
2. Nair, P.M. & Vining, L.C. (1965) BBA, 96, 318.
3. Nair, P.M. & Vining, L.C. (1964) Can. J. Biochem, 42, 1515.
4. Subba Rao, P.V. & Vaidyanathan, C.S. (1967) ABB, 118, 388.

GLUTATHIONE PEROXIDASE

(Glutathione: hydrogen-peroxide oxidoreductase)

2 Glutathione + H_2O_2 = oxidized glutathione + $2H_2O$

Ref.

Molecular weight

source	value	conditions	
Bovine blood	83,800 [4]	pH 7.0	(1)

The enzyme contains 4 gm atoms of selenium per 84,000 daltons. (6)

Specificity and Michaelis constants

Glutathione peroxidase is active with a variety of peroxides such as H_2O_2, cumene peroxide and lipid peroxides (1,3). The enzyme (bovine blood, purified 5000 fold) can utilize the following: glutathione (1.00); mercaptoacetate methyl ester (0.28); γ-L-glutamyl-L-cysteine methyl ester (0.26); N-acetyl-L-cysteine methylester (0.10); N-acetyl-L-cysteine amide (0.09); β-L-aspartyl-L-cysteinylglycine (0.08); L-cysteinylglycine (0.07); 4-mercaptomethylimidazole (0.06); L-cysteine-ethylester (0.04); N-acetyl-L-cysteinylglycine ethyl ester (0.04); L-cysteine methyl ester (0.04); N-acetyl-L-cysteine (0.03); cysteamine-HCl (0.03); L-cysteineamide (0.03); N-acetyl-L-cysteinyl-glycine (0.03); mercaptoacetate (0.02); 3-mercaptopropionate methyl ester (0.02); DL-homocysteine (0.02); N-acetyl-DL-homocysteinyl-glycine (0.02); cysteine-HCl (0.02); N-acetyl-L-cysteinyl-L-histi-dinamide (0.02); γ-glutamyl-L-cysteineamide (0.01); 3-mercaptopropion-ate (0.01); mercaptoethanol (0.01) and ergothioneine (0.01). The following exhibited negligible activities: 2-mercaptopropionate; penicillinamine; 2-mercaptoimidazole and α-methyl-γ-L-glutamyl-L-cysteineamide. (2)

The kinetics of the enzyme from pigs' blood (purified 2500 x) are complex. With cumene peroxide as the variable substrate, nonlinear Lineweaver-Burke plots were obtained. With H_2O_2 or lipid peroxides as the variable substrate, K_m values in the range 1-10μM were obtained. Linear plots were obtained with reduced glutathione as the variable substrate. (K_m = 3 mM). (3)

The kinetics with the enzyme from bovine blood are discussed in Ref. 4. The enzyme from pig erythrocytes can utilize steroid hydroperoxides as substrates. (5)

Inhibitors

source	inhibitor	type	K_i (M)	conditions	
Pigs' blood	ATP[a]	C(reduced glutathione)	2.9×10^{-3}	pH 7, Tris	(3)

[a] several nucleotides were inhibitory and of these the pyrimidine nucleotides were the most effective. The inhibitory powers of the adenosine nucleotides increased in the order adenosine, AMP, ADP, ATP, adenosine 5'-tetraphosphate. NADP was a better inhibitor than NAD; Pi was a weak inhibitor. It is thought that the enzyme possesses an allosteric site for nucleotides.

Abbreviations

glutathione L-γ-glutamyl-L-cysteinylglycine.

References

1. Flohé , L., Eisele, B. & Wendel, A. (1971) Hoppe-Seyler's Z. Physiol. Chem., 352, 151.
2. Flohé , L., Günzler, W., Jung, G., Schaich, E. & Schneider, F. (1971) Hoppe-Seyler's Z. Physiol. Chem., 352, 159.
3. Little, C., Olinescu, R., Reid, K.G. & O'Brien; P.J. (1970) JBC, 245, 3632.
4. Flohé , L., Loschen, G., Günzler, W.A. & Eichele, E. (1972) Hoppe-Seyler's Z. Physiol. Chem., 353, 987.
5. Little, C. (1972) BBA, 284, 375.
6. Flohé , L., Günzler, W.A. & Schock, H.H. (1973) FEBS lett., 32, 132.

PROTOCATECHUATE 4,5-DIOXYGENASE

(Protocatechuate: oxygen 4,5-oxidoreductase (decyclizing))

Protocatechuate + O_2 =

2-hydroxy-4-carboxymuconate semialdehyde

Ref.

Molecular weight

source	value	conditions	
Pseudomonad sp.	150,000	Sephadex G 200, pH 7.2	(1)

The enzyme contains one gm atom of iron per 150,000 daltons. When the iron was removed, the resulting apoenzyme was inactive; reactivation took place when Fe^{2+} (but no other metal ion tested) was added. The enzyme has no light absorption band above 290 nm. (1)

Specific activity

Pseudomonad sp. (46 x) 160 protocatechuate (pH 7.0, Pi, 24°) (1)

Specificity and Michaelis constants

source	substrate	K_m (M)	conditions	
Pseudomonad sp.	protocatechuate[a]	8×10^{-5}	pH 7.0, Pi, 24°	(1)
	O_2	5.4×10^{-5}	pH 7.0, Pi, 24°	(1)

(a) the following compounds were inactive as substrates: catechol; 4-methylcatechol; 3-methylcatechol; protocatechuic aldehyde; pyrogallol; 2-aminophenol; DOPA; caffeate; 3,4-dihydroxyacetophenone; 2,3-dihydroxybenzoate; 3,4-dihydroxyphenylacetate; 3,4-dihydroxymandelate; benzoate; 4-hydroxybenzoate or phthalate.

Inhibitors

The enzyme was inhibited by a number of 2-dihydroxyphenyl compounds including catechol. The inhibition was competitive (protocatechuate). (1)

Light absorption data

The transformation of protocatechuate to 2-hydroxy-4-carboxymuconate semialdehyde is accompanied by a decrease in absorption at 250 nm (molar extinction coefficient = 6060 $M^{-1}cm^{-1}$ at pH 7.0). (1)

References

1. Ono, K., Nozaki, M. & Hayaishi, O. (1970) BBA, 220, 224.

2,5-DIHYDROXYPYRIDINE 5,6-DIOXYGENASE

(2,5-Dihydroxypyridine: oxygen 5,6-oxidoreductase (decyclizing))

2,5-Dihydroxypyridine + O_2 + H_2O = maleamate + formate

Ref.

Molecular weight

source	value	conditions	Ref.
Pseudomonas putida	242,000 [6]	sucrose gradient in the presence of dithiothreitol. In the absence of dithiothreitol dissociation occurred.	(1)

Specific activity

P. putida	(18 x)	38	2,5-dihydroxypyridine (pH 8.0, Pi, 25°)	(1)

Specificity

A single enzyme is responsible for the formation of maleamate and formate from 2,5-dihydroxypyridine. N-Formylmaleamate, a possible product of the ring cleavage of 2,5-dihydroxypyridine, is not hydrolyzed by the enzyme. The enzyme is highly specific for 2,5-dihydroxypyridine and the following compounds were inactive: 2,3-dihydroxypyridine; 2,4-dihydroxypyridine; 2,6-dihydroxypyridine; 2-hydroxypyridine; 3-hydroxypyridine; 4-hydroxypyridine; dipicolinate; picolinate; nicotinate; 6-hydroxynicotinate; pyridoxal; pyridoxamine hydrochloride; catechol and 4-hydroxybenzoate. (1,2)

The enzyme requires Fe^{2+} for activity. (1)

Light absorption data

The molar extinction coefficient of 2,5-dihydroxypyridine at 320 nm = 5200 $M^{-1}cm^{-1}$ (at pH 8.0). (1)

Abbreviations

maleamic acid	maleic acid monoamide

References

1. Gauthier, J.J. & Rittenberg, S.C. (1971) JBC, 246, 3737.
2. Gauthier, J.J. & Rittenberg, S.C. (1971) JBC, 246, 3743.

3,4-DIHYDROXYPHENYLACETATE 2,3-DIOXYGENASE

(3,4-Dihydroxyphenylacetate: oxygen 2,3-oxidoreductase (decyclizing))

3,4-Dihydroxyphenylacetate + O_2 =
2-hydroxy-5-carboxymethylmuconate semialdehyde

Ref.

Molecular weight

source	value	conditions	
Pseudomonas ovalis	135,000	pH 7.5	(1,2)

The enzyme contains 5 g-atoms of ferrous iron per 135,000 daltons; it has a weak light absorption band at 414 nm. The iron is required for enzymic activity. (1,2)

Specific activity

P. ovalis	(44 x)	75	3,4-dihydroxyphenylacetate (pH 7.8, Tris, 20°)	(1,2)

Specificity

The enzyme was active with 3,4-dihydroxyphenylacetate (1.00); 3,4-dihydroxyphenylpropionate (0.029); 3,4-dihydroxybenzoate (0.0008) and catechol (0.00006) but not with dopamine. (1,2)

Light absorption data

2-Hydroxy-5-carboxymethylmuconate semialdehyde has an absorption band at 380 nm (molar extinction coefficient = 38,000 $M^{-1}cm^{-1}$ at pH 7.8). (1,2)

References

1. Kita, H. (1965) J. Biochem (Tokyo), 58, 116.
2. Kita, H. & Senoh, S. (1970) Methods in Enzymology 17A, 645.

CYSTEAMINE DIOXYGENASE

(Cysteamine: oxygen oxidoreductase)

Cysteamine + O_2 = hypotaurine

Ref.

Molecular weight

source	value	conditions	Ref.
Horse kidney	83,000	pH 7.6; amino acid composition	(1)

The enzyme contains 1 gm atom of nonhaem iron per 83,000 daltons. It has no light absorption band above 280 nm. (1)

Specific activity

Horse kidney (1200 x)	3-3.5	cysteamine (pH 7.6, Pi, 38°)	(1,2)

Specificity

The enzyme (horse kidney) is highly specific for cysteamine which could not be replaced by cysteine, cysteine methylester, cysteine ethylester or glutathione. (1)

Cysteamine dioxygenase does not require any added cofactor at low substrate concentrations (10µM). At high substrate concentrations, however, there is a requirement for one of the following compounds: sulphide ion; methylene blue; hydroxylamine; sulphur or selenium. The addition of a cofactor prevents the inhibition of the enzyme by high substrate concentrations. The enzyme cannot utilize artificial dyes in the place of molecular oxygen. (3,4,5)

References

1. Cavallini, D., deMarco, C., Scandurra, R., Dupré, S. & Graziani, M.T. (1966) JBC, 241, 3189.
2. Cavallini, D., Scandurra, R. & Dupré, S. (1971) Methods in Enzymology, 17B, 479.
3. Wood, J.L. & Cavallini, D. (1967) ABB, 119, 368.
4. Cavallini, D., Scandurra, R. & deMarco, C. (1965) BJ, 96, 781.
5. Cavallini, D., Scandurra, R. & Monacelli, F. (1966) BBRC, 24, 185.

3,4-DIHYDROXY-9,10-SECOANDROSTA-1,3,5(10)-TRIENE-9, 17-DIONE 4,5-DIOXYGENASE

(3,4-Dihydroxy-9,10-secoandrosta-1,3,5(10)-triene-9,17-dione: oxygen 4,5-oxidoreductase (decyclizing))

3,4-Dihydroxy-9,10-secoandrosta-1,3,5(10)-triene-9,17-dione + O_2 =
3-hydroxy-5,9,17-trioxo-4,5: 9,10-disecoandrosta-1(10), 2-dien-4-oate

Ref.

Molecular weight

source	value	conditions	
Nocardia restrictus	280,000	various; amino acid composition	(1)

The enzyme contains 1.1 gm atoms of Fe^{2+} per 280,000 daltons. It contains no haem and it has no light absorption band above 280 nm. (1)

Specific activity

N. restrictus (45 x)	22	3,4-dihydroxy-9,10-secoandrosta-1,3,5(10)-triene-9,17-dione (pH 7.5, Pi, 20°)	(1)

Specificity and Michaelis constants

substrate[a]	V relative	K_m (M)	
3,4-dihydroxy-9,10-secoandrosta-1,3,5(10)-triene-9,17-dione	-	2.5×10^{-6}	(1,2)
3-isopropyl catechol[b]	1.000	3.7×10^{-4}	(1,2)
3-t-butyl-5-methyl catechol	0.824	1.8×10^{-3}	(1,2)
3-methyl catechol	0.074	4.5×10^{-4}	(1,2)
4-methyl catechol	0.043	3.4×10^{-4}	(1,2)
catechol	0.006	1.3×10^{-4}	(1,2)
oxygen[c]	-	1.8×10^{-4}	(1,2)

(a) with the enzyme from N. restrictus (pH 7.5, Pi, 20°). The following compounds were inactive: 3-methyl-5-t-octyl catechol; 4-t-octyl catechol; 4-isopropyl catechol and 3,4-dihydroxyphenylalanine.
(b) product = 2-hydroxy-7-methyl-6-oxo-octan-2,4-dienoate.
(c) with 3-isopropyl catechol as the second substrate.

The mechanism of the reaction catalyzed by steroid dioxygenase is discussed in Ref. 2.

Inhibitors

The enzyme is subject to product inhibition. (2)

Light absorption data

substrate or product	wave length (nm)	molar extinction coefficient ($M^{-1}cm^{-1}$)	
3-isopropylcatechol	273	2,240 (pH 7.5)	(2)
4-isopropylcatechol	278	2,240 (pH 7.5)	(2)
3-hydroxy-5,9,17-trioxo-4,5:9,10-disecoandrosta-1(10),2-dien-4-oate	393	10,000 (pH 7.5)	(1)
2-hydroxy-7-methyl-6-oxo-octan-2,4-dienoate	393	18,500 (pH 7.5)	(2)

References

1. Tai, H.H. & Sih, C.J. (1970) JBC, 245, 5062.
2. Tai, H.H. & Sih, C.J. (1970) JBC, 245, 5072.

LYSINE 2-MONOOXYGENASE

(L-Lysine: oxygen 2-oxidoreductase (decarboxylating))

L-Lysine + O_2 = 5-amino-*n*-valeramide + CO_2 + H_2O

Ref.

Molecular weight

source	value	conditions	
Pseudomonas fluorescens	191,000	pH 7.5; amino acid composition	(1,2)

The enzyme contains 2 moles of FAD per 191,000 daltons. The inactive apoenzyme was reactivated by FAD but not by FMN or riboflavin. (2)

Specificity and kinetic properties

source	substrate[a]	V(μmoles/min/mg)	K_m (mM)
P. fluorescens (purified 100 x; see Ref. 3)	L-lysine	52.0	0.62
	DL-2,7-diaminoheptanoate	6.4	1.35
	L-ornithine[b]	4.6	14.5
	L-arginine	4.5	1.8
	DL-5-hydroxylysine	1.4	4.1
	L-*S*-aminoethylcysteine	0.25	0.37
	oxygen (with L-lysine)	-	0.44

(a) conditions: pH 9.5, glycine, 24°. The following were inactive: DL-2,4-diaminobutyrate; DL-2,9-diaminononanoate and D-lysine. DL-2,8-diaminooctanoate was oxidized. (4)

(b) in the reaction L-ornithine + O_2 + H_2O = 2-oxo-5-amino-*n*-valerate + NH_3 + H_2O_2. Ornithine is a competitive inhibitor of lysine oxidation.

Lysine monooxygenase does not require exogenous reducing agents such as reduced pyridine nucleotides, reduced pteridine derivatives or ascorbic acid. Under anaerobic conditions, the enzyme-bound FAD is reduced by the substrate L-lysine which is itself converted to 2-oxo-6-aminocaproate (Δ'-piperidine 2-carboxylate) without the formation of CO_2. (5)

References

1. Nakazawa, T. (1971) Methods in Enzymology, 17B, 154.
2. Takeda, H., Yamamoto, S., Kojima, Y. & Hayaishi, O. (1969) JBC, 244, 2935.
3. Takeda, H. & Hayaishi, O. (1966) JBC, 241, 2733.
4. Nakazawa, T., Hori, K. & Hayaishi, O. (1972) JBC, 247, 3439.
5. Yamamoto, S., Nakazawa, T. & Hayaishi, O. (1972) JBC, 247, 3434.

TRYPTOPHAN 2-MONOOXYGENASE

(L-Tryptophan: oxygen 2-oxidoreductase (decarboxylating))

L-Tryptophan + O_2 = indole-3-acetamide + CO_2 + H_2O

Ref.

Specificity and Michaelis constants

source	substrate	K_m(M)	conditions	
Pseudomonas savastanoi (purified 89 x)	L-tryptophan[a]	2.7×10^{-4}	pH 7.4, Tris, 25°	(1,2)

(a) L-tryptophan (1.00) could be replaced by DL-5-hydroxy-tryptophan (0.17); L-phenylalanine (0.06) and L-tyrosine (0.02). The following compounds exhibited less than 1% the activity of L-tryptophan: D-tryptophan; L-alanine; L-glutamate; glycine; L-histidine; L-methionine; L-serine; L-valine; L-proline; indole-3-acetate; tryptamine; indole-3-acetonitrile; indole-3-pyruvate and indole-3-acetaldehyde.

The enzyme did not require any added cofactor (e.g. pyridoxal phosphate, sulphydryl compounds or metal ions). (2)

The mechanism of the reaction catalyzed by tryptophan oxidase is discussed in Ref. 3.

Inhibitors

The enzyme (P. savastanoi) was inhibited by indole-3-acetamide and indole-3-acetate. Phosphate buffers were also inhibitory. (1)

References

1. Kosuge, T. (1970) Methods in Enzymology, 17A, 446.
2. Kosuge, T., Heskett, M.G. & Wilson, E.E. (1966) JBC, 241, 3738.
3. Hutzinger, O. & Kosuge, T. (1967) BBA, 136, 389.

LYSINE 6-MONOOXYGENASE

(Peptidyllysine: oxygen 6-oxidoreductase (deaminating))

Peptidyllysine + O_2 =

peptidyl-2-aminoadipate-5-semialdehyde + NH_3 + H_2O

Ref.

Molecular properties

source	value	conditions	
Chick embryo cartilage	170,000	gel filtration using Bio-Gel A, pH 7.7	(1)

Lysine 6-monooxygenase contains tightly bound copper which is essential for oxidase activity. When added to the inactive apo-enzyme, Cu^{2+} (1.00), Fe^{2+} (0.67), Co^{2+} (0.57)- and to a lesser extent Mn^{2+}, Cd^{2+} or Zn^{2+} (but not Ni^{2+}, Ca^{2+}, Mg^{2+} or Fe^{3+})- reactivation occurred. (1)

Enzymic properties

The enzyme (purified 440 x from chick embryo cartilage) converts specific lysyl residues in collagen (or elastin) into 2-amino-adipate 5-semialdehyde (allysyl) residues which then condense (probably non-enzymically) to form covalent links. (1,2)

Inhibitors

Lysine 6-monooxygenase is inhibited by the lathyrogen 3-amino-propionitrile fumarate (BAPN). The inhibition is irreversible and very low levels of BAPN are required for effective inhibition (both *in vivo* and *in vitro*). (1,2)

References

1. Siegel, R.C., Pinnell, S.R. & Martin, G.R. (1970) B, 9, 4486.
2. Pinnell, S.R. & Martin, G.R. (1968) PNAS, 61, 708.

PROLINE, 2-OXOGLUTARATE DIOXYGENASE

(Prolyl-glycyl-peptide, 2-oxoglutarate: oxygen oxidoreductase)

Prolyl-glycyl-containing peptide + 2-oxoglutarate + O_2 =

4-hydroxyprolyl-glycyl-containing peptide + succinate + CO_2

Ref.

Molecular weight

source	value	conditions	Ref.
Chick	230,000 [4]	various; amino acid composition	(2)
embryo	248,000 [2]	pH 7.8; gel electrophoresis (SDS)	(3)
	200,000	electron microscopy	(4)

Specific activity

Chick embryo (1600 x; purified by affinity chromatography) 0.34 µmoles 4-hydroxyproline formed per min per mg protein with (Pro-Gly-Pro)n (molecular weight 2700) as substrate (pH 7.8, Tris, 25°) (2)

Specificity and Michaelis constants

The enzyme catalyzes the hydroxylation of proline in synthetic poly-tripeptides with the structure (Gly-X-L-Pro)n where X = proline, alanine or probably any of a variety of other amino acids except glycine. The enzyme does not catalyze the hydroxylation of free proline, proline in the tripeptide Gly-Pro-Pro or in poly-L-proline or of hydroxylysine in protocollagen. (2,3)

The hydroxylase has an absolute requirement for molecular oxygen, Fe^{2+}, 2-oxoglutarate and a reducing agent which can be ascorbate, other enediols or tetrahydropteridines (but not reduced NAD(P) or mercaptoethanol). Ascorbate was the most effective reducing agent. (3,5)

Specificity and Michaelis constants continued

source	cofactor	V relative	K_m (M)	conditions	
Chick embryo	2-oxoglutarate[a]	-	3×10^{-6}	pH 7.5, Tris, 30°	(5)
	Fe^{2+}[b]	-	3×10^{-5}	pH 7.5, Tris, 30°	(5)
	ascorbate[c]	-	1×10^{-3}	pH 7.5, Tris, 30°	(5)
	O_2	-	3×10^{-5}	pH 7.5, Tris, 30°	(5)
	$(Pro\text{-}Pro\text{-}Gly)_5$	1.00	3.5×10^{-4}	pH 7.8, Tris, 25°	(6)
	$(Pro\text{-}Pro\text{-}Gly)_{10}$	1.00	3×10^{-5}	pH 7.8, Tris, 25°	(6)
	$(Pro\text{-}Pro\text{-}Gly)_{15}$	1.05	3×10^{-6}	pH 7.8, Tris, 25°	(6)
	$(Pro\text{-}Pro\text{-}Gly)_{20}$	1.05	2×10^{-6}	pH 7.8, Tris, 25°	(6)

(a) succinate; glutamate; pyruvate; oxaloacetate; glutarate; 3-oxoglutarate; fumarate; malonate; oxomalonate; citrate; malate; proline and hydroxyproline were inactive.

(b) Mg^{2+} and Mn^{2+} had low activities but Cd^{2+}, Cu^{2+}, Hg^{2+}, Ni^{2+}, Zn^{2+}, Co^{3+}, Cr^{3+} or Mo^{6+} were inactive.

(c) the Km of ascorbate was dependent on the concentration of 2-oxoglutarate and Fe^{2+}. The value of 1×10^{-3}M was obtained at 10^{-5}M 2-oxoglutarate and 10^{-5}M Fe^{2+}.

Protocollagen hydroxylase has also been purified from pig uterus. (7)

Inhibitors

Malonate, succinate, oxalate and especially oxomalonate were inhibitory. (5)

References

1. Rosenbloom, J. & Prockop, D.J. (1969) In Repair and Regeneration. The Scientific Basis for Surgical Practice (Eds. Dunphy, J.E. & van Winkle, W.) McGraw-Hill: New York, p 117.
2. Berk, R.A. & Prockop, D.J. (1973) JBC, 248, 1175.
3. Pänkäläinen, M., Aro, H., Simons, K. & Kivirikko, K.I. (1970) BBA, 221, 559.
4. Olsen, B.R., Jimenez, S.A., Kivirikko, K.I. & Prockop, D.J. (1970) JBC, 245, 2649.
5. Hutton, J.J., Tappel, A.L. & Udenfriend, S. (1967) ABB, 118, 231.
6. Kivirikko, K.I., Kishida, Y., Sakakibara, S. & Prockop, J. (1972) BBA, 271, 347.
7. Tang-Kao, K.-Y., Treadwell, C.R., Previll, J.M. & McGavack, T.H. (1968) BBA, 151, 568.

THYMIDINE, 2-OXOGLUTARATE DIOXYGENASE

(Thymidine, 2-oxoglutarate: oxygen oxidoreductase)

Thymidine + 2-oxoglutarate + O_2 = ribothymidine + succinate + CO_2

Ref.

Molecular weight

source	value	conditions	Ref.
Neurospora crassa	47,000	Sephadex G 100, pH 6.8	(1)

Specificity and Michaelis constants

source	substrate	V relative	K_m (mM)	Ref.
N. crassa (purified)	thymidine[a]	1.00	0.09	(1)
	deoxyuridine	2.50	0.19	(1)
	5-bromodeoxyuridine	1.00	0.15	(1)
	5-hydroxymethyldeoxyuridine	0.29	0.29	(1)
	6-azathymidine	0.50	0.29	(1)
	thymidylate	0.03	0.07	(1)
	Fe^{2+}[b]	-	0.45	(1)
	2-oxoglutarate[c]	-	0.25	(1)

(a) the following were inactive: deoxycytidine; deoxyguanosine; deoxyadenosine; deoxyinosine; 1-methyluracil; 1-ethyluracil and deoxyribose. (Conditions: pH 7.2, Pi, 37°).

(b) the enzyme has an absolute requirement for Fe^{2+} which could not be replaced by Mn^{2+}, Co^{2+}, Ni^{2+}, Cu^{2+} or Zn^{2+}.

(c) the following were inactive: pyruvate; 2-oxobutyrate; 2-oxovalerate; oxaloacetate; 2-oxoadipate; 3-oxoadipate; 2-oxopimelate; glutarate; glutamate; iminodiacetate; diglycollate or thiodiglycollate.

The dioxygenase (N. crassa) was stimulated by ascorbate and catalase (EC 1.11.1.6). (1)

The mechanism of the reaction has been investigated. (2)

Inhibitors

The enzyme (N. crassa) was inhibited by high concentrations of the substrate thymidine and also by various oxoacids. Ribothymidine was not inhibitory. (1)

References

1. Bankel, L., Lindstedt, G. & Lindstedt, S. (1972) JBC, 247, 6128.
2. Holme, E., Lindstedt, G., Lindstedt, S. & Tofft, M., (1971) JBC, 246, 3314.

ANTHRANILATE 2,3-DIOXYGENASE (DEAMINATING)

(Anthranilate, reduced NADP: oxygen oxidoreductase
(2,3-hydroxylating and deaminating))

Anthranilate + reduced NADP + O_2 =
2,3-dihydroxybenzoate + NADP + NH_3

Ref.

Specificity and Michaelis constants

source	substrate	K_m (mM)	conditions	
Aspergillus niger (purified 74 x)	anthranilate[a]	0.15	pH 8.2, Tris, 30°	(1)
	reduced NADP[b]	0.16	pH 8.2, Tris, 30°	(1)

(a) the following compounds were inactive: 3-hydroxyanthranilate; benzoate; salicylate; 3-hydroxybenzoate; 4-hydroxybenzoate, 3-aminobenzoate; 4-aminobenzoate; methylanthranilate and ethylanthranilate.

(b) reduced NADP (1.00) could be replaced by reduced NAD (0.02). FAD, FMN or tetrahydrofolate were inactive.

An enzyme (EC 1.14.12.1) has been isolated from *Pseudomonas fluorescens* which catalyzes the conversion of anthranilate to catechol:

anthranilate + reduced NAD(P) + O_2 + $2H_2O$ = catechol + NAD(P) + NH_3 + CO_2.

This enzyme is composed of 2 protein fractions. On its own each protein fraction is inactive. (2,3)

References

1. Subba Rao, P.V., Sreeleela, N.S., Kumar, R.P. & Vaidyanathan, C.S. (1970) *Methods in Enzymology*, *17A*, 510.
2. Kobayashi, S. & Hayaishi, O. (1970) *Methods in Enzymology*, *17A* 505.
3. Taniuchi, H., Hatanaka, M., Kuno, S., Hayaishi, O., Nakajima, M. & Kirihara, N. (1964) *JBC*, *239*, 2204.

METHYLHYDROXYPYRIDINE-CARBOXYLATE DIOXYGENASE

(2-Methyl-3-hydroxypyridine-5-carboxylate, reduced NAD(P): oxygen oxidoreductase (decyclizing))

2-Methyl-3-hydroxypyridine-5-carboxylate + reduced NAD(P) + O_2 = 2-(*N*-acetamidomethylene) succinate + NAD(P)

Ref.

Molecular weight

source	value	conditions	
Pseudomonas sp.	166,000	pH 7.8	(1)

The enzyme contains 2 moles of FAD per 166,000 daltons; it may contain essential metal ions. FMN could not replace FAD. (1)

Specific activity

The enzyme (*Pseudomonas* sp.) has been purified 130 fold to apparent homogeneity. (1)

Specificity and kinetic constants

source	substrate[a]	K_m (M)	
Pseudomonas sp.	reduced NAD	1.0×10^{-4}	(1)
	2-methyl-3-hydroxy-pyridine-5-carboxylate	4.8×10^{-5}	(1)

(a) 2-methyl-3-hydroxypyridine-5-carboxylate (1.00) could be replaced by 3-pyridoxate (0.03) but not by *N*-methylnicotinate; *N*-methylnicotinamide; pyridoxine; pyridoxamine; 3-hydroxy-4-methylbenzoate; 6-methylnicotinate; 2-methyl-3-hydroxypyridine-4,5-dicarboxylate; nicotinate or nicotinamide. Reduced NADP was almost as effective as reduced NAD but 2,6-dichloroindophenol could not replace oxygen (conditions: pH 8.0, Pi, 25°).

The mechanism of the reaction has been investigated. (1)

The enzyme was inhibited by 5-pyridoxate (K_i = 60 μM) and 6-methylnicotinate (K_i = 200 μM). The inhibition was competitive (2-methyl-3-hydroxypyridine-5-carboxylate). 2-Methyl-3-hydroxypyridine-4,5-dicarboxylate; nicotinate and nicotinamide were also inhibitory. (1)

Light absorption data

2-Methyl-3-hydroxypyridine-5-carboxylate has an absorption band at 340 nm (molar extinction coefficient = 17,500 $M^{-1}cm^{-1}$). 2-(*N*-acetamidomethylene)succinate does not absorb at 340 nm. (1)

References

1. Sparrow, L.G., Ho, P.P.K., Sundaram, T.K., Zach, D., Nyns, E.J. & Snell, E.E. (1969) JBC, 244, 2590.

4-HYDROXYBENZOATE 3-MONOOXYGENASE

(4-Hydroxybenzoate, reduced NADP: oxygen oxidoreductase (3-hydroxylating))

4-Hydroxybenzoate + reduced NADP + O_2 = protocatechuate + NADP + H_2O

Ref.

Molecular properties

source	molecular weight	specific activity (at °C (a))	
Pseudomonas desmolytica (133 x)	68,000	40 (-)	(1)
P. fluorescens (163 x)	65,000	47.3 (25°)	(2)
P. putida, strain A-3.12 (90 x)	83,600	15 (23°)	(3)
P. putida, strain M-6 (75 x)	93,622	25.3 (23°)	(4)

(a) pH 8.0, Tris.

The enzyme contains 1 mole FAD per mole protein. A stable enzyme-substrate complex containing 1 mole each of FAD, 4-hydroxybenzoate and enzyme protein (P. desmolytica) has been prepared. (1)

Specificity and kinetic properties

source	substrate	K_m (μM)	conditions	
P. putida	4-hydroxybenzoate[a]	21.3	pH 8.0, Tris, 23°	(3)
(strain A-3.12)	reduced NADP[b]	22.7	pH 8.0, Tris, 23°	(3)

(a) 4-hydroxybenzoate (1.00) could be replaced by 4-toluate (0.003); 3-bromo-4-hydroxybenzoate (0.03); 2,4-dihydroxybenzoate (0.01) or benzene sulphonate (0.003) but not by benzoate; salicylate; 3-hydroxybenzoate; by 4-fluoro-, 4-iodo-, 4-chloro-, 4-amino-, or 4-nitrobenzoate; or by 2,3-dihydroxy-, 2,5-dihydroxy-, 3,4-dihydroxy-, or 3,5-dihydroxybenzoate; or by 2,4,6-trihydroxybenzoate; 4-methoxybenzoate; 4-methylaminobenzoate; 4-hydroxyphenylacetate; phenylacetate; 4-hydroxyphenoxyacetate; phenoxyacetate; 4-cresol; 4-phenylsulphate; 4-toluenesulphonate or catechol.

(b) reduced NAD was inactive.

The hydroxylases isolated from other bacterial sources (see above) are also highly specific for reduced NADP and 4-hydroxybenzoate. Different aromatic compounds vary in their effects on the hydroxylases. Thus, 3,4-dihydroxybenzoate; 2,4-dihydroxybenzoate and benzoate increase the rate of hydroxylation of 4-hydroxybenzoate (with the enzyme from P. fluorescens) and of these 3 "effectors" only 2,4-dihydroxybenzoate is hydroxylated. On the other hand, the enzyme from P. putida (strain A-3.12) was inhibited by a number of aromatic compounds, including benzoate. (5,3)

The mechanism of the reaction has been studied in detail. (4,5)

References

1. Yano, K., Higashi, N. & Arima, K. (1969) BBRC, 34, 1.
2. Howell, L.G., Spector, T. & Massey, V. (1972) JBC, 247, 4340.
3. Hosokawa, K. & Stanier, R.Y. (1966) JBC, 241, 2453.
4. Hesp, B., Calvin, M. & Hosokawa, K. (1969) JBC, 244, 5644.
5. Spector, T. & Massey, V. (1972) JBC, 247, 4679; 5632; 7123.

MELILOTATE 3-MONOOXYGENASE

(3-(2-Hydroxyphenyl)-propionate, reduced NAD:
oxygen oxidoreductase (3-hydroxylating))

3-(2-Hydroxyphenyl)-propionate + reduced NAD + O_2 =
3-(2,3-dihydroxyphenyl)-propionate + NAD + H_2O

Ref.

Equilibrium constant

The reaction is irreversible [F] with the enzyme from Arthrobacter sp. as the catalyst. (1)

Molecular properties

source	value	conditions	
Arthrobacter sp.	65,000	sucrose density gradient; Sephadex G 100, pH 7.5	(2)

The enzyme contains bound FAD which is removed during purification.

Specific activity

Arthrobacter sp. (230 x) 12.5 melilotate (pH 7.4, Pi, 30°) (2)

Specificity and Michaelis constants

source	substrate	K_m (μM)	conditions	
Arthrobacter sp.	melilotate[(a)]	86	pH 7.3, Pi, 30°	(1)
	FAD[(b)]	0.8	pH 7.4, Pi, 30°	(1,2)
	reduced NAD[(c)]	91	pH 7.3, Pi, 30°	(1)

(a) melilotate (1.00) could be replaced by 3-(3-hydroxyphenyl)-propionate (0.20) and phenylpropionate (0.015) but not by 3-(4-hydroxyphenyl)-propionate; 2-,3- or 4-methoxyphenylpropionate; 2- or 4-hydroxyphenylacetate; 2-,3- or 4-hydroxybenzoate; phenylacetate; benzoate; cinnamate; phenylalanine; L-tyrosine; 2,4-dihydroxyphenylpropionate; 4- or 7-hydroxycoumarin; or 2-, 3- or 4-hydroxycinnamate.

(b) FAD (1.00) could be replaced by FMN (0.27) or riboflavin (0.18).

(c) reduced NAD (1.00) could be replaced by reduced NADP (0.48) but not by cysteine, glutathione, ascorbate, mercaptoethanol or 2-amino-4-hydroxy-6,7-dimethyltetrahydropteridine.

Abbreviations

melilotate 3-(2-hydroxyphenyl)-propionate or 2-hydroxyphenylpropionate

References

1. Levy, C.C. & Frost, P. (1966) JBC, 241, 997.
2. Levy, C.C. (1967) JBC, 242, 747.
   Strickland, S. & Massey, V. (1973) JBC, 248, 2944; 2953.

IMIDAZOLEACETATE 4-MONOOXYGENASE

(Imidazoleacetate, reduced NAD:
oxygen oxidoreductase (hydroxylating))

Imidazoleacetate + reduced NAD + O_2 =
imidazoloneacetate + NAD + H_2O

Ref.

Molecular weight

source	value	conditions	
Pseudomonas sp.	90,000	pH 7.5; Sephadex G 200, pH 7.2	(1)

The enzyme contains 1 mole of FAD per 90,000 daltons. Metal components were not detected. (1)

Specific activity

Pseudomonas sp. (360 x) 25 imidazoleacetate (pH 9.0, Tris, 24°) (1)

Specificity and kinetic properties

source	substrate	K_m (μM)	conditions	
Pseudomonas sp.	imidazoleacetate[(a)]	300	pH 9.0, Tris, 24°	(1)
	diethyldithiocarbamate	10	pH 9.0, Tris, 24°	(1)
	reduced NAD[(b)]	20	pH 9.0, Tris, 24°	(1)
	O_2	20	pH 9.0, Tris, 24°	(1)

(a) imidazoleacetate (1.00) could be replaced by diethyldithiocarbamate (1.10); imidazolepropionate (0.10) or imidazolelactate (0.04) but not by histidine; histamine; imidazole; urocanate or parabanate. These compounds were also inactive as inhibitors.

(b) β-reduced NAD (1.00) could be replaced by β-reduced NADP (0.13) but not by α-reduced NADP.

The enzyme was inhibited by phenylacetate, indole acetate and nicotinate. These compounds were not substrates. (1)

The product of the reaction, imidazolone acetate, decomposes non-enzymically to formiminoaspartate. (1)

Light absorption data

Imidazoleacetate has an absorption band at 215 nm (molar extinction coefficient = 5,000 $M^{-1}cm^{-1}$). (1)

References

1. Maki, Y., Yamamoto, S., Nozaki, M. & Hayaishi, O. (1969) JBC, 244, 2942, 2951.

ORCINOL 2-MONOOXYGENASE

(Orcinol, reduced NAD: oxygen oxidoreductase (2-hydroxylating))

Orcinol + reduced NAD + O_2 = 2,3,5-trihydroxytoluene + NAD + H_2O

Ref.

Molecular weight

source	value	conditions	
Pseudomonas putida	65,000	gel filtration and ultracentrifugation	(1)

The enzyme contains 1 mole FAD per 65,000 daltons; it probably does not contain any bound metal ions. (1)

Specific activity

P. putida (33 x)	14	orcinol (pH 6.8, Pi, 30°)	(1)

The product of the reaction, 2,3,5-trihydroxytoluene, oxidizes non-enzymically to a quinone. (1)

References

1. Ohta, Y. & Ribbons, D.W. (1970) FEBS letters, 11, 189.

PHENOL 2-MONOOXYGENASE

(Phenol, reduced NADP: oxygen oxidoreductase (2-hydroxylating))

Phenol + reduced NADP + O_2 = catechol + NADP + H_2O

Ref.

Molecular properties

source	value	conditions	
Trichosporon cutaneum[a]	148,000	Sephadex G 200, pH 7.6	(1)

(a) the enzyme contains 1 mole FAD per 148,000 daltons. FMN and riboflavin were inactive in the place of FAD.

Specific activity

T. cutaneum (purified 28 x)	8.3	phenol (pH 7.6, Pi, 22°)	(1)

Specificity and catalytic properties

source	substrate	K_m (μM)	conditions	
T. cutaneum	phenol[a]	18	pH 7.6, Pi, 22°	(1)
	reduced NADP	71	pH 7.6, Pi, 22°	(1)
	O_2	53	pH 7.6, Pi, 22°	(1)

(a) phenol (1.00) could be replaced by resorcinol (1.02); catechol (0.58); quinol (0.90); 2-aminophenol (0.11); 3-aminophenol (0.66); 4-aminophenol (1.01); 2-fluorophenol (0.29); 3-fluorophenol (0.66); 4-fluorophenol (0.93); 2-chlorophenol (0.10); 3-chlorophenol (0.26); 4-chlorophenol (0.29); 2-methylphenol (0.07); 3-methylphenol (0.10) and 4-methylphenol (0.16). The following were poor substrates: 2-, 3- and 4-carboxyphenol; benzoate; 3-hydroxybenzaldehyde; 3-hydroxybenzyl alcohol and 4-*sec*-butylphenol.

Inhibitors

The enzyme was inhibited by concentrations of phenol exceeding 250 μM. Other inhibitors include Cl- and ammonium sulphate. (1)

References

1. Neujahr, H.Y. & Gaal, A. (1973) EJB, 35, 386.

2,6-DIHYDROXYPYRIDINE 3-MONOOXYGENASE

(2,6-Dihydroxypyridine, reduced NAD:
oxygen oxidoreductase (3-hydroxylating))

2,6-Dihydroxypyridine + reduced NAD + O_2 =
2,3,6-trihydroxypyridine + NAD + H_2O

Ref.

Molecular weight

source	value	conditions	
Arthrobacter oxydans	89,000	sucrose density gradient	(1)

The enzyme has a flavin prosthetic group. (1)

Specificity

The enzyme (A. oxydans, purified 50 x) is highly specific for 2,6-dihydroxypyridine which could not be replaced by 2-, 3- or 4-hydroxypyridine; 2,5-dihydroxypyridine; 2,6-dipicolinate; 4-hydroxybenzoate; nicotinate or by nicotine. Reduced NAD (1.00) could be replaced by reduced NADP (0.33) but not by ascorbate. Under anaerobic conditions, phenazine methosulphate; pyocyanin perchlorate; tetramethyl-4-phenylene diamine hydrochloride or 2,3,5-triphenyl-2H-tetrazolium chloride were all inactive as electron acceptors. Methylene blue and 2,6-dichlorophenol indophenol were, however, reduced anaerobically. (1)

The final product of the oxidation of 2,6-dihydroxypyridine is a blue pigment; the immediate product of the reaction is thought to be 2,3,6-trihydroxypyridine which rapidly oxidizes in the presence of oxygen. The blue pigment is a dipyridyl derivative of unknown structure. In crude extracts of A. oxydans this blue pigment is not formed and instead 2,3,6-trihydroxypyridine is converted to maleamate (also see EC 1.13.11.9). (1,2)

The enzyme does not require cations such as Fe^{2+}, Fe^{3+}, Mg^{2+}, Mn^{2+}, Li^{+} or Ca^{2+} for activity. Further, metal chelating agents such as EDTA or o-phenanthroline were non inhibitory. (1)

References

1. Holmes, P.E. & Rittenberg, S.C. (1972) JBC, 247, 7622.
2. Holmes, P.E., Rittenberg, S.C. & Knackmuss, H.J. (1972) JBC, 247, 7628.

ALKANE 1-MONOOXYGENASE

(Alkane, reduced-rubredoxin: oxygen 1-oxidoreductase)

Octane + reduced rubredoxin + O_2 =

1-octanol + oxidized rubredoxin + H_2O

Ref.

Molecular properties

The hydroxylation of fatty acids and hydrocarbons in *Pseudomonas oleovorans* is catalyzed by an enzyme system consisting of 3 components: rubredoxin, rubredoxin-NAD reductase (EC 1.6.7.2) and alkane 1-monooxygenase. The 3 components have been separated. Rubredoxin (*P. oleovorans*) is a non-haem iron protein with 2 iron-binding sites; it has a molecular weight of 19,141 daltons, and it has been sequenced (Ref. 4). The alkane 1-monooxygenase (purified 10 x) has a molecular weight of 2 x 10^6 daltons; it contains 4.2 gm atoms of iron per mole but little FAD or haem. The iron is required for enzymic activity. (1,2)

An alkane hydroxylating system is present in *Corynebacterium* sp. (7E 1C). This consists of 2 components which have been separated: one fraction is particulate and contains cytochrome P-450; the other is a soluble flavoprotein. (3)

Catalytic properties

The hydroxylase system of *P. oleovorans* catalyzes the following type of reaction:

Octane + reduced NAD(P) + O_2 = 1-octanol + NAD(P) + H_2O

Of a number of substances hydroxylated, octane was the most active: *n*-alkanes (C_6 to C_{14}); cyclohexane; methylcyclohexane; dimethylhexane; fatty acids (C_6 to C_{12}) and certain glyceride derivatives of fatty acids. Benzene; toluene; xylene; oleate; diolein; tricaproin; triolein and benzphetamine were not oxidized. The following relative velocities and Michaelis constants (at pH 7.4) have been obtained: octane (relative velocity = 1.00, Km = 0.77 mM); hexane (0.75; 6mM); decane (0.80, 0.58 mM); hexanoate (0.25, 22 mM); heptanoate (0.59; 5.2 mM); nonanoate (0.25; 0.69 mM) and laurate (0.13, 0.032 mM). (1,2)

References

1. McKenna, E.J. & Coon, M.J. (1970) *JBC*, *245*, 3882.
2. Peterson, J.A., Kusunose, M., Kusunose, E. & Coon, M.J. (1967) *JBC*, *242*, 4334.
3. Cardini, G. & Jurtshuk, P. (1970) *JBC*, *245*, 2789.
4. Benson, A., Tomoda, K., Chang, J., Matsueda, G., Lode, E.T., Coon, M.J. & Yasunobu, K.T. (1971) *BBRC*, *42*, 640.

TRYPTOPHAN 5-MONOOXYGENASE

(L-Tryptophan, tetrahydropteridine: oxygen oxidoreductase (5-hydroxylating))

L-Tryptophan + tetrahydropteridine + O_2 =

5-hydroxy-L-tryptophan + dihydropteridine + H_2O

Ref.

Specificity and catalytic properties

source	substrate	K_m	conditions[a]	
Rabbit hind brain (purified 10 x)	tryptophan	290 μM	6,7-DMPH$_4$ = 330 μM	(1)
	6,7-DMPH$_4$	130 μM	tryptophan = 700 μM	(1)
	O_2	20%	6,7-DMPH$_4$ = 330 μM	(1)
	tryptophan[b]	50 μM	BH$_4$ = 83 μM	(1)
	BH$_4$	30 μM	tryptophan = 100 μM	(1)
	O_2[c]	2.5%	BH$_4$ = 140 μM	(1)

(a) pH 7.4, Tris, 37°; with the use of a fluorometric assay method. The reaction was stimulated by reduced NADP, by Fe^{2+} or catalase.

(b) high concentrations of tryptophan were inhibitory with BH$_4$, but not 6,7-DMPH$_4$, as the co-factor.

(c) high O_2 concentrations were not inhibitory.

The low activity and the insensitivity and inconvenience of the assay methods available have precluded detailed studies of tryptophan 5-monooxygenase. When a solution of the product of the reaction, 5-hydroxytryptophan, is excited at 295 nm, fluorescence occurs at 538 nm. This phenomenon has been utilized in an assay method for the enzyme (Ref. 1). Other assay methods are discussed in Refs. 1 and 2.

Light absorption data and abbreviations

tetrahydropterin	abbreviation	Δε 340 nm[a]	
2-amino-4-hydroxy-6-methyl-tetrahydropteridine	6-MPH$_4$	4,500	(3)
2-amino-4-hydroxy-6,7-di-methyltetrahydropteridine	6,7-DMPH$_4$	4,080	(3)
2-amino-4-hydroxy-7-methyltetrahydropteridine	7-MPH$_4$	3,850	(3)
2-amino-4-hydroxy-tetrahydropteridine	PH$_4$	3,900	(3)
tetrahydrobiopterin	BH$_4$	-	

(a) molar absorbance observed at 340 nm upon complete oxidation of the tetrahydro- to the 7,8-dihydropterin.

References

1. Friedman, P.A., Kappelman, A.H. & Kaufman, S. (1972) JBC, 247, 4165.
2. Ichiyama, A., Nakamura, S., Nishizuka, Y. & Hayaishi, O. (1970) JBC, 245, 1699.
3. Shiman, R., Akino, M. & Kaufman, S. (1971) JBC, 246, 1330.

HAEM OXYGENASE (DECYCLIZING)

(Haem, hydrogen-donor: oxygen oxidoreductase

(α-methene-oxidizing, hydroxylating))

Haem + $3AH_2$ + $3O_2$ = biliverdin + Fe^{2+} + CO + 3A + $3H_2O$

Ref.

Specificity and Michaelis constants

substrate[a]	V relative	K_m (μM)	
protohaemin IX	1.00	5.0	(1)
mesohaemin IX	0.75 - 0.80	5.1	(1)
deuterohaemin IX	0.24 - 0.29	4.7	(1)
methaemoglobin	0.31 - 0.44	4.9	(1)
α chains of haemoglobin	0.25 - 0.29	4.7	(1)
β chains of haemoglobin	0.26 - 0.30	4.5	(1)

(a) with the enzyme in rat spleen microsomes (pH 7.4, Pi). The following were poor substrates: coprohaemin I (0.11 - 0.15); haemoglobin-haptoglobin 1-1 (0.06 - 0.10) and haemoglobin-haptoglobin 1-2 (0.06 - 0.10). The following were inactive: myoglobin, oxy-haemoglobin, carboxy-haemoglobin, protoporphyrin IX, mesoporphyrin IX, deuteroporphyrin IX, coproporphyrin I, uroporphyrin I and uroporphyrin III.

Haem oxygenase is a component of a microsomal enzyme system which catalyzes the oxidation of haem compounds to bilirubin. Cytochrome P-450 may be involved in the reaction catalyzed by haem oxygenase which is inhibited by CO. Reduced NADP (AH, above) could not be replaced by reduced NAD, ascorbate or glutathione. The enzyme did not require added metal ions. It was inhibited by KCN but not by substances (e.g. aminopyrine, hexobarbital) that inhibit the microsomal oxidation of certain drugs. (1)

Light absorption properties

Bilirubin (formed by the action of biliverdin reductase, EC 1.3.1.24, on biliverdin) has an absorption band at 468 nm (molar extinction coefficient = 27,700-31,700 $M^{-1}cm^{-1}$ at pH 7.4). (1)

References

1. Tenhunen, R., Marver, H.S. & Schmid, R. (1969) JBC, 244, 6388.

SQUALENE MONOOXYGENASE (2,3-EPOXIDIZING)

(Squalene, hydrogen-donor: oxygen oxidoreductase (2,3-epoxidizing))

Squalene + AH_2 + O_2 = 2,3-oxidosqualene + A + H_2O

Ref.

Squalene epoxidase activity of rat liver requires 3 components for activity: a protein component (this has been purified 84 x from liver supernatant; it has a molecular weight of 44,000 daltons); a heat stable component (e.g. phosphatidylserine, phosphatidylglycerol or phosphatidylinositol; certain other phospholipids were also active) and washed microsomes. The supernatant protein component did not bind squalene nor another substrate, dihydrosqualene nor 2,3-oxidosqualene. It could not be replaced by other proteins tested (including bovine serum albumin and human high density apolipoprotein). In addition to the 3 components detailed above, the epoxidase requires FAD (FMN was inactive) and reduced NADP (reduced NAD was less active) for activity. The enzyme does not require any metal ions for activity; it was not inhibited by metal chelating agents or CO. (1,2)

References

1. Tai, H.H. & Bloch, K. (1972) JBC, 247, 3767.
2. Yamamoto, S. & Bloch, K. (1970) JBC, 245, 1670.

SUPEROXIDE DISMUTASE

(Superoxide: superoxide oxidoreductase)

$O_2^- + O_2^- + 2H^+ = O_2 + H_2O_2$

Ref.

Molecular properties

source (purification factor)	molecular weight	composition(a)	Ref.
Bovine heart (570 x)	32,600 [2]	$2Zn^{2+}$; $2Cu^{2+}$; amino acid	(2)
Bovine erythrocytes (2750 x)	32,500 [2]	$2Zn^{2+}$; $2Cu^{2+}$; amino acid	(2,3)
Green pea (650 x)	31,500	$2Zn^{2+}$; $2Cu^{2+}$; amino acid	(4)
Neurospora crassa (60 x)	31,000 [2]	$2Zn^{2+}$; $2Cu^{2+}$; amino acid	(5)
Escherichia coli B (2100 x)	39,500 [2]	$2Mn^{2+}$; amino acid	(6)
Streptococcus mutans (2520 x)	40,250 [2]	$2Mn^{2+}$; amino acid	(7)
Saccharomyces cerevisiae (83 x)	32,700 [2]	$2Zn^{2+}$; $2Cu^{2+}$; amino acid	(8)

(a) metal content expressed as gm atoms metal per molecular weight dismutase.

Catalytic properties

The following proteins have superoxide dismutase activity: haemocuprein (bovine erythrocytes); erythrocuprein (human erythrocytes); cerebrocuprein (human brain) and hepatocupreine (bovine and equine liver). The enzyme is highly effective in destroying the toxic superoxide radical. At low concentrations it prevents the reduction of ferricytochrome c and tetranitromethane by superoxide radicals. (2,3)

The mechanism of the reaction catalyzed by superoxide dismutase is discussed in Ref. 9 and the preparation of solutions containing stable superoxide radicals in Ref. 3.

References

1. Fridovich, I. (1972) Acc. Chem. Rec, 5, 321.
2. Keele, B.B., McCord, J.M. & Fridovich, I. (1971) JBC, 246, 2875.
3. McCord, J.M. & Fridovich, I. (1969) JBC, 244, 6049.
4. Sawada, Y., Ohyama, T. & Yamazaki, I. (1972) BBA, 268, 305.
5. Misra, H.P. & Fridovich, I. (1972) JBC, 247, 3410.
6. Keele, B.B., McCord, J.M. & Fridovich, I. (1970) JBC, 245, 6176.
7. Vance, P.G., Keele, B.B. & Rajagopalan, K.V. (1972) JBC, 247, 4782.
8. Goscin, S.A. & Fridovich, I. (1972) BBA, 289, 276.
9. Rotilio, G. (1973) Biochem Soc. Trans., 1, 50.
   Weisiger, R.A. & Fridovich, I. (1973) JBC, 248, 3582.
   Fee, J.A. (1973) JBC, 248, 4229.

CDP-4-KETO-6-DEOXY-D-GLUCOSE REDUCTASE

(CDP-4-keto-3,6-dideoxy-D-glucose:
NAD(P) 3-oxidoreductase)

CDP-4-keto-3,6-dideoxy-D-glucose + NAD(P) + H_2O =
CDP-4-keto-6-deoxy-D-glucose + reduced NAD(P)

Ref.

Equilibrium constant

The reaction is reversible with the enzyme from *Pasteurella pseudotuberculosis* as the catalyst. (1)

Molecular properties

source		value	conditions	
P. pseudotuberculosis	E_1	61,000 [1]	Sephadex G 100 (thin layer); gel electrophoresis (SDS)	(1)
	E_3	41,000 [1]		

The enzyme consists of two protein fractions, E_1 and E_3, both of which are required for activity. Fraction E_1 contains tightly bound pyridoxamine 5'-phosphate. When the cofactor is removed (by ion exchange chromatography) an inactive protein results. Pyridoxamine could not replace pyridoxamine 5'-phosphate. CDP-4-Keto-6-deoxy-D-glucose binds weakly to E_1; the binding is greatly enhanced in the presence of pyridoxamine-5'-phosphate. Reduced NAD had no effect on the binding. (1)

Specific activity

E_1(at saturating E_3)	0.11	CDP-4-keto-6-deoxy-D-glucose (pH 7.5, Pi, 37° with 1 x 10^{-5}M pyridoxamine-5'-phosphate).	(1)
E_3(at saturating E_1)	0.31		

Kinetic properties

source	substrate	K_m(M)	conditions	
P. pseudotuberculosis	CDP-4-keto-6-deoxy-D-glucose	1.5 x 10^{-4}	pH 7.4, Pi, 25°	(2)

The enzyme does not require divalent metal ions for activity. Reduced NAD and reduced NADP were about equally active. (2)

The reductase has also been studied in *Salmonella typhimurium*. (2,3)

References

1. Gonzales-Porgue, P. & Strominger, J.L. (1972) JBC, 247, 6748.
2. Pape, H. & Strominger, J.L. (1969) JBC, 244, 3598.
3. Matsuhashi, S., Matsuhashi, M. & Strominger, J.L. (1966) JBC, 241, 4267.

RIBONUCLEOSIDE-DIPHOSPHATE REDUCTASE

(2'-Deoxyribonucleoside-diphosphate: oxidized-thioredoxin 2'-oxidoreductase)

2'-Deoxyribonucleoside diphosphate + oxidized thioredoxin + H_2O = ribonucleoside diphosphate + reduced thioredoxin

Ref.

Molecular properties

source	component	value	conditions	Ref.
Escherichia coli	B_1 (a)	160,000 [2]	ultracentrifugation with or without guanidine HCl; gel electrophoresis (SDS); amino acid analysis	(1)
	B_2 (b)	78,000 [2]	as for B_1. Also peptide mapping	(1,2)

(a) purified by affinity chromatography on dATP-Sepharose.

(b) B_2 contains 2 gm atoms of non-haem iron per 78,000 daltons. The iron is essential for enzymic activity.

The reductase consists of one molecule each of component B_1 (or $\alpha\alpha'$) and component B_2 (or β_2). On its own each component is inactive. The interaction between B_1 and B_2 requires Mg^{2+} and is enhanced by the positive allosteric effectors dTTP, dGTP and ATP. The negative allosteric effector dATP causes the B_1 - B_2 complex to aggregate to a large, inactive protein. All of these effectors bind specifically to the B_1 component. (1,2)

Specificity and catalytic properties

In the presence of the appropriate allosteric effector, ribonucleoside-diphosphate reductase catalyzes the reduction of all 4 ribonucleoside diphosphates (CDP, UDP, ADP and GDP) about equally well. The enzyme requires Mg^{2+} for activity but under certain conditions monovalent ions (Na^+, K^+ etc.) are also active. Thioredoxin could be replaced by dithioerythritol. This enzyme does not require cobamide coenzyme. (2,3,4)

The relationships between nucleotides as substrates, activators and allosteric effectors are very complex. There are 4 binding sites, all on component B_1, for nucleoside triphosphates: two h-sites with a high affinity for dATP (K_D = 0.03 μM); these sites also bind ATP, dGTP (K_D = 0.08 μM) and dTTP (K_D = 0.3 μM) and two l-sites which bind only dATP (K_D = 0.5 μM) and ATP. Sigmoidal saturation curves are obtained with ATP (K_D ∿ 10 μM). The influence of the effectors on the catalytic properties is summarized below. (2,3,4)

Ref.

Specificity and catalytic properties continued

nucleotide (μM)	effect on catalytic activity	effect on substrate specificity (for base)
ATP (2000)	stimulation	pyrimidines
dATP (1)	stimulation	pyrimidines
dATP (100)	inhibition	purines and pyrimidines
dGTP (10)	stimulation	purines
dTTP (10)	stimulation	purines and pyrimidines

Ribonucleoside-diphosphate reductase has also been studied in Novikoff ascites rat tumour. With this enzyme, the reduction of CDP and UDP was inhibited by dTTP, dUTP, dGTP and dATP; the reduction of GDP was inhibited by dGTP and dATP and the reduction of ADP was inhibited by dATP. dCTP was a poor inhibitor. (5)

Highly purified ribonucleoside-triphosphate reductase (see EC 1.17.4.2) from *Lactobacillus leichmannii* has low activity with ribonucleoside diphosphates.

A partially purified coenzyme B_{12}-dependent ribonucleotide reductase from *Rhizobium meliloti* can utilize ribonucleoside di- and triphosphates as substrates. (6)

Thioredoxin

see EC 1.17.4.2

References

1. Thelander, L. (1973) JBC, 248, 4591.
2. Brown, N.C., Eliasson, R., Reichard, P. & Thelander, L. (1969) EJB, 9, 512; 561.
3. Brown, N.C. & Reichard, P. (1969) JMB, 46, 39.
4. Larsson, A. & Reichard, P. (1966) JBC, 241, 2533; 2540.
5. Moore, E.C. & Hurlbert, R.B. (1966) JBC, 241, 4802.
6. Cowles, J.R. & Evans, H.J. (1968) ABB, 127, 770.

RIBONUCLEOSIDE-TRIPHOSPHATE REDUCTASE

(2'-Deoxyribonucleoside-triphosphate: oxidized-thioredoxin 2'-oxidoreductase)

2'-Deoxyribonucleoside triphosphate + oxidized thioredoxin + H_2O = ribonucleoside triphosphate + reduced thioredoxin

Ref.

Molecular properties

source	value	conditions	Ref.
Lactobacillus leichmannii	110,000	pH 7.0, the buffer contained mercaptoethanol; amino acid composition.	(2)

Specificity and catalytic properties

The specificity and the catalytic properties of the reductase (L. leichmannii) are complex and they depend on the presence of ATP and Mg^{2+}. (2,3)

substrate	relative velocity[a] A	B	C	Km[a] A	B	C
CTP	1.00	0.05	0.33	1.7 mM	9.5 mM	-
UTP	0.08	0.05	0.28	-	-	-
GTP	0.22	0.48	2.56	-	-	-
ATP (as ATP-8-^{14}C)	0.43	0.49	0.60	-	2.3 mM	-
CDP	0.04	0.00	0.03	-	-	-
CMP	0.002	0.00	-	-	-	-
ADP	0.34	0.41	0.07	-	-	-
UDP	0.01	0.008	0.03	-	-	-
GDP	0.004	0.02	0.01	-	-	-
DBC coenzyme[b]	-	-	-	0.58 μM	-	-

(a) different conditions were used:
A = the reaction mixture contained DBC coenzyme, dihydrolipoate, ATP, Mg^{2+} and Tris, pH 7.5 (37°).
B = as for A but minus ATP.
C = as for A but minus ATP and Mg^{2+}.

(b) DBC coenzyme, which is required for activity, could be partially replaced by benzimidazolylcobamide coenzyme but not by methylcobalamin, cyanocobalamin or hydroxycobalamin.

The enzyme (L. leichmannii) can use reduced thioredoxin or dihydrolipoate as the hydrogen donor but the following were inactive: reduced NAD, reduced NADP and 2-mercaptoethanol. It requires Mg^{2+} (Mn^{2+} and Ca^{2+} were also active) for full activity; the activity obtained in the absence of Mg^{2+} (about 30%) was not inhibited by EDTA. With CTP as the substrate, ATP was required for full activity (see table above); dATP also activated but dTTP and especially GTP and UTP were much less effective. (2)

Ref.

Specificity and catalytic properties continued

A partially purified coenzyme B_{12}-dependent ribonucleotide reductase from *Rhizobium meliloti* has requirements similar to the enzyme from *L. leichmannii* except that ribonucleoside diphosphates are used in preference to the corresponding triphosphates. (4)

Thioredoxin

Thioredoxin is a single chain protein of molecular weight about 12,000. Oxidized thioredoxin has a single disulphide bond; this bond is cleaved in the presence of thioredoxin reductase (EC 1.6.4.5) to yield reduced thioredoxin. Thioredoxin from *Escherichia coli* has been sequenced. (5)

Abbreviations

DBC coenzyme 5,6-dimethylbenzimidazolylcobamide coenzyme

References

1. Barker, H.A. (1972) Ann. Rev. Biochem, 41, 72.
2. Goulian, M. & Beck, W.S. (1966) JBC, 241, 4233.
3. Abeles, R.H. & Beck, W.S. (1967) JBC, 242, 3589.
4. Cowles, J.R. & Evans, H.J. (1968) ABB, 127, 770.
5. Holmgren, A. (1968) EJB, 6, 475.

TETRAHYDROPTEROYLGLUTAMATE METHYLTRANSFERASE

(5-Methyltetrahydropteroyl-L-glutamate: L-homocysteine S-methyltransferase)

5-Methyltetrahydropteroylglutamate + L-homocysteine = tetrahydropteroylglutamate + L-methionine

Ref.

Molecular properties

source	value	conditions	Ref.
Escherichia coli B[a]	150,000	Sephadex G 200, pH 7.4	(3)

[a] the enzyme, which is composed of subunits, contains 1 mole of cobalamin (probably dimethylbenzimidazole-cobalamin; see ref. 4) per 150,000 daltons. When the cobalamin is removed the resulting apo-enzyme has a more expanded, less dense structure than the holoenzyme.

The enzyme from pig kidney (purified 1,800 x) also contains a cobalamin prosthetic group. (6)

Specific activity

E. coli (860 x)	8.8 L-homocysteine (pH 7.0, Pi, 37°)	(4)

Specificity and Michaelis constants

Homocysteine methyltransferase catalyzes a complex reaction. First, the methyl group of 5-methyl-FH_4 is transferred to enzyme bound cobalamin. This reaction requires catalytic amounts of S-adenosylmethionine and a reducing system (e.g. EC 1.6.99.3). Second, the methyl group is transferred from the methylated cobalamin to homocysteine to yield methionine. The bound cobalamin, therefore, functions as a methyl-carrier prosthetic group. (2,3,4,5,6)

source	substrate	K_m (M)	conditions	
E. coli	DL-homocysteine	3×10^{-3}	pH 7.0, Pi, 37°	(4)
	5-methyl-FH_4	6.7×10^{-4}	pH 7.0, Pi, 37°	(4)
	S-adenosylmethionine	7.5×10^{-7}	pH 7.0, Pi, 37°	(4)
E. coli B	homocysteine	1.6×10^{-5}	pH 7.4, Pi, 37°	(5)
	5-methyl-FH_4	6.0×10^{-5}	pH 7.4, Pi, 37°	(5)
	S-adenosylmethionine	1.6×10^{-6}	pH 7.4, Pi, 37°	(5)

Abbreviations

FH_4	5,6,7,8-tetrahydrofolate or 5,6,7,8-tetrahydropteroylglutamate

References

1. Buchanan, J.M., Elford, H.L., Loughlin, R.E., McDougall, B.M. & Rosenthal, S. (1964) Ann. N.Y. Acad. Sci., 112 (2), 756.
2. Baker, H.A. (1972) Ann. Rev. Biochem, 41, 75.
3. Taylor, R.T. (1971) BBA, 242, 355.
4. Stavrianopoulos, J. & Jaenicke, L. (1967) EJB, 3, 95.
5. Taylor, R.T. & Weissbach, H. (1971) Methods in Enzymology, 17B, 379.
6. Mangum, J.H. & North, J.A. (1971) B, 10, 3765.

TETRAHYDROPTEROYLTRIGLUTAMATE METHYLTRANSFERASE

(5-Methyltetrahydropteroyltri-L-glutamate:
L-homocysteine S-methyltransferase)

5-Methyltetrahydropteroyltriglutamate + L-homocysteine =
tetrahydropteroyltriglutamate + L-methionine

Ref.

Molecular properties

source	value	conditions	Ref.
Escherichia coli K12	84,000 [2-4]	pH 7.8; amino acid composition	(1)
Baker's yeast	75,000	gel filtration	(2)

The methyltransferase has a typical protein absorption spectrum with no absorption of visible light. (1,2)

Specific activity

E. coli K12 (20 x)	0.18	L-homocysteine (pH 7.8, Pi, 37°)	(1)

Specificity and Michaelis constants

source	substrate	K_m(M)	conditions	Ref.
E. coli K12	5-CH_3-H_4-pteroylglu$_3$	4.7×10^{-6}	pH 7.8, Pi, 37°	(1)
Yeast (purified	L-homocysteine	2.2×10^{-5}	pH 7.0, Pi, 37°	(2)
48 x)	5-CH_3-H_4-pteroylglu$_3$	4×10^{-4}	pH 7.0, Pi, 37°	(2)

The enzyme from E.coli K12 could not utilize 5-methyltetrahydropteroylglutamate or 5-methyltetrahydropteroyldiglutamate in the place of 5-CH_3-H_4-pteroylglu$_3$. Cysteine, mercaptoethanol and dithiothreitol could not replace homocysteine as methyl acceptor. The enzyme requires Pi and is stimulated by Mg^{2+} and dithiothreitol. It neither contains nor requires vitamin B_{12}. (1)

The enzyme (yeast) is also highly specific for its substrates. It requires Pi but not Mg^{2+} or vitamin B_{12} for activity. (2)

Inhibitors

Pteroyl-α-glutamylglutamate, pteroyl-γ-glutamyl-γ-glutamyl-glutamate and 5-methyltetrahydropteroyl-α-glutamylglutamate are competitive inhibitors (5-CH_3-H_4-pteroylglu$_3$). The following compounds were non-inhibitory: γ-glutamylglutamate, pteroate and pteroylglutamate. The data refer to the enzyme from E.coli. (1)

Abbreviations

5-CH_3-H_4-pteroylglu$_3$ 5-methyltetrahydropteroyltriglutamate

References

1. Whitfield, C.D., Steers, E. J. & Weissbach, H. (1970) JBC, 245, 390.
2. Burton, E. & Sakami, W. (1971) Methods in Enzymology, 17B, 388.

FATTY-ACID METHYLTRANSFERASE

(*S*-Adenosyl-L-methionine: fatty acid-*O*-methyltransferase)

S-Adenosyl-L-methionine + a fatty acid = *S*-adenosyl-L-homocysteine + a fatty acid methyl ester

Ref.

Specificity and Michaelis constants

source	substrate	K_m (M)	conditions	
Mycobacterium phlei (acetone powder extract)	*S*-adenosyl-methionine[a]	2.5×10^{-5}	pH 7.0, Pi, 30°	(1)
	oleic acid[b]	1.3×10^{-3}	pH 7.0, Pi, 30°	(1)

(a) the following were inactive: methionine, serine, formate and *S*-adenosylethionine.

(b) oleic acid was the most effective acceptor substrate. The following fatty acids were also active: lauric, myristic, palmitic, stearic, arachidic, myristoleic, palmitoleic and linoleic.

Fatty-acid methyltransferase does not require divalent metal ions for activity. (1)

Inhibitors

The enzyme (*M. phlei*) was inhibited by Ca^{2+} (but not Mg^{2+}) and *S*-adenosylhomocysteine. (1)

References

1. Akamatsu, Y. & Law, J.H. (1970) JBC, 245, 709.

TRIMETHYLSULPHONIUM-TETRAHYDROFOLATE METHYLTRANSFERASE

(Trimethylsulphonium-chloride: tetrahydrofolate N-methyltransferase)

Trimethylsulphonium chloride + tetrahydrofolate = dimethylsulphide + 5-methyltetrahydrofolate

ref.

Molecular properties

source	value	conditions	
Pseudomonas sp.	100,000	sucrose gradient	(1)

The light absorption spectrum of the enzyme revealed a small shoulder at 313 nm. The enzyme contained no vitamin B_{12} derivative. (1)

Specificity and Michaelis constants

source	substrate	K_m(M)	conditions	
Pseudomonas sp. (purified 28 x)	trimethylsulphonium chloride[a]	5.8×10^{-3}	pH 7.8, Tris, 35°	(1)
	tetrahydrofolate[b]	1.0×10^{-3}	pH 7.8, Tris, 35°	(1)

(a) S-adenosylmethionine and dimethyl-β-propiothetin were inactive.

(b) 2-mercaptoethanol was inactive.

The methyltransferase has a requirement for a reducing agent such as 2-mercaptoethanol or ascorbic acid for full activity. High concentrations of 2-mercaptoethanol were inhibitory. (1)

References

1. Wagner, C., Lusty, S.M., Kung, H-F. & Rogers, N.L. (1967) JBC, 242, 1287.

GLYCINE METHYLTRANSFERASE

(S-Adenosyl-L-methionine: glycine N-methyltransferase)

S-Adenosyl-L-methionine + glycine =

S-adenosyl-L-homocysteine + sarcosine

Ref.

Molecular properties

source	value	conditions	
Rabbit liver	123,500 [3 or 4]	various; amino acid composition	(1)

Glycine methyltransferase contains 4 residues sialic acid, 2 residues hexose and 12 acetyl groups per 123,500 daltons. (1)

Specific activity

Rabbit liver (114 x) 0.67 glycine (pH 8.6, Tris, 37°) (1)

Specificity and kinetic constants

source	substrate	K_m (mM)	conditions	
Rabbit liver	S-adenosylmethionine	0.1	pH 8.6, Tris, 37°	(1)
	glycine[(a)]	2.2	pH 8.6, Tris, 37°	(1)

(a) all other compounds tested were inactive in the place of glycine: glygly; glycinamide; ethanolamine; aminomethylphosphonate; thioglycollate and the L-amino acids. The reaction stops at the monomethyl stage: neither dimethyl glycine nor betaine is formed. Thus sarcosine was not methylated, instead it was a poor inhibitor.

Glycine methyltransferase (rabbit liver) was inhibited by S-adenosylhomocysteine (C(S-adenosylmethionine, Ki = 0.035 mM) and acetate (Ki = 4.4 mM). (1)

The enzyme does not require any divalent metal ion or sulphydryl compound for activity. (1)

Abbreviations

sarcosine N-methylglycine

References

1. Heady, J.E. & Kerr, S.J. (1973) JBC, 248, 69.

GLUTAMATE METHYLTRANSFERASE

(Methylamine: L-glutamate N-methyltransferase)

Methylamine + L-glutamate = NH_3 + N-methylglutamate

Ref.

Equilibrium constant

The reaction is reversible with the enzyme from Pseudomonas MA as the catalyst. The forward reaction is favoured. (1,3)

Molecular properties

source	value	conditions	
Pseudomonas MA	350,000 [12]	pH 7.4. In the presence of L-glutamate: in its absence dissociation occurred	(1)

The enzyme contains bound FMN. The apoenzyme was partially reactivated by FMN but not by FAD, riboflavin or NAD. (1)

Specific activity

Pseudomonas MA (20 x)	0.14	L-glutamate (pH 8.3, tricine, 30°)	(1)

Specificity and Michaelis constants

source	substrate	K_m (M)	conditions	
Pseudomonas MA	methylamine	9.2×10^{-2}	pH 8.5, Tris, 30°	(2,3)
(crude extract)	L-glutamate	6.8×10^{-3}	pH 8.5, Tris, 30°	(2,3)

With the purified methyltransferase, L-glutamate (1.00) could be replaced by L-aspartate (0.023) but not by any of the other amino acids; by DL-2-hydroxyglutarate; D-glutamate; 2-methyl aspartate or by sarcosine. The enzyme was less specific for methylamine (1.00) which could be replaced by ethylamine (0.46); propylamine (0.26); butylamine (0.10); ethanolamine (0.39) or cysteamine (0.77) but not by isopropylamine or dimethylamine. Similarly, in the reverse reaction, ammonia could be replaced by several amines. (1,2)

The mechanism of the reaction is discussed in ref. 3. The enzyme does not require added metal ions for activity.

Inhibitors

Ammonia is a competitive inhibitor of methylamine demethylation (Ki = 4.9×10^{-2}M). High concentrations of Tris buffers were also inhibitory. (2,3)

References

1. Pollock, R.J. & Hersh, L.B. (1971) JBC, 246, 4737.
2. Shaw, W.V. & Stadtman, E.R. (1970) Methods in Enzymology, 17A, 868.
3. Shaw, W.V., Tsai, L. & Stadtman, E.R. (1966) JBC, 241, 935.

CARNOSINE N-METHYLTRANSFERASE

(S-Adenosyl-L-methionine: carnosine N-methyltransferase)

S-Adenosyl-L-methionine + carnosine =

S-adenosylhomocysteine + anserine

Ref.

Specificity and Michaelis constants

source	substrate	K_m (M)	conditions	
Chick pectoral muscle (purified 7.5 x)	S-adenosylmethionine	9×10^{-5}	pH 8.0, Tris, 37°	(1)
	carnosine[a]	4×10^{-3}	pH 8.0, Tris, 37°	(1)

[a] the following were inactive: homocarnosine, histidine, histamine, urocanate and imidazole.

The enzyme did not require metal ions for activity. (1)

Abbreviations

carnosine	β-alanyl-L-histidine
anserine	β-alanyl-1-methyl-L-histidine

References

1. McManus, I.R. (1962) JBC, 237, 1207.

PROTEIN O-METHYLTRANSFERASE

(S-Adenosyl-L-methionine: protein O-methyltransferase)

S-Adenosyl-L-methionine + protein-COOH =
S-adenosylhomocysteine + protein-$COOCH_3$ — Ref.

Molecular weight

source	value	conditions	Ref.
Calf thymus	35,000 [1]	Sephadex G 100, pH 7.4; amino acid composition	(1)

Specific activity

Calf thymus (900 x)	2.7×10^{-3}	S-adenosylmethionine (acceptor protein = F-P-100. Conditions = pH 6.0, citrate-Pi, 37°)	(1)

Specificity and Michaelis constants (a)

methyl donor	K_m (M)	methyl acceptor	V relative	Ref.
S-adenosylmethionine	1.05×10^{-6}	F-P-100	1.00	(1)
S-adenosylmethionine	1.54×10^{-6}	histone type II-A	0.28	(1)

(a) with the enzyme from calf thymus. The conditions were: pH 6.0, citrate-Pi, 37°.

The enzyme requires free protein carboxyl groups for activity and it could utilize the following as acceptor proteins: F-P-100 (1.00); ribonuclease A, pancreatic (0.45); ovalbumin (0.40); γ-globulin (0.36); histone F_1, lysine rich (0.29); histone F_2, slightly lysine rich (0.25); histone F_3, arginine rich (0.33); lysozyme (0.16); pepsin (0.08); human serum albumin (0.008); RNA (0.005); DNA (0.001) and bovine serum albumin (0.0008) but not polylysine or polyglutamic acid. (1,2)

The enzyme does not require any known cofactor. (1)

A protein (arginine) methyltransferase (protein methylase I) has been isolated from calf thymus. This enzyme has been purified 34 x. (3)

Abbreviations

F-P-100 — methylacceptor protein isolated from calf thymus. The amino acid composition of this protein has been determined. (1)

References

1. Kim, S. & Paik, W.K. (1970) JBC, 245, 1806.
2. Kim, S. & Paik, W.K. (1971) B, 10, 3141.
3. Paik, W.K. & Kim, S. (1968) JBC, 243, 2108.

IODOPHENOL METHYLTRANSFERASE

(S-Adenosyl-L-methionine: 2-iodophenol O-methyltransferase)

S-Adenosyl-L-methionine + 2-iodophenol =

S-adenosylhomocysteine + 2-iodophenol methyl ester

Ref.

Specificity and Michaelis constants

source	substrate	K_m (M)	conditions	Ref.
Rat liver (purified 200 x)	DIB[a]	3.4×10^{-6}	pH 6, Pi, 37°	(1)
	S-adenosylmethionine	1.5×10^{-6}	pH 6, Pi, 37°	(1)

(a) at high concentrations DIB was inhibitory. 3,5,3',5'-Tetraiodothyroacetate was active in the place of DIB at pH 7.5; 3,5,3'-triiodothyroacetate, thyroxine and 3,5,3'-triiodothyronine were poor substrates and 4-hydroxybenzoate and thyroacetate were inactive.

The enzyme does not require divalent ions for activity. (1)

Abbreviations

DIB	3,5-diiodo-4-hydroxybenzoate
thyroacetate	4-(4'-hydroxyphenoxy) phenylacetate

References

1. Tomita, K., Cha, C-J M. & Lardy, H.A. (1964) JBC, 239, 1202.

NORADRENALINE N-METHYLTRANSFERASE

(S-Adenosyl-L-methionine: phenylethanolamine N-methyltransferase)

S-Adenosyl-L-methionine + noradrenaline
= S-adenosylhomocysteine + adrenaline

Ref.

Molecular properties

source	value	conditions	
Bovine adrenals (medulla)	38,000	Sephadex G 100, pH 6.8; sucrose gradient	(2)

The absorption spectrum of the enzyme revealed no peak above 280 nm. (2)

Specific activity

Bovine adrenals (60 x) 4×10^{-2} 3-O-methylnoradrenaline (pH 7.9, Pi, 37°) (2)

Specificity and kinetic properties

The enzyme has an absolute requirement for an hydroxyl group at the β position of the side chain of a variety of phenylethylamines but it is much less specific for other substituents attached to the aromatic ring. The purified enzyme does not require any added cofactor. The reaction mechanism is complex and kinetic constants have not been obtained. Substrate dependence curves were sigmoidal. (2)

The methyltransferase has also been purified from the adrenal glands of *Macaca mulatta* (monkey). This enzyme has similar specificity properties to the bovine enzyme. (3)

Inhibitors

The enzyme was strongly inhibited by noradrenaline and adrenaline and also by a number of amines not structurally related to phenylethanolamine, e.g. tranylcypromine and 2-cyclohexylcyclopropaneamine. (4,5)

References

1. Molinoff, P.B. & Axelrod, J. (1971) Ann Rev Biochem, 40, 473.
2. Connett, R.J. & Kirshner, N. (1970) JBC, 245, 329.
3. Axelrod, J. (1962) JBC, 237, 1657.
4. Fuller, R.W., & Hunt, J.M. (1967) Life Sci. 6, 1107.
5. Krakoff, L.R. & Axelrod, J. (1967) Biochem. Pharmacol. 16, 1384.
   Pohorecky, L.A. & Baliga, G.S. (1973) ABB, 156, 703.

DNA METHYLTRANSFERASE

(S-Adenosyl-L-methionine: DNA(cytosine-5-)-methyltransferase)

S-Adenosyl-L-methionine + DNA = S-adenosyl-L-homocysteine + DNA containing 5-methylcytosine (and 6-methylaminopurine)

Ref.

Equilibrium constant

The reaction was irreversible [F] with the enzyme from E. coli as catalyst. (2)

Molecular properties

DNA methyltransferase from Escherichia coli B exists in several forms each having 2 nonidentical subunits: β(molecular weight = 60,000 daltons) and γ(55,000 daltons). Both molecular weights were obtained by gel electrophoresis (SDS). Freshly isolated enzyme has the structure $\beta_1\gamma_1$; after storage at neutral pH and low salt the structure $\beta_3\gamma_1$ results. (1)

Specific activity

E. coli B 2×10^{-5} fd 101 RF.0 (pH 6.0, Pi, 37°; in the presence of dithiothreitol) (1)

Specificity and kinetic properties

The DNA methyltransferase from E. coli B can methylate a wide variety of DNAs (except DNA produced by E. coli B). Adenosylmethionine could not be replaced by S-adenosyl homocysteine; S-adenosylethionine or by 5'-methylthioadenosine. S-Adenosylethionine and 5'-methylthioadenosine were inhibitory, S-adenosylhomocysteine was not (but see Ref. 2). The enzyme did not require divalent metal ions for activity. However, low concentrations of Mg^{2+} or Mn^{2+} activated but high concentrations were inhibitory. (1)

DNA methyltransferase from rat spleen (purified 200 x) differs from the bacterial enzymes in that it is able to methylate homologous DNA. This enzyme did not require divalent metal ions for activity. (3)

source	substrate	K_m (μM)	Ref.
Rat spleen	S-adenosylmethionine[a]	4.5	(3)
	DNA from calf thymus[b]	44	(3)
E. coli (purified 400 x)	S-adenosylmethionine[c]	33	(2)

(a) S-adenosylethionine was inactive (pH 7.8, Tris, 37°).

(b) the Km values obtained with DNAs from several other sources were similar.

(c) pH 8.0, Tris, 37°.

Abbreviations

fd 101 RF.0 — double-stranded, circular replicative form of DNA produced by phage fd 101 isolated from a nonmodifying (methylating) strain of *E. coli*.

References

1. Lautenberger, J.A. & Linn, S. (1972) JBC, 247, 6176.
2. Gold, M. & Hurwitz, J. (1964) JBC, 239, 3858.
3. Kalousek, F. & Morris, N.R. (1969) JBC, 244, 1157.

O-DEMETHYLPUROMYCIN METHYLTRANSFERASE

(S-Adenosyl-L-methionine: O-demethyl-puromycin O-methyltransferase)

S-Adenosyl-L-methionine + O-demethylpuromycin = S-adenosyl-L-homocysteine + puromycin

Ref.

Molecular weight

source	value	conditions	
Streptomyces alboniger	68,000	sucrose density gradient	(1)

Specificity and Michaelis constants

source	substrate	K_m (M)	conditions	
S. alboniger (purified 30 x)	S-adenosylmethionine	1×10^{-5}	pH 7.4, Pi, 38°	(1)
	O-demethylpuromycin(a)	2.1×10^{-4}	pH 7.4, Pi, 38°	(1)

(a) the following were inactive: L-tyrosine, L-tyrosine ethyl ester, L-tyrosine amide, catechol and adrenaline.

No metal or other cofactor requirement could be shown for the enzyme. (1)

Abbreviations

Puromycin 6-dimethylamino-9[3'-(4-methoxy-L-phenylalanyl-amino)-β-D-ribofuranosyl]purine

References

1. Rao, M.M., Rebello, P.F. & Pogell, B.M. (1969) JBC, 244, 112.

Δ^{24}-STEROL METHYLTRANSFERASE

(S-Adenosyl-L-methionine: zymosterol methyltransferase)

S-Adenosyl-L-methionine + 5α-cholesta-8,24-dien-3β-ol =

S-adenosyl-L-homocysteine +

24-methylene-5-α-cholest-8-en-3β-ol

Ref.

Specificity and Michaelis constants

source	substrate	K_m(M)	conditions	
Yeast microsomes (purified 670 x)	zymosterol	6.25×10^{-5}	pH 7.5, Tris, 37°	(1)

The enzyme requires Mg^{2+} and glutathione for maximum activity. High concentrations of glutathione were inhibitory. (1)

Abbreviations

zymosterol 5α-cholesta-8,24-dien-3β-ol.

References

1. Moore, J.T. & Gaylor, J.L. (1969) JBC, 244, 6334.

o-DIHYDRIC PHENOL METHYLTRANSFERASE

(S-Adenosyl-L-methionine: 5,7,3',4'-tetra-hydroxyflavone 3'-O-methyltransferase)

S-Adenosyl-L-methionine + 5,7,3',4'-tetra-hydroxyflavone = S-adenosylhomocysteine + 5,7,4'-trihydroxy-3'-methoxyflavone

Ref.

Molecular weight

source	value	conditions	
Petroselinum hortense (parsley)	48,000	Sephadex G 100, pH 7.5	(1)

Different molecular forms of the enzyme have been obtained on gel filtration. (1)

Specificity and Michaelis constants

source	substrate	V relative	K_m (M)
P. hortense (purified 82 x)	luteolin-7-glucoside	1.00	3.1×10^{-5}
	luteolin[a]	0.88	4.6×10^{-5}
	eriodictyol	0.15	1.2×10^{-3}
	caffeate[b]	1.48	1.6×10^{-3}
	S-adenosylmethionine	-	1.5×10^{-4}

[a] luteolin (1.00) could be replaced by quercetin (0.40) or proto-catechuate (0.40) but not by diosmetin, chrysoeriol, apigenin, ferulate isoferulate or p-cumarate. (Conditions: pH 9.3, glycine, 30°; ref. 1)

[b] the product was ferulate

The enzyme requires Mg^{2+} for activity (1)

Abbreviations

luteolin	5,7,3',4'-tetrahydroxyflavone
eriodictyol	5,7,3',4'-tetrahydroxyflavanone
caffeate	3,4-dihydroxycinnamate
ferulate	3-methoxy-4-hydroxycinnamate

References

1. Ebel, J. Hahlbrock, K. & Grisebach, H. (1972) BBA, 268, 313.

HISTIDINE METHYLTRANSFERASE

(S-Adenosyl-L-methionine: L-histidine α-N-methyltransferase)

S-Adenosyl-L-methionine + L-histidine =

S-adenosylhomocysteine + α-N-methyl-L-histidine

Ref.

Molecular weight

source	value	conditions	
Neurospora crassa	270,000 - 300,000	Bio-Gel A-0.5m, pH 8.5	(1)

Specific activity

N. crassa (500 x) 0.0434 dimethylhistidine (pH 8.5, NH_4Cl, 25°) (1)

Specificity

The enzyme (N. crassa) utilized dimethylhistidine (1.00), methylhistidine (0.94) and histidine (0.24) as methyl acceptors. In each case the final product was hercynine. The following were inactive as methyl acceptors or as inhibitors: nicotinamide; guanidinoacetate; tyramine; 2,6-diaminopurine; sarcosine; dimethylglycine; β-alanine; 4-aminobutyrate; 2-amino-3-DL-hydroxybutyrate; the protein amino acids; cysteate or S-methylcysteine. The following compounds could not replace S-adenosylmethionine as methyl donor: L-methionine; S-methylmethionine; S-methylcysteine; dimethylthetin; betaine; choline; choline sulphate ester; dimethylaminoethanol; trimethylamine; carnitine; ergothioneine; hercynine; dimethylhistidine and methylhistidine. The enzyme does not require metal ions for activity. (1)

Inhibitors

source	inhibitor	type	K_i (M)	conditions	
N.crassa	Zn^{2+}	C(dimethyl-histidine)	3.9×10^{-5}	pH 8.5, NH_4Cl, 25°	(1)
	hercynine	C(dimethyl-histidine)	-	-	(1)
	S-adenosyl-homocysteine	UC(dimethyl-histidine)	-	-	(1)

The enzyme (N. crassa) was inhibited by imidazolepropionate (but not by a number of other imidazoles tested). (1)

Abbreviations

methylhistidine	α-N-methyl-L-histidine
dimethylhistidine	α-N,N-dimethyl-L-histidine
hercynine	α-N,N,N-trimethyl-L-histidine

References

1. Ishikawa, Y. & Melville, D.B. (1970) JBC, 245, 5967.

METHIONYL-tRNA FORMYLTRANSFERASE

(10-Formyltetrahydrofolate: L-methionyl-tRNA N-formyltransferase)

10-Formyltetrahydrofolate + L-methionyl-tRNA =
tetrahydrofolate + N-formylmethionyl-tRNA

Ref.

Molecular weight

source	value	conditions	
Escherichia coli B	25,000	Yphantis	(1)

Specificity and Michaelis constants

source	substrate	K_m (μM)	
E. coli B	methionyl-tRNA (F) [(a)]	10	(1,2,3)
(purified 1500 x)	10-formyltetrahydrofolate	13.3	(1,2,3)

(a) methionyl-tRNA (M) was inactive (pH 7.4, Tris, 37°). Other substrates were E. coli B aminoacylethionyl-tRNA and norleucyl-tRNA.

The enzyme requires Mg^{2+} for full activity. (1)

References

1. Dickerman, H.W., Steers, E., Redfield, B.G. & Weissbach, H. (1967) JBC, 242, 1522.
2. Dickerman, H.W. & Weissbach, H. (1968) Methods in Enzymology, 12B, 681.
3. Trupin, J., Dickerman, H., Nirenberg, M. & Weissbach, H. (1966) BBRC, 24, 50.

GLYCINE SYNTHASE

(5,10-Methylenetetrahydrofolate: ammonia hydroxymethyltransferase (carboxylating, reducing))

5,10-Methylenetetrahydrofolate + CO_2 + NH_3 + reduced hydrogen-carrier protein = tetrahydrofolate + glycine + oxidized hydrogen-carrier protein + H_2O

Ref.

Equilibrium constant

The reaction is reversible. (1)

Molecular and kinetic properties

2 Proteins have been isolated from rat liver mitochondria which together catalyze the reaction:

L-serine + CO_2 + NH_3 + $2H^+$ = 2 glycine + H_2O

This is the sum of 2 reactions:

L-serine + tetrahydrofolate = glycine + 5,10-methylene tetrahydrofolate

and

5,10-methylene tetrahydrofolate + CO_2 + NH_3 + $2H^+$ = glycine + tetrahydrofolate + H_2O

1 of the proteins has been highly purified (34 x); it has a molecular weight of 17,000 (Sephadex G 100; pH 7.4). It has 1 thiol group per mole and it functions as a hydrogen carrier protein. The other protein may be a pyridoxal phosphate enzyme; it has not been extensively purified. The overall reaction proceeds best anaerobically and requires pyridoxal phosphate. The Km for the hydrogen-carrier protein is 91 μM (pH 8.0, Tris, 37°). (1)

A glycine cleavage system has also been isolated from *Peptococcus glycinophilus*. This system consists of 4 protein components: P_1, a pyridoxal phosphate protein (molecular weight = 125,000 daltons); P_2, a heat stable and low molecular weight protein which has similar properties to the hydrogen-carrier protein of rat liver mitochondria; P_3, a flavoprotein which could be replaced by lipoamide dehydrogenase (EC 1.6.4.3) and P_4, an uncharacterized protein. (2,3)

With glycine synthase from *P. glycinophilus*, the following Km values have been obtained (pH 7.0, Pi, 37°): glycine, 32 mM; HCO_3^-, 31 mM; pyridoxal phosphate, 4.6 μM and hydrogen carrier protein, 1.3 mg ml^{-1}. (2)

References

1. Motokawa, Y. & Kikuchi, G. (1969) ABB, 135, 402.
2. Klein, S.M. & Sagers, R.D. (1966) JBC, 247, 197; 206.
3. Klein, S.M. & Sagers, R.D. (1967) JBC, 242, 297; 301.
   Robinson, J.R., Klein, S.M. & Sagers, R.D. (1973) JBC, 248, 5319.

CHLORAMPHENICOL ACETYLTRANSFERASE

(Acetyl-CoA: chloramphenicol 3-O-acetyltransferase)

Acetyl-CoA + chloramphenicol = CoA + chloramphenicol 3-acetate

Ref.

Molecular weight

source	value	conditions	Ref.
Staphylococcus aureus [a]	78,000	sucrose density gradient	(1)
Escherichia coli [a]	78,000	sucrose density gradient	(1)

(a) a chloramphenicol resistant strain.

Specific activity

source	value	conditions	Ref.
E. coli (200 x)	153	chloramphenicol (pH 7.8, Tris, 37°)	(1)

Specificity and Michaelis constants

source	substrate	K_m (μM)	conditions	Ref.
S. aureus	chloramphenicol	2.7	pH 7.8, Tris, 37°	(1)
	dibromoacetyl analogue	22	pH 7.8, Tris, 37°	(1)
	1-phenyl analogue	68	pH 7.8, Tris, 37°	(1)
E. coli	chloramphenicol	6.1	pH 7.8, Tris, 37°	(1)
	dibromoacetyl analogue	3.7	pH 7.8, Tris, 37°	(1)
	1-phenyl analogue	21	pH 7.8, Tris, 37°	(1)

Abbreviations

chloramphenicol — D(-)-threo-2,2-dichloro-N-[β-hydroxy-α-(hydroxymethyl)-4-nitrophenethyl] acetamide

References

1. Shaw, W.V. & Brodsky, R.F. (1968) J. Bact., 95, 28.

GLYCINE ACETYLTRANSFERASE

(Acetyl-CoA: glycine C-acetyltransferase)

Acetyl-CoA + glycine = CoA + 2-amino-3-oxobutyrate

Ref.

Equilibrium constant

The reaction is irreversible [F] since 2-amino-3-oxobutyrate is highly unstable at neutral pH:-

2-amino-3-oxobutyrate = aminoacetone + CO_2 (1)

Specificity and Michaelis constants

source	substrate	K_m (M)	conditions	
Arthrobacter sp. (purified 70 x)	acetyl-CoA[a]	4.3×10^{-5}	pH 7.5, Pi	(1)
	glycine	7×10^{-2}	pH 7.5, Pi	(1)

(a) acetyl-CoA (1.00) could be replaced by propionyl-CoA (0.85) or butyryl-CoA (0.023).

The acetyl transferase may require pyridoxal phosphate for activity. It does not require added divalent metal ions. (1)

References

1. McGilvray, D. & Morris, J.G. (1971) Methods in Enzymology, 17B, 585.

SERINE ACETYLTRANSFERASE

(Acetyl-CoA: L-serine O-acetyltransferase)

Acetyl-CoA + L-serine = O-acetyl-L-serine + CoA

Ref.

Molecular weight

source	value	conditions	Ref.
Salmonella typhimurium[a]	160,000	pH 7.6; amino acid composition	(1)

(a) the enzyme is associated with O-acetylserine sulphydrolase (EC 4.2.99.8) to form the L-cysteine synthetase complex. The complex is composed of 1 molecule of serine acetyltransferase, 2 molecules of O-acetylserine sulphydrolase and 4 molecules of pyridoxal phosphate. The molecular weight of the complex is 309,000 daltons.

Specific activity

S. typhimurium (300 x) 200 L-serine (pH 7.6, Tris, 25°) (1)

Specificity and Michaelis constants

source	substrate	K_m (mM)	conditions	Ref.
S. typhimurium	L-serine[a]	0.7	pH 7.6, Tris, 25°	(1,3)
	acetyl-CoA[b]	0.12	pH 7.6, Tris, 25°	(1,3)

(a) L-serine could not be replaced by any of the other protein amino acids or by L-cystine; D-serine; DL-homoserine; cycloserine; ethanolamine; L-serine amide; L-serylglycine; O-phospho-L-serine; DL-serine-O-methylester; DL-serine hydroxyamate; phosphatidylserine; DL-β-phenylserine; β-2-thienylserine; allylglycine; D-glucosamine; D-galactosamine; isopropylthiogalactopyranoside; cystathionine or taurine.
(b) acetyl-CoA (1.0) could be replaced by formyl-CoA (0.4) but not by malonyl-CoA or acyl carrier protein.

Serine acetyltransferase has also been purified from Phaseolus vulgaris (50 x). (2)

Inhibitors

The acetyltransferase from S. typhimurium was inhibited by L-cysteine (Ki = 0.6 μM; NC(L-serine), C(acetyl-CoA)) and that from P. vulgaris by CoA and L-cysteine. (1,2,3)

References

1. Kredich, N.M., Becker, M.A. & Tomkins, G.M. (1969) JBC, 244, 2428.
2. Smith, I.K. & Thompson, J.F. (1971) BBA, 227, 288.
3. Kredich, N.M. & Tomkins, G.M. (1966) JBC, 241, 4955.

HOMOSERINE ACETYLTRANSFERASE

(Acetyl-CoA: L-homoserine O-acetyltransferase)

Acetyl-CoA + L-homoserine = CoA + O-acetyl-L-homoserine

Ref.

Specificity and kinetic properties

source	substrate[a]	K_m (mM)	conditions	
Neurospora crassa (purified 60 x)	O-acetylhomoserine[b]	2	pH 7.5, Pi, 30°	(1,2)
	homoserine	1	pH 7.5, Pi, 30°	(1,2)

(a) in the exchange reaction:

[homoserine]* + O-acetylhomoserine = O-acetyl[homoserine]* + homoserine

where * denotes a radioactive amino acid.

(b) O-succinylhomoserine was inactive.

In the acetyltransferase reaction, acetyl-CoA could not be replaced by acetyl phosphate, acetyl carnitine or acyl carrier protein from E. coli. (1)

Homoserine acetyltransferase isolated from Saccharomyces cerevisiae was inhibited by S-adenosylmethionine (but not by methionine); the enzyme isolated from N. crassa was not inhibited by L-methionine or S-adenosylmethionine, separately or together. (1,3)

References

1. Nagai, S. & Kerr, D. (1971) Methods in Enzymology, 17B, 442.
2. Nagai, S. & Flavin, M. (1967) JBC, 242, 3884.
3. de Robichon-Szulmajster, H. & Cherest, H. (1967) BBRC, 28, 256.

GLUTAMATE ACETYLTRANSFERASE

(N^2-Acetyl-L-ornithine: L-glutamate N-acetyltransferase)

N^2-Acetyl-L-ornithine + L-glutamate =

L-ornithine + N-acetyl-L-glutamate

Ref.

Equilibrium constant

$$\frac{[\text{ornithine}]\ [N\text{-acetylglutamate}]}{[\text{acetylornithine}]\ [\text{glutamate}]} = 0.47 \text{ (pH 7.5, Tris, 37°)}$$ (1,2)

Specificity and catalytic properties

source	substrate	K_m (mM)	
Chlamydomonas reinhardti (purified 60 x)	glutamate[a]	13	(1)
	acetylornithine[b]	5.5	(2)

(a) L-glutamate (or L-ornithine) could not be replaced by L-aspartate, 2-amino-5-hydroxyvalerate or L-lysine (conditions: pH 7.5, Tris, 37°).

(b) N^2-acetylornithine (1.00) could be replaced by N^2-propionylornithine (0.40) but not by N^2-butyrylornithine.

The enzyme (C. reinhardti) also catalyses the hydrolysis of N^2-acetylornithine (to acetate and ornithine) and of N^2-butyrylornithine (to butyrate and ornithine). The ratio of the rate of the transfer reaction to the rate of the hydrolytic reaction with N^2-acetylornithine is 100 : 1. (1,2)

The transferase does not require any added cofactor (e.g. CoA or divalent metal ions). (2)

References

1. Dénes, G. (1970) Methods in Enzymology, 17A, 273.
2. Staub, M. & Dénes, G. (1966) BBA, 128, 82.

δ-AMINOLAEVULINATE SYNTHASE

(Succinyl-CoA: glycine C-succinyltransferase (decarboxylating))

Succinyl-CoA + glycine = aminolaevulinate + CoA + CO_2

Ref.

Molecular weight

source	value	conditions	Ref.
Micrococcus denitrificans	68,000 [1]	gel filtration and gel electrophoresis (SDS)	(2)
Rhodopseudomonas spheroides	57,000 [1]	gel electrophoresis (SDS)	(3)
Rat liver	295,000	pH 7.5, in 0.25M NaCl	(4)
	>500,000	pH 7.5, no NaCl	(4)

Specific activity

R. spheroides (1300 x)	2.1	succinyl-CoA (pH 7.5, Tris, 37°)	(3)
Rabbit reticulocytes (4400 x)	0.005	succinyl-CoA (pH 7.4, Tris, 37°)	(6)

Specificity and kinetic properties

source	substrate	K_m (M)	conditions	Ref.
M. denitrificans (purified 788 x)	succinyl-CoA [a]	1.0×10^{-5}	pH 7.0, Tris, 37°	(2)
	glycine	1.2×10^{-2}	pH 7.0, Tris, 37°	(2)
	pyridoxal phosphate	1.1×10^{-5}	pH 7.0, Tris, 37°	(2)
Rat liver (purified 200 x)	succinyl-CoA [c]	7×10^{-5}	pH 7.5, Tris, 37°	(4)
	glycine [b]	1×10^{-2}	pH 7.5, Tris, 37°	(4)
	pyridoxal phosphate	3×10^{-6}	pH 7.5, Tris, 37°	(4)

(a) acetyl-CoA was inactive.
(b) alanine and serine were inactive.
(c) succinyl-CoA (1.00) could be replaced by acetyl-CoA (0.01) but not by propionyl-CoA.

The kinetic properties of the enzyme from R. spheroides are discussed in Refs. 3 and 5. With this enzyme, plots of the concentration of pyridoxal phosphate versus the reaction rate gave "S" shaped curves. The mechanism of the reaction is discussed in Ref. 4 and the properties of the enzyme isolated from rabbit reticulocytes in Ref. 6. All the δ-aminolaevulinate synthases studied to date require pyridoxal phosphate for activity.

Inhibitors

The enzyme (R. spheroides) is subject to feedback inhibition by haemin. Protoporphyrin IX and Mg-protoporphyrin were good inhibitors; haemoglobin and myoglobin were poor inhibitors and cytochrome c was non-inhibitory. (Ref. 3,5). The enzyme from rat liver was also inhibited by haemin (Ki = 2×10^{-5}M) and by a number of metalloporphyrins, porphyrins, bilirubin and by high concentrations of pyridoxal phosphate (Ref. 4). The enzyme from Spirillum itersonii was not inhibited by haemin (Ref. 7).

References

1. Jordan, P.M. & Shemin, D. (1972) The Enzymes, 7, 339.
2. Tait, G.H. (1973) BJ, 131, 389.
3. Warnick, G.R. & Burnham, B.F. (1971) JBC, 246, 6880.
4. Scholnick, P.L., Hammaker, L.E. & Marver, H.S. (1972) JBC, 247, 4126, 4132.
5. Yubisui, T. & Yoneyama, Y. (1972) ABB, 150, 77.
6. Aoki, Y., Wada, O., Urata, G., Takaku, F. & Nakao, K. (1971) BBRC, 42, 568.
7. Ho, Y.K. & Lascelles, J. (1971) ABB, 144, 734.

ACYL CARRIER PROTEIN ACETYLTRANSFERASE

(Acetyl-CoA: acyl carrier protein S-acetyltransferase)

Acetyl-CoA + ACP = CoA + acetyl-ACP

Ref.

Equilibrium constant

$$\frac{[\text{CoA}]\,[\text{acetyl-ACP}]}{[\text{acetyl-CoA}][\text{ACP}]} = 2.09 \text{ (pH 6.5, Pi, } 38^{o}\text{)}$$ (1,4)

Specificity and kinetic properties

The acetyltransferase has been purified 232 x from *Escherichia coli*. Acetyl-CoA (1.00) could be replaced by propionyl-CoA (0.23); butyryl-CoA (0.10) or hexanoyl-CoA (0.05) but not by malonyl-CoA. Acetyl pantetheine was also active. ACP could be replaced by pantetheine but not by mercaptoethanol. (1,3,4)

A stable acetyl-enzyme intermediate has been reported. (4)

Fatty acid synthetase

ACP acetyltransferase is the first enzyme of the fatty acid synthetase complex of *E. coli*. The complex catalyzes, in 7 discrete steps, the addition of 2-carbon units to acetyl-ACP to finally form palmitoyl-ACP. The first step is the formation of butyryl-ACP from acetyl-CoA and malonyl-CoA:

reaction	catalyst
Acetyl-CoA + ACP = CoA + acetyl-ACP	EC 2.3.1.38
Malonyl-CoA + ACP = CoA + malonyl-ACP	EC 2.3.1.39
Acetyl-ACP + malonyl-ACP = 3-oxobutyryl-ACP + CO_2 + ACP	EC 2.3.1.41
3-Oxobutyryl-ACP + reduced NADP = D-3-hydroxybutyryl-ACP + NADP	EC 1.1.1.100
D-3-Hydroxybutyryl-ACP = *trans*-crotonoyl-ACP + H_2O	EC 4.2.1.58
trans-Crotonoyl-ACP + reduced NADP = butyryl-ACP + NADP	EC 1.3.1.10

Fatty acid synthetase continued

The overall reaction may be represented by:-

Acetyl-CoA + 7 malonyl-CoA + 14 reduced NADP =
palmitate + 14 NADP + 8 CoA + $7CO_2$ + $6H_2O$

The enzyme components of the complex of E. coli (and several plants) are present in the cytosol and they have been separated. The enzyme components from yeast and from animal sources form stable complexes that have not been separated. (1,2)

The fatty acid synthetase of bakers' yeast has been crystallized. (1,5)

Fatty acid synthetase from Mycobacterium phlei has been purified 340 x; this complex is stimulated by certain mycobacterial polysaccharides. (6)

Fatty acid synthetase has also been isolated from Corynebacterium diphtheria. (7)

Abbreviations

ACP acyl carrier protein (see EC 2.3.1.39)

References

1. Vagelos, P.R. (1973) The Enzymes, 8A, 155.
2. Ginsburg, A. & Stadtman, E.R. (1970) Ann. Rev. Biochem, 39, 449.
3. Alberts, A.W., Majerus, P.W. & Vagelos, P.R. (1969) Methods in Enzymology, 14, 50.
4. Williamson, I.P. & Wakil, S.J. (1966) JBC, 241, 2326.
5. Lynen, F. (1969) Perspectives in Biology and Medicine, 12, 204.
6. Vance, D.E., Mitsuhashi, O. & Block, K. (1973) JBC, 248, 2303.
7. Knoche, H.W. & Koths, K.E. (1973) JBC, 248, 3517.
   Volpe, J.J. & Vagelos, P.R. (1973) Ann. Rev. Biochem., 42, 33.

ACYL CARRIER PROTEIN MALONYLTRANSFERASE

(Malonyl-CoA: acyl carrier protein S-malonyltransferase)

Malonyl-CoA + ACP = CoA + malonyl-ACP

Ref.

Equilibrium constant

$$\frac{[\text{CoA}]\ [\text{malonyl-ACP}]}{[\text{malonyl-CoA}]\ [\text{ACP}]} = 1.8 \times 10^{-2} \text{ (pH 6.5, Pi, } 25^{\circ}\text{)}$$ (3)

A value of 2.33 has also been reported. (6)

Molecular weight

source	value	conditions	Ref.
Escherichia coli	37,000 [1]	Sephadex G 100, pH 7.4; gel electrophoresis (SDS); amino acid composition	(1,3)

ACP-malonyltransferase is a component of the fatty acid synthetase complex (see EC 2.3.1.38).

Specific activity

E. coli	(4800 x)	1850	malonyl-CoA	(1)

Specificity and kinetic properties

substrate(a)	second substrate	V relative	K_m (M)	Ref.
ACP	malonyl-CoA	1.00	5.4×10^{-5}	(3)
pantetheine	malonyl-CoA	2.18	3.2×10^{-2}	(3)
N-acetylcysteamine	malonyl-CoA	0.50	4.8×10^{-1}	(3)
CoA	malonyl-ACP	5.14	2.8×10^{-5}	(3)
pantetheine	malonyl-ACP	3.02	6.0×10^{-5}	(3)
N-(N-acetyl-β-alanyl)-cysteamine	malonyl-ACP	6.68	3.3×10^{-2}	(3)
N-acetylcysteamine	malonyl-ACP	25.00	1.7×10^{-1}	(3)
malonyl-CoA	ACP	-	2.5×10^{-5}	(3)

(a) with the enzyme from E.coli (purified 950 x; conditions: pH 6.5, Pi, 25°).

The enzyme (E. coli) is specific for malonyl-CoA which could not be replaced by acetyl-CoA. During catalysis, a covalent bond is formed between malonate and the enzyme protein; an enzyme serine residue is involved. The intermediate is stable and can be isolated and it can donate its malonyl group to ACP or CoA. (1,3)

Malonyl transferase (E. coli) was inhibited by acetyl-CoA (C(malonyl-CoA); Ki = 115 μM); CoA (C(ACP); Ki = 45 μM) and malonyl-ACP (C (malonyl-CoA); Ki = 16 μM). (3)

ACP (acyl carrier protein)

ACPs from different sources (E. coli, avocado, spinach, yeast etc.) are proteins of low molecular weights (9,500 - 16,000). ACP possesses a single 4'-phosphopantetheine residue which is covalently bound in a phosphodiester linkage to a seryl hydroxyl group of the protein. It is the sulphydryl group of this pantetheine residue which is the functional group of ACP. (1,2)

The amino acid compositions of several ACPs are known as is the amino acid sequence of E. coli ACP. (1,2,4)

ACP has not been obtained from animal fatty acid synthetase. However, these complexes also contain functional 4'-phosphopantetheine covalently linked to a seryl residue. (1,5)

References

1. Vagelos, P.R. (1973) The Enzymes, 8A, 176.
2. Prescott, D.J. & Vagelos, P.R. (1972) Advances in Enzymology, 36, 269.
3. Joshi, V.C. & Wakil, S.J. (1971) ABB, 143, 493.
4. Vanaman, T.C., Wakil, S.J. & Hill, R.L. (1968) JBC, 243, 6420.
5. Roncari, D.A.K., Bradshaw, R.A. & Vagelos, P.R. (1972) JBC, 247, 6234.
6. Alberts, A.W., Majerus, P.W. & Vagelos, P.R. (1969) Methods in Enzymology, 14, 53.

3-OXOACYL-ACYL CARRIER PROTEIN SYNTHASE

(Acyl-acyl carrier protein: malonyl-acyl carrier protein *C*-acyltransferase (decarboxylating))

Acyl-ACP + malonyl-ACP = 3-oxoacyl-ACP + CO_2 + ACP

Ref.

Molecular weight

source	value	conditions	Ref.
Escherichia coli	66,000 [2]	pH 7.0; gel electrophoresis (SDS); amino acid composition	(3,4)

3-Oxoacyl-ACP synthase is a component of the fatty acid synthetase complex (see EC 2.3.1.38). A stable acetyl-enzyme complex has been described. (1,3)

Specific activity

			Ref.
E. coli (700 x)	14	acetyl-ACP (second substrate = malonyl-ACP; pH 7.0, Pi, 25°)	(4)

Specificity and kinetic properties

substrate[(a)]	V relative	K_m (µM)	Ref.
acetyl-ACP	1.00	0.52	(1,2)
decanoyl-ACP	1.00	0.33	(1,2)
dodecanoyl-ACP	0.35	0.27	(1,2)
tetradecanoyl-ACP	0.11	0.28	(1,2)
cis-3-decenoyl-ACP	0.68	0.71	(1,2)
cis-5-dodecenoyl-ACP	0.61	0.20	(1,2)
cis-9-hexadecenoyl-ACP	0.13	0.37	(1,2)

(a) with the enzyme from *E. coli* (conditions = pH 7.0, Pi, 25°; with malonyl-ACP). Hexadecanoyl-ACP and *cis*-11-octadecenoyl-ACP were inactive.

The enzyme catalyzes the transfer of the malonyl group to a number of saturated or unsaturated fatty acyl-ACPs but not to hexadecanoyl-ACP (palmitoyl-ACP), the predominating saturated fatty acid in *E. coli*. The enzyme shows an absolute specificity for the ACP portion of the substrate and thioesters of CoA or pantetheine are completely inactive. (1)

Specificity and kinetic properties continued

In addition to the acyl transfer reaction, the enzyme also catalyzes the following reactions:

acyl-CoA + ACP = acyl-ACP + CoA (1)

and

malonyl-ACP = acetyl-ACP + CO_2 (1)

Abbreviations

ACP acyl carrier protein (see EC 2.3.1.39)

References

1. Vagelos, P.R. (1973) The Enzymes, 8A, 188.
2. Prescott, D.J. & Vagelos, P.R. (1972) Advances in Enzymology, 36, 295.
3. Greenspan, M.D., Alberts, A.W. & Vagelos, P.R. (1969) JBC, 244, 6477.
4. Prescott, D.J. & Vagelos, P.R. (1970) JBC, 245, 5484.
5. Alberts, A.W., Majerus, P.W. & Vagelos, P.R. (1969) Methods in Enzymology, 14, 57.

γ-GLUTAMYLTRANSFERASE

((γ-Glutamyl)-peptide: amino-acid γ-glutamyltransferase)

(γ-Glutamyl)-peptide + an amino acid =
peptide + γ-L-glutamyl-amino acid

Ref.

Equilibrium constant

The reaction (transfer) is reversible. (1)

Molecular weight

source	value	conditions	
Swine kidney, cortex	80,000[(a)]	sucrose gradient	(1)
Kidney bean, fruit	180,000	gel filtration	(2)

(a) a sedimentation constant of 8.5S is reported in Ref. 3. The isolation procedure reported in Ref. 1 involves digestion with ficin, that reported in Ref. 3 does not. Proteolytic digestion may effect the molecular and catalytic properties of the enzyme (see Ref. 2). The enzyme may be a glycoprotein (Ref. 3, but also see Ref. 1).

Specific activity

Swine kidney	(1880 x)	750	reduced glutathione (with glygly; pH 8.0, Tris, 37°)	(1)
Kidney bean	(360 x)	0.052	γ-glutamylaniline (with S-methyl-L-cysteine; pH 9.5, Tris, 37°)	(2)

Specificity and catalytic properties

γ-Glutamyltransferase catalyzes both hydrolysis and transfer. The enzyme isolated from kidney bean, for instance, catalyzes the following reactions:

γ-glutamylaniline + S-methyl-L-cysteine =
γ-glutamyl-S-methyl-L-cysteine + aniline

2 γ-glutamylaniline = γ-glutamyl-γ-glutamylaniline + aniline

γ-glutamylaniline + H_2O = glutamic acid + aniline (2)

At pH 9.5 and with the enzyme from kidney bean as catalyst, the ratio of the transfer activity to the hydrolytic activity was 1.5 but at pH 7.5, very little transfer activity was obtained. The enzyme showed little activity towards the α-L- or γ-D-glutamyl moiety. The enzyme from swine kidney had similar properties (Refs. 2 & 3; but see Ref. 1).

Further kinetic properties of the enzyme from swine kidney are reported in Ref. 3 and of that from kidney bean in Ref. 2. The mechanism of the reaction is discussed in Ref. 2.

Abbreviations

Glutathione (reduced) L-γ-glutamyl-L-cysteinylglycine

References

1. Leibach, F.H. & Binkley, F. (1968) ABB, 127, 292.
2. Goore, M.Y. & Thompson, J.F. (1967) BBA, 132, 15.
3. Orlowski, M. & Meister, A. (1965) JBC, 240, 338.
   Folk, J.E. & Chung, S.I. (1973) Advances in Enzymology, 38, 109.

γ-L-GLUTAMYLCYCLOTRANSFERASE

((γ-L-Glutamyl)-L-amino-acid γ-L-glutamyltransferase (cyclizing))

(γ-L-Glutamyl)-L-amino acid = L-pyrrolidone carboxylate + L-amino acid

Ref.

Molecular weight

source	value	conditions	
Pig liver	22,400	pH 6.5; amino acid composition	(1)
Human brain	$S_{20,w}$ = 1.2 S	-	(2)

The cyclotransferases from pig liver and human brain occur in at least 2 forms and that from sheep brain in 5 forms. The different forms have similar molecular weights but different isoelectric points. (1,3)

Specific activity

Pig liver (3700 x)	147	γ-L-glutamyl-γ-L-glutamyl-α-naphthylamide (pH 8.0, Tris, 37°)	(1)
Human brain (2500 x)	250	γ-L-glutamyl-γ-L-glutamyl-α-naphthylamide (pH 8.0, Tris, 37°)	(3)

Specificity and catalytic properties

γ-Glutamyl cyclotransferase catalyzes the transfer of the γ-carboxyl group of terminal L-glutamyl residues from linkage with α-amino groups of amino acids and peptides to linkage with the amino group of the same glutamyl residue. The best substrates have 2 γ-glutamyl residues. Triglutamyl analogues were poor substrates and γ-glutamyl-α-naphthylamide was inactive. The following were good substrates: γ-glutamyl-γ-glutamyl-4-nitroanilide (Km = 0.4 mM with the enzyme from human brain); -α-naphthylamide; -alanine; -glutamine; -glycine; and -2-aminobutyrate. γ-Glutamyl-γ-glutamyl-phenylalanine; -tyrosine; -leucine, -valine and also glutathione and opthalmic acid were only slightly active. Compounds containing D-glutamyl residues were inactive. The enzyme does not require any added cofactor or activator. (1,2,3)

References

1. Adamson, E.D., Connell, G.E. & Szewczuk, A. (1970) Methods in Enzymology, 19, 789.
2. Orlowski, M. & Meister, A. (1970) Methods in Enzymology, 17A, 863.
3. Orlowski, M., Richman, P.G. & Meister, A. (1969) B, 8, 1048.
   Orlowski, M. & Meister, A. (1973) JBC, 248, 2836.
   Bodnaryk, R.P. & McGirr, L. (1973) BBA, 315, 352.

ARGINYLTRANSFERASE

(L-Arginyl-tRNA: protein arginyltransferase)

L-Arginyl-tRNA + protein = tRNA + arginyl-protein

Ref.

Specific activity

Rabbit liver (7000 x) 0.135 arginyl-tRNA (acceptor = bovine serum albumin; pH 9.0, Tris, 37°) (1)

Specificity and catalytic properties

Arginyltransferase (rabbit liver) is concerned with the modification of proteins rather than with their synthesis *de novo* since the reaction does not require ribosomes, template nucleic acids, Mg^{2+} or GTP. It does, however, require the presence of a thiol compound (e.g. dithiothreitol) and a monovalent cation (e.g. K^+). (1)

The transfer reaction is highly specific with respect to the protein acceptor which must possess either aspartic or glutamic acid at its amino terminus. The following were good acceptors: serum albumin (human, bovine and rabbit); thyroglobulin (bovine); Bence-Jones protein ("con" and "pot") and trypsin inhibitor (soybean). (2)

The enzyme is also highly specific for arginyl-tRNA which could not be replaced by any of 14 other amino acyl-tRNAs tested. (1)

An enzyme which transfers leucine or phenylalanine from the respective amino acyl-tRNA to the amino terminal arginine of a protein acceptor has been purified from *Escherichia coli*. Good acceptors were α_{s1}-casein and β-casein. The transferase was about equally active with leucyl-tRNA and phenylalanyl-tRNA but valyl-tRNA was inactive. (3)

References

1. Soffer, R.L. (1970) JBC, 245, 731.
2. Soffer, R.L. (1971) JBC, 246, 1481; 1602.
3. Leibowitz, M.J. & Soffer, R.L. (1971) JBC, 246, 4431; 5207.
   Soffer, R.L. (1973) JBC, 248, 2918.

UDP-ACETYLMURAMOYLPENTAPEPTIDE LYSINE N^6-ALANYLTRANSFERASE

(L-Alanyl-tRNA: UDP-N-acetylmuramoyl-L-alanyl-D-glutamyl-L-lysyl-D-alanyl-D-alanine N^6-alanyltransferase)

L-Alanyl-tRNA + UDP-N-acetylmuramoyl-L-Ala-D-Glu-L-Lys-D-Ala-D-alanine = tRNA + UDP-N-acetylmuramoyl-L-Ala-D-Glu-[(N^6-L-Ala)-L-Lys]-D-Ala-D-alanine

Ref.

Equilibrium constant

With the enzyme from *Lactobacillus viridescens*, the reaction is essentially irreversible [F]. (1)

Molecular weight

source	value	conditions	
L. viridescens	40,000 [1]	gel electrophoresis (SDS)	(1)

Specific activity

L. viridescens (531 x) 0.28 L-alanyl-tRNA; acceptor = UDP-N-acetylmuramoyl pentapeptide (conditions = pH 7.2, Tris, 30°). (1)

Specificity and Michaelis constants

source	substrate	K_m(M)	conditions	
L. viridescens	UDP-N-acetylmuramoyl pentapeptide	5×10^{-8}	pH 7.2, Tris, 30°	(1)

UDP-N-acetylmuramoylpentapeptide (1.00) could be replaced by UDP-acetylmuramoyl-L-Ala-D-Glu-*meso*-2,6-diaminopimelate-D-Ala-D-Ala (0.15) but not by phosphoacetylmuramoyl-L-Ala-D-Glu-L-Lys-D-Ala-D-Ala; acetylmuramoyl-L-Ala-D-Glu-L-Lys-D-Ala-D-Ala or UDP-acetylmuramoyl-L-Ala-D-Glu-L-Lys. (1)

tRNA preparations from a number of bacteria could substitute for homologous tRNA from *L. viridescens* in the transfer of both L-alanine and L-serine to the uridine nucleotide. The enzyme was able to utilize the following amino acid tRNAs in the following order of activity: L-alanyl-tRNA; L-seryl-tRNA; L-cysteinyl-tRNA and glycyl-tRNA. (1)

The enzyme did not require added monovalent or divalent cations for activity. (1)

Abbreviations

UDP-N-acetylmuramoylpentapeptide UDP-N-acetylmuramoyl-L-Ala-D-Glu-L-Lys-D-Ala-D-Ala

References

1. Plapp, R. & Strominger, J.L. (1970) JBC, 245, 3675.

1,3-β-OLIGOGLUCAN PHOSPHORYLASE

(1,3-β-Oligoglucan: orthophosphate glucosyltransferase)

$$(1,3\text{-}\beta\text{-D-Glucosyl})_n + Pi =$$
$$(1,3\text{-}\beta\text{-D-glucosyl})_{n-1} + \alpha\text{-D-glucose 1-phosphate}$$

Ref.

Equilibrium constant

$$\frac{[\text{laminaritriose}]\ [\text{glucose 1-phosphate}]}{[\text{laminaritetraose}]\ [Pi]} = \text{about } 0.2 \quad (\text{pH } 6.5, \text{imidazole}, 37^o)$$ (1)

Specificity and Michaelis constants

substrate[a]	second substrate	V relative	K_m (mM)
glucose	glucose 1-phosphate	1.00	40
laminaribiose[b]	glucose 1-phosphate	1.31	4
laminaritriose	glucose 1-phosphate	1.21	4.5
laminaritetraose	glucose 1-phosphate	1.21	3
laminaripentaose	glucose 1-phosphate	0.86	2
glucose 1-phosphate[c]	laminaribiose	-	1.8
Pi[d]	laminaritriose	-	2

(a) with the enzyme from *Euglena gracilis* (purified 10 x). Conditions: pH 7.2, imidazole, 30°. EDTA and mercaptoethanol were required for full activity (Ref. 2).
(b) laminaribiose (1.00) could also be replaced by arbutin (0.98); salicin (0.98); β-phenylglucoside (0.78); cellobiose (0.72); gentiobiose (0.37) but not by laminarin; maltose; L-xylose; D-xylose; raffinose; sucrose; isomaltose; melibiose; sorbitol; 2-deoxyglucose; trehalose; fructose or α-methylglucoside.
(c) α-glucose 1-phosphate could not be replaced by α-mannose 1-phosphate, α-galactose 1-phosphate, β-glucose 1-phosphate or fructose 1,6-diphosphate.
(d) arsenate was active.

A laminaribiose phosphorylase which catalyzes the same reaction as 1,3-β-oligoglucan phosphorylase but which possesses different kinetic properties has also been isolated from *E. gracilis*. The 2 enzymes have been separated. The laminaribiose phosphorylase is more active with smaller substrates than 1,3-β-oligoglucan phosphorylase. It does not require a sulphydryl compound for activity. (1,2,3)

Abbreviations

laminaribiose 3-*O*-β-D-glucopyranosyl-D-glucose

References

1. Goldemberg, S.H., Maréchal, L.R. & deSouza, B.C. (1966) *JBC*, *241*, 45.
2. Maréchal, L.R. (1967) *BBA*, *146*, 417.
3. Maréchal, L.R. (1967) *BBA*, *146*, 431.

1,3-β-GLUCAN SYNTHASE

(UDPglucose: 1,3-β-D-glucan 3-β-glucosyltransferase)

UDPglucose + $(1,3\text{-}\beta\text{-D-glucosyl})_n$ = UDP + $(1,3\text{-}\beta\text{-D-glucosyl})_{n+1}$

Ref.

Specificity and Michaelis constants

source	substrate	K_m (mM)	conditions	Ref.
Euglena gracilis (particulate)	UDP-α-glucose[a]	0.6	pH 7.5, glygly, 23°	(1)

(a) the following were inactive: ADPglucose; dADPglucose; UDP-β-glucose and TDPglucose. The acceptor substrate was endogenous. The enzyme did not require added activators or cofactors.

Inhibitors

UDP was an inhibitor (C(UDPglucose); Ki = 0.1 mM). Phosphoenolpyruvate and high salt concentrations were also inhibitory. (1)

References

1. Maréchal, L.R. & Goldemberg, S.H. (1964) JBC, 239, 3163.

UDP-ACETYLGLUCOSAMINE-STEROID ACETYLGLUCOSAMINYLTRANSFERASE

(UDP-2-acetamido-2-deoxy-D-glucose: 17α-hydroxysteroid-3-D-glucuronoside 17α-acetamidodeoxyglucosyltransferase)

UDP-2-acetamido-2-deoxy-D-glucose + 17α-hydroxysteroid-3-D-glucuronoside = UDP + 17α-(2-acetamido-2-deoxy-D-glucosyloxy)-steroid-3-D-glucuronoside

Ref.

Specificity and Michaelis constants

substrate(a)	K_m (M)	
UDP-N-acetylglucosamine	6.84×10^{-5}	(1)
17α-oestradiol-3-glucuronoside	1.68×10^{-4}	(1)

(a) with the enzyme from rabbit liver (microsomes). The conditions were: pH 8.0, Pi, 37°.

The enzyme is highly specific for the 17α-hydroxyl group in phenolic steroids which are attached through position 3 to glucuronic acid. It is inactive with free steroids. Thus, the monoglucuronosides of 17α-oestradiol; 17β-methyl-17α-oestradiol; 17-epioestriol and 16,17-epioestriol were active as acceptor substrates. The monoglucuronosides of the following steroids were inactive: oestrone; 17β-oestradiol; 15α-hydroxyoestrone; oestriol; 16-epioestriol; tetrahydrocortisone; tetrahydrocortisol; 17α-hydroxypregnenolone; epitestosterone; 3α, 17α-androstenediol; 3β, 17α-androstenediol; diethylstilboestrol; oestradiol-17β-glucuronoside and oestradiol-17α-N-acetylglucosaminide. (1)

References

1. Collins, D.C., Jirku, H. & Layne, D.S. (1968) JBC, 243, 2928.

UDP-ACETYLGALACTOSAMINE-PROTEIN ACETYLGALACTOSAMINYLTRANSFERASE

(UDP-2-acetamido-2-deoxy-D-galactose:
protein acetamidodeoxygalactosyltransferase)

UDP-2-acetamido-2-deoxy-D-galactose + protein =
UDP + 2-acetamido-2-deoxy-D-galactose-protein

Ref.

Molecular weight

source	value	conditions	
Bovine submaxillary gland	100,000-200,000	Sephadex G 200, pH 7.2. The buffer contained Triton X-100.	(1)

Specificity and Michaelis constants

The enzyme (bovine submaxillary gland, purified 70 x) transfers N-acetylgalactosamine to a serine or threonine residue of the acceptor protein substrate. It requires Mn^{2+} for activity; Mn^{2+} (1.00) could be replaced by Co^{2+} (0.74), Ni^{2+} (0.30) or Cd^{2+} (0.30) but not by Ca^{2+}, Mg^{2+}, Cu^{2+}, Fe^{2+}, Zn^{2+} or Pb^{2+}. (1)

substrate	K_m (M)	conditions	
UDP-N-acetylgalactosamine	6×10^{-5}	Tris, pH 7.2, 37°	(1)
E receptor[a]	9.5×10^{-4}	Tris, pH 7.2, 37°	(1)
P receptor	5.0×10^{-3}	Tris, pH 7.2, 37°	(1)

(a) the acceptor substrates used were bovine submaxillary gland glycoprotein with sialic acid removed followed by the removal of the N-acetylhexosamine units either by hexosaminidase (to provide E receptor) or by periodate hydrolysis treatment (to provide P receptor). The Km values are expressed as hexosamine. The enzyme is highly specific for its acceptor protein substrate.

Some of the properties of the transferase isolated from sheep submaxillary gland have been reported. (2)

References

1. Hagopian, A. & Eylar, E.H. (1969) ABB, 129, 515.
2. McGuire, E.J. & Roseman, S. (1967) JBC, 242, 3745.

UDP-GALACTOSE-LIPOPOLYSACCHARIDE GALACTOSYLTRANSFERASE

(UDPgalactose-lipopolysaccharide galactosyltransferase)

UDPgalactose + a lipopolysaccharide =

UDP + a D-galactosyl-lipopolysaccharide

Ref.

Molecular weight

source	value	conditions	
Salmonella typhimurium	20,000[a]	sucrose density gradient; amino acid composition	(1)

(a) lipid free enzyme; this material had 80% the enzyme activity of the native lipoprotein. Multiple forms have been observed. The enzyme may contain carbohydrate.

Specificity and catalytic properties

source	substrate	K_m (μM)	conditions	
S. typhimurium (purified 6000 x)	UDPgalactose[a]	74	pH 8.5, Tris, 37°	(1)

(a) ADPgalactose; UDPglucose; TDPglucose and galactose 1-phosphate were inactive.

The galactosyl transferase (S. typhimurium) is specific for lipopolysaccharides lacking the 3-α-galactosyl residue. It requires Mg^{2+} (partially replaced by Mn^{2+} or Ca^{2+} but not by Zn^{2+}) and a phospholipid (e.g. phosphatidylethanolamine) for activity. Several non-substrate lipopolysaccharides were active inhibitors. (1)

The galactosyltransferase has also been purified from the slime mould, Dictyostelium discoideum. This enzyme required K^{+} for full activity but there was no requirement for Mg^{2+} or Mn^{2+}. (2)

References

1. Endo, A. & Rothfield, L. (1969) B, 8, 3501; 3508.
2. Sussman, M. & Osborn, M.J. (1964) PNAS, 52, 81.

CELLODEXTRIN PHOSPHORYLASE

(1,4-β-Oligoglucan: orthophosphate α-glucosyltransferase)

$(1,4\text{-}\beta\text{-D-Glucosyl})_n + Pi =$

$(1,4\text{-}\beta\text{-D-glucosyl})_{n-1} + \alpha\text{-D-glucose 1-phosphate}$

Ref.

Equilibrium constant

$$\frac{[\text{cellobiose}]\,[\alpha\text{-D-glucose 1-phosphate}]}{[\text{cellotriose}]\,[Pi]} = 0.4 \quad (\text{pH } 7.5,\ 37°)$$ (1)

Specificity and Michaelis constants

substrate[a]	second substrate	K_m (mM)
cellobiose[b]	glucose 1-phosphate	1.2
glucose 1-phosphate	cellobiose	4.7
cellotriose, cellotetraose or cellopentaose	Pi	1.0
cellohexaose[c]	Pi	0.37
Pi	cellotriose	0.13
	cellohexaose	0.26

(a) with the enzyme from *Clostridium thermocellum* (purified 450 x). The conditions were: pH 7.5, Tris, 37° (Ref. 1).

(b) cellobiose (1.00) could be replaced by cellotriose (1.06); cellotetraose (1.03); cellopentaose (0.76); cellohexaose (0.48); cellobiitol (0.18); laminaritriose (0.74); 4-*O*-β-glucosyl-D-altrose (0.29); 4-*O*-β-glucosyl-2-deoxy-D-glucose (0.60); 4-*O*-β-glucosyl-D-mannose (0.54) or 4-*O*-β-glucosyl-D-xylose (0.34). A large number of other sugars and their derivatives were inactive as glucosyl acceptors.

(c) cellobiose, cellulose, laminaritriose, melezitose and raffinose were inactive.

The enzyme requires a sulphydryl compound or Na_2SO_3 for activity. Nonsulphur reducing compounds were inactive. There was no requirement for a divalent metal ion. (1)

References

1. Sheth, K. & Alexander, J.K. (1969) JBC, 244, 457.

UDP-GALACTOSE-COLLAGEN GALACTOSYLTRANSFERASE

(UDPgalactose: 5-hydroxylysine-collagen galactosyltransferase)

UDPgalactose + 5-hydroxylysine-collagen =

UDP + O-D-galactosyl-5-hydroxylysine-collagen

Specificity and catalytic properties

source	substrate	K_m (M)	conditions	Ref.
Rat kidney cortex (purified 10 x)	UDPgalactose[a]	3.1×10^{-4}	pH 6.8, Tris, 37°	(1)

(a) UDPgalactose could not be replaced by ADPgalactose, GDPgalactose, galactose 1-phosphate or galactose. CDPgalactose and TDPgalactose had less than 10% the activity obtained with UDPgalactose.

The enzyme from rat kidney could utilize all native collagens tested (vertebrate or invertebrate) as galactosyl acceptors. Glomerular basement membrane (bovine) became active only after removal of its disaccharide units (resulting in the exposure of hydroxylysine residues). The affinity of the enzyme was limited to high molecular weight acceptors containing free hydroxylysine - hydroxylysine itself or small peptides containing hydroxylysine were inactive as acceptor substrates as were N-acetylated derivatives of active acceptors. (1)

The enzyme requires Mn^{2+} for activity; Mg^{2+} was also active but several other cations tested were inactive. The following compounds were inhibitory: ATP, UTP, UDP and sucrose. (1)

References

1. Spiro, M.J. & Spiro, R.G. (1971) JBC, 246, 4910.

UDP-GLUCURONATE-OESTRADIOL GLUCURONOSYLTRANSFERASE

(UDPglucuronate: 17β-oestradiol 3-glucuronosyltransferase)

UDPglucuronate + 17 β-oestradiol =

UDP + 17β-oestradiol 3-D-glucuronoside

Ref.

Specificity and Michaelis constants

source	substrate	K_m (M)	conditions	Ref.
Pig small intestine	17β-oestradiol [a]	7.4×10^{-6}	pH 8.0, Tris, 37°	(1)
(purified 5-10 x)	UDPglucuronate	8.4×10^{-5}	pH 8.0, Tris, 37°	(1)

(a) the following compounds were active in the place of 17β-oestradiol (1.00): oestrone (0.78); oestriol (0.04); 17α-oestradiol (0.04); dehydroepiandrosterone (0.006) and testosterone (0.002). The following were inactive: bilirubin and 4-nitrophenol.

Although the enzyme was not inhibited by EDTA, Mg^{2+} was required for maximum activity. (1)

Inhibitors

inhibitor [a]	type	K_i (M)
UDP	NC (UDPglucuronate)	2.05×10^{-4}
ATP	C (UDPglucuronate)	4.8×10^{-3}
oestrone	C (17β-oestradiol)	7.9×10^{-6}

(a) with the enzyme from pig small intestine (pH 8.0, Tris, 37°; Ref. 1).

References

1. Rao, G.S. & Breuer, H. (1969) JBC, 244, 5521.

UDP-GLUCURONATE-OESTRIOL 16α-GLUCURONOSYLTRANSFERASE

(UDPglucuronate: oestriol 16α-glucuronosyltransferase)

UDPglucuronate + oestriol = UDP + oestriol 16α-mono-D-glucuronoside

Ref.

Specificity and Michaelis constants

source	substrate	K_m (M)	conditions	Ref.
Human liver	UDPglucuronate	1×10^{-4}	pH 8.0, Tris, 37°	(1)
(purified 100 x)	oestriol[(a)]	1.33×10^{-5}	pH 8.0, Tris, 37°	(1)

(a) oestriol (1.00) could be replaced by 17β-oestradiol (0.06); 17α-oestradiol (0.026); oestrone (0.023) and cortisol (0.01) but not by aldosterone or 17α-hydroxyprogesterone.

The enzyme requires Mg^{2+} for activity; Mn^{2+} or Fe^{2+} were equally effective. (1)

Inhibitors

source	inhibitor	type	K_i (μM)	conditions	Ref.
Human	UDP[(a)]	NC(oestriol)	13	pH 8.0, Tris, 37°	(1)
liver	UDP	C(UDPglucuronate)	4.4	pH 8.0, Tris, 37°	(1)
	17-epioestriol	C(oestriol)	0.8	pH 8.0, Tris, 37°	(1)
	oestrone[(b)]	NC(oestriol)	5	pH 8.0, Tris, 37°	(1)

(a) oestriol 16α-mono-D-glucuronoside, the other product of the reaction, did not inhibit.

(b) several other steroids tested were also inhibitory.

References

1. Rao, G.S., Rao, M.L. & Breuer, H. (1970) BJ, 118, 625.

TREHALOSE PHOSPHORYLASE

(α,α-Trehalose: orthophosphate β-glucosyltransferase)

α,α-Trehalose + Pi = D-glucose + β-D-glucose 1-phosphate

Ref.

Equilibrium constant

$$\frac{[\text{glucose}]\ [\text{glucose 1-phosphate}]}{[\text{trehalose}]\ [\text{Pi}]} = 0.24 \text{ (pH 7.0, imidazole, 37°)}$$ (1)

Molecular weight

source	value	conditions	Ref.
Euglena gracilis	344,000	sucrose density gradient	(1)

Specificity and Michaelis constants

source	substrate[a]	V relative	K_m (mM)
E. gracilis (purified 75 x)	trehalose	1.0	33
	Pi	1.0	9.4
	glucose 1-phosphate	2.6	6
	glucose	2.6	32

(a) conditions: pH 7.0, imidazole 37° (Ref. 1).

The following compounds were inactive in the place of D-glucose: D-mannose; D-galactose; D-fructose; D-2-deoxyglucose; maltose; laminaribiose; laminaribiosyl β-1,4-glucose and D-glucose 6-phosphate. D-6-Deoxyglucose possessed 93% and xylose 23% the activity of D-glucose. α-Glucose 1-phosphate; α-xylose 1-phosphate; α-mannose 1-phosphate, D-fructose 1-phosphate and α-galactose 1-phosphate could not replace β-glucose 1-phosphate nor could trehalose be replaced by maltose, sucrose, lactose, laminaribiose, cellobiose or melibiose.

References

1. Maréchal, L.R. & Belocopitow, E. (1972) JBC, 247, 3223.

4-FUCOSYLTRANSFERASE

(GDP-L-fucose: β-2-acetamido-2-deoxy-D-glucosaccharide 4-α-L-fucosyltransferase)

GDP-L-fucose + 2-acetamido-2-deoxy-β-D-glucosaccharide = GDP + (1,4)-α-fucosyl-2-acetamido-2-deoxy-D-glucosaccharide

Ref.

Specificity and kinetic properties

The enzyme (human milk, purified 16 x) could not utilize L-fucose or L-fucose 1-phosphate in the place of GDP-fucose. Lacto-N-fucopentaose I or lacto-N-tetraose could serve as acceptor substrates but the following were inactive: lacto-N-pentaose II; lactose; fucosyllactose; glucosamine and glucose. Certain glycoproteins were also acceptor substrates. The enzyme requires Mn^{2+} or Mg^{2+} for full activity; other divalent ions tested did not activate the reaction. The assay conditions were: pH 6.8, Tris, 37°. (1)

References

1. Jarkovsky, Z., Marcus, D.M. & Grollman, A.P. (1970) B, 9, 1123.

UDP-GLUCOSE-CERAMIDE GLUCOSYLTRANSFERASE

(UDPglucose: *N*-acylsphingosine glucosyltransferase)

UDPglucose + ceramide = glucosylceramide + UDP

Ref.

Specificity and Michaelis constants

source	substrate	K_m (M)	conditions	
Embryonic	UDPglucose[a]	1.2×10^{-4}	pH 8.1, bicine, 30°	(1)
chicken brain	ceramide[b]	8×10^{-5}	pH 8.1, bicine, 30°	(1)
(membrane bound)	sphingosine	5×10^{-3}	pH 8.1, bicine, 30°	(1)

(a) UDPglucose (1.00) could be replaced by CDPglucose (0.06) but not by UDPgalactose; ADPglucose; GDPglucose; glucose or glucose 1-phosphate.

(b) ceramide (1.00) could be replaced by sphingosine (0.25) or dihydrosphingosine (0.25).

The enzyme has an absolute requirement for detergent and optimal activity was obtained with a mixture of Cutscum and Triton X-100. Metal ions such as Mg^{2+} or Mn^{2+} did not stimulate the reaction. (1)

References

1. Basu, S., Kaufman, B. & Roseman, S. (1973) JBC, 248, 1388.

UDP-GLUCOSE-APIGENIN β-GLUCOSYLTRANSFERASE

(UDPglucose: apigenin β-D-glucosyltransferase)

UDPglucose + apigenin = UDP + 7-O-(β-D-glucosyl)-apigenin

Ref.

Molecular weight

source	value	conditions	
Petroselinum hortense (parsley)	50,000	Sephadex G 100	(1)

Specificity and Michaelis constants

source	substrate	V relative	K_m (M)	conditions	
Parsley (purified 89 x)	UDPglucose[a]	1.00	1.2×10^{-4}	pH 7.5, Tris, 30°	(1)
	TDPglucose	0.29	2.6×10^{-4}	pH 7.5, Tris, 30°	(1)
	CDPglucose	0.07	3.6×10^{-3}	pH 7.5, Tris, 30°	(1)
	apigenin[b]	1.00	2.7×10^{-6}	pH 7.5, Tris, 30°	(1)
	naringenin	0.45	1.0×10^{-5}	pH 7.5, Tris, 30°	(1)
	luteolin	2.10	1.5×10^{-6}	pH 7.5, Tris, 30°	(1)

(a) the following were inactive: GDPglucose; ADPglucose; UDPxylose; UDPglucuronate and α-D-glucose 1-phosphate.

(b) the following flavenoids were also active (the figures in the brackets are V(relative) with apigenin = 1.00): chrysoeriol (0.63); acacetin (0.24); 7,4'-dihydroxyflavone (0.20); 5,7-dihydroxyflavanone (0.15); quercetin (0.16); kaempferid (0.17); fisetin (0.14); kaempferol (0.11); galangin (0.09); 3,7,4'-trihydroxyflavone (0.08); datiscetin (0.04) and dihydrokaempferol (0.01). The following compounds were inactive: cyanidin; daidzein; biochanin A; p-coumarate; 4-hydroxyacetophenone; resacetophenone; phloroacetophenone and phloroglucinol. (The formulae of the above compounds will be found in Ref. 1).

The enzyme did not require added divalent metal ions for activity. (1)

Abbreviations

apigenin 4',5,7-trihydroxyflavone

References

1. Sutter, A, Ortmann, R. & Grisebach, H. (1972) BBA, 258, 71.

UDP-APIOSE-FLAVONE APIOSYLTRANSFERASE

(UDP-D-apiose: 7-O-(β-D-glucosyl)-apigenin apiofuranosyltransferase)

UDPapiose + 7-O-(β-D-glucosyl)-apigenin =

UDP + 7-O(β-D-apiofuranosyl (1 → 2) β-D-glucosyl)-apigenin

Ref.

Molecular weight

source	value	conditions	
Petroselinum hortense (parsley)	50,000	Sephadex G 100	(1)

Specificity and Michaelis constants

source	substrate	K_m(M)	conditions	
Parsley (purified 123 x)	apigenin-7-glucoside	6.6×10^{-5}	pH 7.0, Tris, 30°	(1)

(a) apigenin-7-glucoside (1.00) could be replaced by biochanin-A-7-glucoside (1.00); formononetin-7-glucoside (0.66); chrysoeriol-7-glucoside (0.47); luteolin-7-glucoside (0.36); naringenin-7-glucoside (0.27); apigenin-7-glucuronide (0.25); 4-vinylphenol-glucoside (0.25) and 4-nitrophenol-β-glucoside (0.07). 4-Nitrophenol-α-glucoside; kaempferol-3- and -7-glucosides and flavenoid aglycones could not function as acceptors.

Apiosyltransferase (parsley) is specific for UDPapiose which could not be replaced by UDPglucose, UDPgalactose, UDPglucuronate, UDPxylose or TDPglucose. The enzyme did not require any added cofactor for activity; it was inhibited by a number of nucleotides of which UDP was the most potent. ATP did not inhibit. (1)

Abbreviations

apigenin	4',5,7-trihydroxyflavone
apiin	7-O-(β-D-apiofuranosyl (1 → 2) β-D-glucosyl)-apigenin
D-apiose	3-C-(hydroxymethyl)-D-glyceroaldotetraose

References

1. Ortmann, R., Sutter, A. & Grisebach, H. (1972) BBA, 289, 293.

UDP-XYLOSE-PROTEIN XYLOSYLTRANSFERASE

(UDPxylose: serine-protein xylosyltransferase)

UDPxylose + serine-protein = UDP + O-D-xylosyl-serine-protein

Ref.

Molecular weight

source	value	conditions	
Embryonic chick epiphyseal cartilage	115,000 and 175,000	Sephadex G 200, 1M NaCl (pH 7.5)	(1)

Specificity and Michaelis constants

source	substrate	K_m(M)	conditions	
Embryonic chick epiphyseal cartilage (purified 50 x)	UDPxylose	2.5×10^{-5}	pH 6.5, MES, 37°	(1)
	PG-SD	240 μg ml^{-1}	pH 6.5, MES, 37°	(1)

Xylosyltransferase catalyses the formation of a glycosidic bond between xylose and the hydroxyl group of a serine residue of the acceptor protein substrate. The best acceptor found was a protein fraction (PG-SD) obtained by Smith degradation (removal of chondroitin sulphate chains from the protein core) of bovine chondromucoprotein. (1)

References

1. Stoolmiller, A.C., Horwitz, A.L. & Dorfman, A. (1972) JBC, 247, 3525.

UDP-N-ACETYLGALACTOSAMINE-TRIGLYCOSYLCERAMIDE N-ACETYLGALACTOSAMINYL TRANSFERASE

(UDP-2-acetamido-2-deoxy-D-galactose: triglycosylceramide 2-acetamido-2-deoxy-β-D-galactosyltransferase)

UDP-N-acetylgalactosamine + triglycosylceramide = UDP + galactosaminyl-triglycosylceramide

Ref.

Specificity and catalytic properties

source	substrate	K_m (mM)	conditions	
Embryonic chicken brain (particulate)	UDP-N-acetyl-galactosamine[a]	0.2	pH 6.5, MES, 37°	(1)
	triglycosylceramide[b]	1.7	pH 6.5, MES, 37°	(1)

(a) UDPglucose and UDP N-acetylglucosamine were inactive.

(b) from pig heart. Triglycosylceramide preparations from human serum and erythrocytes and from rabbit erythrocytes were also active acceptor substrates. Triglycosylceramide (1.0) could be replaced by haematoside (1.0); disialganglioside (1.0); lac-n Tet-ceramide (0.2); lac Tet-ceramide (0.2); globoside (0.1); lactosylceramide (0.1) or galactosylceramide (0.01).

The enzyme requires Mn^{2+} (Co^{2+} had 46% and Zn^{2+}, Ca^{2+}, Mg^{2+}, Cd^{2+} and Ni^{2+} had 1-10% the activity of Mn^{2+}) and a detergent for activity. EDTA was inhibitory. (1)

Abbreviations

triglycosylceramide (or Globo Tri-cer)

O-α-galactosyl(1,4)-O-β-galactosyl(1,4)-O-β-glucosyl-(1,1)-ceramide

References

1. Chien, J-L., Williams, T. & Basu, S. (1973) JBC, 248, 1778.

GALACTINOL-SUCROSE GALACTOSYLTRANSFERASE

(Galactinol : sucrose 6-galactosyltransferase)

Galactinol + sucrose = raffinose + *myo*-inositol

Ref.

Molecular weight

source	value	conditions	
Vicia faba (broad bean, seed)	90,000	glycerol density gradient; Sephadex G 200, pH 7.5	(1)

Multiple forms have been observed. (1)

Specificity and catalytic properties

In addition to its galactosyltransferase activity the enzyme catalyzes the following exchange reaction: (1)

raffinose + [^{14}C] sucrose = [^{14}C] raffinose + sucrose

source	substrate	K_m (mM) galactosyl-transferase	K_m (mM) exchange	conditions	
V. faba (purified 400 x)	galactinol	7.0	-	pH 7.2, Tris, 32°	(1)
	raffinose	-	10	pH 7.2, Tris, 32°	(1)
	sucrose[a]	1.0	2.9	pH 7.2, Tris, 32°	(1)
	sucrose[b]	1.4	0.47	pH 7.2, Tris, 32°	(1)

(a) in the presence of 0.02 M donor.

(b) in the presence of 0.2 M donor.

The enzyme is highly specific for sucrose which could not be replaced by raffinose, stachyose, fructose, glucose, galactose, lactose, cellobiose, melibiose or glycerol. In the absence of sucrose, galactinol hydrolyses slowly (releasing galactose). Galactinol could be replaced by 4-nitrophenyl-α-D-galactopyranoside but not by UDP-galactose. The enzyme does not have α-galactosidase activity (EC 3.2.1.22). (1)

The enzyme is inhibited by heavy metal ions (e.g. Mn^{2+}, Zn^{2+} etc.) and it requires a thiol compound (e.g. dithioerythritol) for activity. (1)

Stachyose and verbascose are synthesized by different enzymes: (1)

galactinol + raffinose = stachyose + *myo*-inositol (EC 2.4.1.67)
galactinol + stachyose = verbascose + *myo*-inositol (EC 2.4.1-)

Abbreviations

galactinol	1-*O*-α-D-galactosyl-*myo*-inositol
raffinose	α-*D*-galactosyl-(1,6)-α-D-glucosyl-(1,2)-β-D-fructoside

References

1. Lehle, L. & Tanner, W. (1973) EJB, 38, 103.

PYRIMIDINE - NUCLEOSIDE PHOSPHORYLASE

(Pyrimidine-nucleoside: orthophosphate ribosyltransferase)

Pyrimidine nucleoside + Pi = α-D-ribose 1-phosphate + pyrimidine

Ref.

Equilibrium constant

The reaction is reversible. (1)

Molecular weight

source	value	conditions	Ref.
Bacillus stearothermophilus	78,000	sucrose density gradient	(1)
Escherichia coli	148,000	sucrose density gradient	(1)

Specificity and Michaelis constants

source	substrate	relative velocity	K_m (μM)	Ref.
B. stearothermophilus (purified 112 x)	uridine[a]	1.00	250	(1)
	thymidine	1.39	380	(1)
Haemophilus influenzae (purified 400 x)	uridine[b]	1.00	240	(2)
	5-methyluridine	0.27	70	(2)
	5-bromouridine	0.40	30	(2)
	deoxyuridine	0.12	130	(2)
	thymidine	0.21	110	(2)
	5-bromodeoxyuridine	0.74	100	(2)

(a) uridine (1.00) could also be replaced by deoxyuridine (1.79); 5-bromouridine (0.95); 5-bromodeoxyuridine (0.95) but not by cytidine or deoxycytidine. In the reverse reaction, ribose 1-phosphate (1.00) could be replaced by deoxyribose 1-phosphate (0.94). (Conditions = pH 7.5, Pi, 60°).

(b) the conditions were: pH 7.4, Pi, 37°. The figures under "relative velocity" are V(relative). Purine- and 4-amino-pyrimidine-nucleosides were inactive.

Light absorption data

base	λ(nm)	molar extinction coefficient	Ref.
uracil	290	5700 $M^{-1}cm^{-1}$	(2)
thymine	300	3700 $M^{-1}cm^{-1}$	(2)
5-bromouracil	312	2200 $M^{-1}cm^{-1}$	(2)

References

1. Saunders, P.P., Wilson, B.A. & Saunders, G.F. (1969) JBC, 244, 3691.
2. Scocca, J.J. (1971) JBC, 246, 6606.

ANTHRANILATE PHOSPHORIBOSYLTRANSFERASE

(N-(5'-Phosphoribosyl)-anthranilate:
pyrophosphate phosphoribosyltransferase)

N-(5'-Phospho-D-ribosyl)-anthranilate + PPi =
anthranilate + 5-phospho-α-D-ribose 1-diphosphate

Ref.

Molecular properties

source	value	conditions	
Salmonella typhimurium	150,000 [2]	Sephadex G 150 and G 200, pH 7.4; gel electrophoresis (SDS)	(1,2)

Both molecular forms (monomer and dimer) of the transferase are enzymically active. (1)

Anthranilate phosphoribosyltransferase is an enzyme of the tryptophan pathway (see EC 4.1.3.27) and in many organisms it is associated with anthranilate synthase (EC 4.1.3.27). In S. typhimurium, the aggregated form has the structure I_2II_2 and the unaggregated form II_1 (or II_2). Unless otherwise stated, the properties given here are those of the unaggregated enzyme.

Specific activity

S. typhimurium	(30 x)	1.5	anthranilate (pH 7.4, triethanolamine)	(1)

The enzyme was purified by affinity chromatography on a column of anthranilate coupled to succinylamidohexamethylimino-Sepharose.

Specificity and catalytic properties

source	substrate	K_m (μM)	conditions	
S. typhimurium	anthranilate	8.3 (a)	pH 7.4, triethanolamine	(2)
	PRPP	6.7 (a)	pH 7.4, triethanolamine	(2)
	Mg^{2+} (b)	20	pH 7.4, triethanolamine	(2)

(a) the values are for the aggregated transferase.

(b) the aggregated form required Mg^{2+}. The unaggregated form did not require Mg^{2+} nor was it inhibited by EDTA.

Both aggregated and unaggregated forms of the transferase are inhibited by tryptophan (see EC 4.1.3.27). The transferase requires dithiothreitol for full activity. (2)

Abbreviations

PRPP 5-phospho-α-D-ribose 1-diphosphate

References

1. Marcus, S.L. & Balbinder, E. (1972) BBRC, 47 438.
2. Henderson, E.J., Zalkin, H. & Hwang, L.H. (1970) JBC, 245, 1424.

NICOTINATENUCLEOTIDE-DIMETHYLBENZIMIDAZOLE PHOSPHORIBOSYLTRANSFERASE

(Nicotinatenucleotide: 5,6-dimethylbenzimidazole phosphoribosyltransferase)

β-Nicotinate D-ribonucleotide + 5,6-dimethylbenzimidazole = nicotinate + 1-α-D-ribosyl-5,6-dimethylbenzimidazole 5'-phosphate

Ref.

Equilibrium constant

The reaction is irreversible [F] with the enzyme from *Propionibacterium shermanii* as catalyst. (1)

Specificity and catalytic properties

source	substrate	second substrate	V relative	K_m (μM)	
Clostridium sticklandii (purified 680 x)	β-N_aMN [a]	benzimidazole	1.0	700	(2)
	benzimidazole [b]	β-N_aMN	1.0	500	(2)
	β-N_aMN	adenine	0.2	300	(2)
	adenine [c]	β-N_aMn	0.2	10	(2)

(a) β-N_aMN (1.00) could be replaced by desamido NAD(0.10) but not by NAD or nicotinamide mononucleotide (conditions: pH 8.6, Tris, 37°).

(b) a number of bases could replace benzimidazole e.g. 5,6-dimethylbenzimidazole; 5,6-dichloro- or 5(6)-nitro-benzimidazole.

(c) with adenine as the base the product of the reaction was 7-α-D-ribosyladenine-5'-phosphate (see Ref. 3).

Abbreviations

β-N_aMN β-nicotinate D-ribonucleotide

References

1. Friedmann, H.C. (1965) JBC, 240, 413.
2. Fyfe, J.A. & Friedmann, H.C. (1969) JBC, 244, 1659.
3. Friedmann, H.C. & Fyfe, J.A. (1969) JBC, 244, 1667.

SIALYLTRANSFERASE

(CMP-*N*-acetylneuraminate: D-galactosylglycoprotein
N-acetylneuraminyltransferase)

CMP-*N*-acetylneuraminate + D-galactosylglycoprotein =
CMP + *N*-acetylneuraminyl-D-galactosylglycoprotein

Ref.

Specificity and Michaelis constants

substrate	K_m (mM)
CMP-*N*-acetylneuraminate[a]	0.27
thyroglobulin glycopeptides[a,b]	0.59
CMP-*N*-acetylneuraminate[c]	0.047

(a) with the enzyme from calf thyroid (purified 94 x). Conditions: pH 8.0, Tris, 37°; Ref. 1.

(b) sialic acid free. Km expressed as *N*-acetylneuraminate acceptor site concentration.

(c) with the enzyme from rat liver (purified 75 x). Conditions: pH 7.5, Tris, 37°; Ref. 2.

The most favourable acceptor substrates with the sialyltransferase from calf thyroid were glycopeptides or glycoproteins from which sialic acid had been removed and which contained terminal galactose. Low molecular weight compounds such as galactose, lactose and *N*-acetyllactosamine were inactive as were also galactosylhydroxylysine derivatives. Low activities were obtained with glycopeptides having terminal *N*-acetylglucosamine or mannose. The enzyme did not require added mono- or di-valent metal ions for activity. (1)

Sialyltransferase is also present in human serum. This enzyme could utilize the following as acceptor substrates: lactose (1.00); *N*-acetylneuraminic acid free fetuin (0.81); *N*-acetylneuraminic acid and galactose free fetuin (0.05) and native fetuin (0.04). Native or *N*-acetylneuraminic acid free ovine submaxillary mucin was inactive. The serum enzyme may need a divalent metal ion for activity. (3)

References

1. Spiro, M.J. & Spiro, R.G. (1968) JBC, 243, 6520.
2. Hickman, J., Ashwell, G., Morell, A.G., van den Hamer, C.J.A. & Scheinberg, I.H. (1970) JBC, 245, 759.
3. Kim, Y.S., Perdomo, J., Bella, A. & Nordberg, J. (1971) BBA, 244, 505.
   Bartholomew, B.A., Jourdian, G.W. & Roseman, S. (1973) JBC, 248, 5751; 5763.

ENOYLPYRUVATE TRANSFERASE

(Phosphoenolpyruvate: UDP-2-acetamido-2-deoxy-D-glucose 2-enoyl-1-carboxyethyltransferase)

Phosphoenolpyruvate + UDP-2-acetamido-2-deoxy-D-glucose = Pi + UDP-2-acetamido-2-deoxy-3-enoylpyruvoylglucose

Ref.

Equilibrium constant

The reaction is reversible. (1)

Specificity and Michaelis constants

substrate[a]	K_m (M)	conditions	Ref.
phosphoenolpyruvate[b]	3×10^{-5}	pH 7.4, Tris, 37°	(1)
UDP-N-acetylglucosamine[c]	4.6×10^{-4}	pH 7.4, Tris, 37°	(1)

(a) with the enzyme from Enterobacter cloacae (purified 326 x).

(b) glycerate 2-phosphate, glycerate 3-phosphate and pyruvate were inactive.

(c) UDP-N-acetylgalactosamine and UDPglucose were inactive.

The enzyme requires the presence of a sulphydryl compound but not of monovalent or divalent metal ions for activity. No inhibition was observed in the presence of EDTA. (1)

The characterization and properties of UDP-2-acetamido-2-deoxy-3-enoylpyruvoylglucose (a precursor of UDP-N-acetylmuramic acid) are given in Ref. 1.

References

1. Gunetileke, K.G. & Anwar, R.A. (1968) JBC, 243, 5770.

RIBOFLAVIN SYNTHASE

(6,7-Dimethyl-8-(1'-D-ribityl)lumazine:
6,7-dimethyl-8-(1'-D-ribityl)lumazine 2,3-butanediyltransferase)

2 6,7-Dimethyl-8-(1'-D-ribityl)lumazine =
riboflavin + 4-(1'-D-ribitylamino)-5-amino-2,6-dihydroxypyridine

Ref.

Molecular properties

source	value	
Yeast	75,000[a]	(1,2)

(a) the enzyme contains one mole riboflavin per 75,000 daltons. The riboflavin can be displaced by 6-methyl-7-hydroxy-8-(1'-D-ribityl)lumazine.

Specific activity

Yeast (4000 x)	0.25	DRL(pH 7.0, Pi, 37°. In the presence of 10mM $NaHSO_3$. The enzyme was inactivated by O_2)	(1)

Specificity and kinetic properties

source	substrate	K_m(M)	conditions
yeast	DRL	1×10^{-5}	pH 7.0, Pi, 37°

Of the epimeric forms of pentyl derivatives tested as substrates, only those with hydroxyl groups in the D-configuration in positions 2' and 4' of the side chains are bound to the enzyme. Thus, 6,7-dimethyl-8-(1'-(2'-deoxy-D-ribityl))lumazine is neither a substrate nor inhibitor. Further, lengthening of the sugar side chain to D-glucitol derivatives leads to loss of activity and shortening of the sugar chain to D- and L-threityl and D- and L-erythrityl derivatives provides relatively ineffective competitive inhibitors (Ki approximately 0.5 to 1×10^{-3}M). Of several compounds substituted at the D-ribitol group by simple substituents, only 6,7-dimethyl-8-(1'-D-xylitol)lumazine (Ki = 9×10^{-5}M) is bound to the enzyme. Replacement of the 2 methyl groups at positions 6 and 7 leads to inactive substances (methyl + ethyl; methyl + *n*-propyl; methyl + *n*-butyl; ethyl + ethyl; *n*-propyl + *n*-propyl; phenyl + phenyl or methyl + phenyl) or competitive inhibitors (methyl + *n*-pentyl (Ki = 2×10^{-4}M); 5,6,7,8-tetrahydro-9-(1'-D-ribityl) isoalloxazine (Ki = 1.6×10^{-4}M); 6-methyl-7-hydroxy-8-(1'-D-ribityl)lumazine (Ki = 2×10^{-6}M) and 6,7-dihydroxy-8-(1'-D-ribityl)-lumazine (Ki = 9×10^{-9}M). Riboflavin is a product inhibitor (Ki = 5×10^{-6}M) but FMN and FAD are inactive. The reader is referred to Refs. 1,2 & 4 for further specificity details and also for the mechanism of action of riboflavin synthase.

Riboflavin synthase has also been purified from *Ashbya gossypii* (180 x). This enzyme has similar properties to the enzyme in yeast. (3)

Ref.

Light absorption data

Riboflavin has an absorption maximum at 470 nm (molar extinction coefficient 9600 M^{-1} cm^{-1}). (1)

Abbreviations

DRL 6,7-dimethyl-8-(1'-D-ribityl)lumazine.

References

1. Plaut, G.W.E. & Harvey, R.A. (1971) Methods in Enzymology, 18B, 527.
2. Harvey, R.A. & Plaut, G.W.E. (1966) JBC, 241, 2120.
3. Plaut, G.W.E. (1963) JBC, 238, 2225.
4. Plaut, G.W.E., Beach, R.L. & Aogaichi, T. (1970) B, 9, 771.

GERANYLTRANSFERASE

(Geranyldiphosphate: isopentenyldiphosphate geranyltransferase)

Geranyldiphosphate + isopentenyldiphosphate =

PPi + farnesyldiphosphate

Ref.

Specificity and Michaelis constants

source	substrate	K_m (μM)	conditions	
Pig liver (purified 10-15 x)	geranyl diphosphate	4.0	pH 7.0, imidazole 38°	(1)
	isopentenyl diphosphate	2.0	pH 7.0, imidazole, 38°	(1)

The enzyme required Mn^{2+} (Mg^{2+} was much less effective) as activator. It did not have isopentenyl diphosphate isomerase (EC 5.3.3.2), geranyl diphosphate synthase (EC 2.5.1.1) or geranyl geranyl diphosphate synthase activities. High concentrations of isopentenyl diphosphate were inhibitory. (1)

References

1. Benedict, C.R., Kett, J. & Porter, J.W. (1965) ABB, 110, 611.

GLUTATHIONE S-ALKYLTRANSFERASE

(Alkyl-halide: glutathione S-alkyltransferase)

Methyl iodide + glutathione = S-methylglutathione + HI

Ref.

Specificity and catalytic properties

source	substrate	K_m (mM)	conditions	
Rat liver	methyl iodide[a]	0.37	pH 7.0, Pi, 25°	(1)
(purified 14 x)	glutathione[b]	2.2	pH 7.0, Pi, 25°	(1)

(a) methyl iodide (1.00) could be replaced by methyl bromide (0.40); ethyl iodide (0.20); ethyl bromide (0.13) or 1-iodopropane (0.13). 1-Bromopropane, chloroacetonitrile, iodoacetamide and β-propiolactone were poor substrates and the following were inactive: 2-bromopropane; bromochloromethane; 2-chloroethanol; 2-bromoethanol; chloroacetaldehyde; chloroacetate; chloroacetamide; iodoacetate and 2- and 3-bromopropionate.

(b) glutathione could not be replaced by L-cysteine; L-histidine; 2-mercaptopropionate; 2,3-dimercaptopropan-1-ol (BAL); 2-mercaptoethylamine or 2-mercaptoethanol.

Several alkyl halides and thiol compounds react non-enzymically; thus chloroacetaldehyde reacts with glutathione. The enzyme did not require any added cofactor; it was not inhibited by EDTA. (1)

Glutathione alkyltransferase was inhibited by S-methylglutathione, oxidized glutathione and by several aromatic nitro compounds. (1)

References

1. Johnson, M.K. (1966) BJ, 98, 44.

DIHYDROPTEROATE SYNTHASE

(2-Amino-4-hydroxy-6-hydroxymethyl-7,8-dihydropteridine-diphosphate: 4-aminobenzoate 2-amino-4-hydroxydihydropteridine-6-methenyl-transferase)

2-Amino-4-hydroxy-6-hydroxymethyl-7,8-dihydropteridine diphosphate + 4-aminobenzoate = dihydropteroate + PPi

Ref.

Equilibrium constant

The reaction is essentially irreversible [F] with the enzyme from Escherichia coli as catalyst. (1,2)

Molecular weight

source	value	conditions	
E. coli	52,000	Sephadex G 100, pH 7.0	(1,2)

Specificity and Michaelis constants

source	substrate	K_m(M)	conditions	
E. coli (purified 52 x)	4-aminobenzoate	2.5×10^{-6}	pH 8.6, Tris, 37°	(1,2)

The enzyme requires Mg^{2+} for activity. (1,2,3)

Dihydropteroate synthase is also present in Lactobacillus plantarum. This enzyme can utilize 4-aminobenzoyl glutamate in the place of 4-aminobenzoate. With 4-aminobenzoate or 4-aminobenzoyl glutamate as the variable substrate, sigmoid dependence curves were obtained. (3)

Inhibitors

The enzyme (E. coli) was inhibited by 4-aminobenzoylglutamate (C(4-aminobenzoate); Ki = 10^{-3}M) and by dihydropteroate (C(2-amino-4-hydroxy-6-hydroxymethyl-7,8-dihydropteridine diphosphate)) but not by 4-aminohippurate, 2-aminobenzoate, 3-aminobenzoate, dihydrofolate, tetrahydropteroyltriglutamate, PPi or Pi. (1,2)

Sulphathiozole and other sulphonamides were inhibitory. (4)

References

1. Richey, D.P. & Brown, G.M. (1971) Methods in Enzymology, 18B, 770.
2. Richey, D.P. & Brown, G.M. (1969) JBC, 244, 1582.
3. Shiota, T., Baugh, C.M., Jackson, R. & Dillard, R. (1969) B, 8, 5022.
4. Brown, G.M. (1962) JBC, 237, 536.

PROPYLAMINE TRANSFERASE

(5'-Deoxyadenosyl-(5'), 3-aminopropyl-(1), methylsulphonium salt: putrescine 3-aminopropyltransferase)

5'-Deoxyadenosyl-(5'), 3-aminopropyl-(1), methylsulphonium salt + putrescine = 5'-methylthioadenosine + spermidine

Ref.

Molecular weight

source	value	Ref.
Escherichia coli W	73,000 [2]	(1)

Specificity and kinetic properties

source	substrate	K_m(M)	conditions	Ref.
Rat prostate (purified 80 x)	decarboxylated S-adenosylmethionine	5×10^{-5}	pH 7.2, Pi, 30°	(2)

Propylamine transferases isolated from different sources differ in their specificity towards the amine substrate. Thus, the enzyme from E. coli (purified 2000 x) was most active with putrescine but some propylamine transfer was also observed with spermidine (to form spermine) or with 1,5-diamino-pentane. Rat brain, however, has 2 propylamine transferases - one is specific for putrescine and the other for spermidine. The two enzymes have been separated. (1,3)

Propylamine transferase from E. coli does not require any added cofactor for activity. (1)

Abbreviations

putrescine	1,4-diaminobutane
decarboxylated S-adenosyl methionine	5'-deoxyadenosyl-(5'), 3-aminopropyl-(1),-methylsulphonium salt

References

1. Tabor, H. & Tabor, C.W. (1972) Advances in Enzymology, 36, 215, 232.
2. Jänne, J., Schenone, A. & Williams-Ashman, H.G. (1971) BBRC, 42, 758.
3. Raina, A. & Hannonen, P. (1971) FEBS lett, 16, 1.
   Bowman, W.H., Tabor, C.W. & Tabor, H. (1973) JBC, 248, 2480.

AQUOCOB(I)ALAMIN ADENOSYLTRANSFERASE

(ATP: aquocob(I)alamin Co-adenosyltransferase)

ATP + aquocob(I)alamin + H_2O = adenosylcobalamin + Pi + PPi

Ref.

Equilibrium constant

The reaction is irreversible [F] with the enzyme from *Clostridium tetanomorphum* as the catalyst (Ref. 2 but also see Ref. 1).

Specificity and Michaelis constants

source	substrate	K_m (M)	conditions	
Cl. tetanomorphum (purified 300 x)	ATP (a)	1.6×10^{-5}	pH 8.0, Tris, 37°	(1,2)
	aquocob(I)-alamin (b)	1.0×10^{-5}	pH 8.0, Tris, 37°	(1,2)

(a) ATP (1.00) could be replaced by CTP (1.02), ITP (0.53), UTP (0.39) or GTP (0.22) but not by ADP or AMP or *S*-adenosylmethionine.

(b) other vitamin B_{12} derivatives tested were inactive.

Mn^{2+} was required for full activity; Co^{2+} and Mg^{2+} were less effective. (2)

The enzyme from *Propionibacterium shermanii* (purified 337 x) could not utilize UTP, GTP or ITP in the place of ATP. This enzyme requires the presence of Mn^{2+}, K^+, reduced FAD and a sulphydryl compound for activity. It was active with a number of vitamin B_{12} derivatives. (3)

Inhibitors

The enzyme (*Cl. tetanomorphum*) was inhibited by PPi, tripolyphosphate, trimetaphosphate and high concentrations of ATP but not by Pi. (2)

Light absorption data

The transformation of aquocob(I)alamin to adenosylcobalamin is accompanied by an increase in absorption at 525 nm (molar extinction coefficient = 5,800 $M^{-1}cm^{-1}$). (2)

Abbreviations

aquocob(I)alamin — α-(5,6-dimethylbenzimidazolyl)aquocobamide with the cobalt atom in the +1 oxidation state. Also called vitamin B_{12s}

adenosylcobalamin — also called adenosyl-B_{12}

References

1. Mudd, S.H. (1973) *The Enzymes*, *8A*, 144.
2. Vitols, E., Walker, G.A. & Huennekens, F.M. (1966) *JBC*, *241*, 1455.
3. Brady, R.O., Castanera, E.G. & Barker, H.A. (1962) *JBC*, *237*, 2325.

RUBBER TRANSFERASE

((*cis*-1,4-Isoprene)$_n$ diphosphate: isopentenyl diphosphate *cis*-1,4-isoprenyltransferase)

(*cis*-1,4-Isoprene)$_n$ diphosphate + isopentenyl diphosphate = (*cis*-1,4-isoprene)$_{n+1}$ diphosphate + PPi

Ref.

Molecular weight

source	value	conditions	Ref.
Hevea latex	60,000	gel filtration	(1)

Catalytic properties

Rubber transferase (purified 350 x) utilizes rubber particles as the acceptor substrate. The enzyme is strictly stereospecific and yields exclusively *cis*-polyisoprene from isopentenyl pyrophosphate. The enzyme was activated by neryl pyrophosphate but farnesyl pyrophosphate was inhibitory. Mg^{2+} (which could not be replaced by Mn^{2+}, Co^{2+} or Fe^{2+}) and a sulphydryl compound were essential for activity. Ascorbate had no effect on the reaction. (1)

The preparation of rubber particles has been described. (1,2)

References

1. Archer, B.L. & Cockbain, E.G. (1969) Methods in Enzymology, 15, 476.
2. McMullen, A.I. & McSweeney, G.P. (1966) BJ, 101, 42.

PRESQUALENE SYNTHASE

(Farnesyl diphosphate: farnesyl diphosphate farnesyltransferase)

2 Farnesyl diphosphate = PPi + presqualene diphosphate

Ref.

Molecular and catalytic properties

Squalene synthase has been purified from bakers' yeast (120 x). The enzyme exists in 2 interconvertible forms: a protomeric form of molecular weight 450,000 daltons (by sucrose density gradient centrifugation) and a polymeric form of a very high molecular weight. Both forms catalyze the synthesis of presqualene diphosphate from farnesyl diphosphate about equally well. The polymeric form (but not the protomeric form) also catalyzes the formation of squalene from presqualene diphosphate:-

presqualene diphosphate + reduced NADP = squalene + NADP + PPi

A divalent metal ion (Mg^{2+} or Mn^{2+}) is essential for the formation of presqualene pyrophosphate but not for the formation of squalene. The latter reaction was, however, activated by Mg^{2+}. (1,2)

The enzyme was inhibited by PPi and NADP. The mechanisms of the reactions have been investigated. (1)

substrate	K_m (µM)	conditions	
farnesyl diphosphate[a]	0.44	pH 7.5, Pi, 37°	(1)
reduced NADP[b]	61	pH 7.5, Pi, 37°	(1)
presqualene diphosphate	0.77	pH 7.4, Pi, 37°	(1)

(a) farnesyl diphosphate (1.00) could be replaced by geranylgeranyl diphosphate (0.06) to form lycopersene.

(b) reduced NADP (1.00) could be replaced by reduced NAD (0.37).

References

1. Qureshi, A.A., Beytia, E. & Porter, J.W. (1973) JBC, 248, 1848; 1856.
2. Qureshi, A.A., Beytia, E.D. & Porter, J.W. (1972) BBRC, 48, 1123.

GLUTAMINE-OXOACID AMINOTRANSFERASE

(L-Glutamine: 2-oxoacid aminotransferase)

L-Glutamine + a 2-oxoacid =

2-oxoglutaramate + an L-amino acid

Ref.

Equilibrium constant

$$\frac{[\text{2-oxoglutaramate}]\ [\text{glycine}]}{[\text{L-glutamine}]\ [\text{glyoxylate}]} = 607 \text{ (pH 8.4, Tris, 37°)}$$ (1)

The reaction is in the direction of glutamine utilization since 2-oxoglutaramate cyclizes nonenzymically to 5-hydroxypyrolidone carboxylate. The equilibrium constants for the glutamine-pyruvate and glutamine-2-oxo-4-methiolbutyrate reactions have been determined. (1)

Molecular weight

source	value	conditions	
Rat liver	110,000 [2]	pH 7.2	(1)

The enzyme contains tightly bound pyridoxal 5'-phosphate. (1)

Specific activity

Rat liver (870 x)	5	L-glutamine (second substrate = glyoxylate; pH 8.4, Tris, 37°)	(1,2)

Ref.

source	substrate	second substrate	V relative	K_m (mM)
Rat liver	L-glutamine[a]	2-oxoglutaramate	1.00	2.0
	L-methionine	2-oxoglutaramate	0.04	1.9
	L-methionine sulphoxide	2-oxoglutaramate	0.04	2.0
	L-glutamic acid α-methyl ester	2-oxoglutaramate	0.04	4.4
	2-oxoglutaramate[b]	L-glutamine	1.00	0.2
	2-oxo-4-methiol-butyrate	L-glutamine	0.75	3.2
	3-mercaptopyruvate	L-glutamine	0.57	4.8
	glyoxylate	L-glutamine	0.27	8.0
	3-hydroxypyruvate	L-glutamine	0.18	8.0
	pyruvate	L-glutamine	0.07	10.5

(a) L-glutamine (1.00) could also be replaced by L-ethionine (0.08) and L-homoserine (0.10). Low activities were shown by L-glutamic γ-benzyl ester; L-methionine sulphone; L-phenylalanine; L-asparagine; L-cysteine; L-alanine; L-serine or glycine. Similar results were obtained with glyoxylate or pyruvate as the second substrate. With glyoxylate the following were inactive as amino group donors: L-glutamate; L-aspartate; L-ornithine; L-tyrosine; L-cystine; L-leucine; L-threonine; L-valine; L-isoleucine; L-arginine; L-cysteate; DL-2-amino-4-phosphonobutyrate; L-tryptophan; L-histidine; L-lysine; β-alanine; L-2,4-diaminobutyrate and amino malonate. (Conditions: pH 8.4, Tris, 37°; Ref. 1).

(b) the concentration of 2-oxoglutaramate is expressed as the open-chair form. 2-Oxoglutaramate could be replaced by a number of other 2-oxo-acids tested but not by 2-oxoisovalerate; trimethylpyruvate; 2-oxo-3-hydroxybutyrate or 4-hydroxyphenylpyruvate.

A partially purified preparation of the aminotransferase has been obtained from rat kidney. This preparation also hydrolysed 2-oxo-glutaramate to 2-oxoglutarate and NH_3. (3)

References

1. Cooper, A.J.L. & Meister, A. (1972) B, 11, 661.
2. Cooper, A.J.L. & Meister, A. (1970) Methods in Enzymology, 17A, 1016.
3. Kupchik, H.Z. & Knox, W.E. (1970) Methods in Enzymology, 17A, 951.

DIIODOTYROSINE AMINOTRANSFERASE

(3,5-Diiodo-L-tyrosine: 2-oxoglutarate aminotransferase)

3,5-Diiodo-L-tyrosine + 2-oxoglutarate =

3,5-diiodo-4-hydroxyphenylpyruvate + L-glutamate

Ref.

Equilibrium constant

The reaction is reversible. (1)

Molecular weight

source	value	conditions	
Rat kidney	80,000	Sephadex G 100, pH 7.4	(2)
Rabbit kidney	95,000 [2]	sucrose gradient centrifugation	(1)

Specificity and kinetic properties

source	substrate	V relative	K_m (M)	pH [(a)]
Rat kidney (purified 32 x)	3,5-dibromotyrosine	1.00	3.7×10^{-3}	6.5
	3,5-dichlorotyrosine	1.02	6.1×10^{-3}	6.7
	3,5-diiodotyrosine [(b)]	0.48	1.3×10^{-3}	6.2
	3-iodotyrosine	0.70	6.5×10^{-3}	7.4
	2-oxoglutarate [(c)]	-	3.7×10^{-4}	6.2
	pyridoxal phosphate	-	4.1×10^{-6}	6.2

(a) Pi, 37° (Ref. 2)

(b) 3,5-diiodo-L-tyrosine (1.00) could be replaced by the thyroid hormones; L-tyrosine (0.04); L-phenylalanine (0.04); L-tryptophan (0.04) or by 4-fluoro-DL-phenylalanine (0.03) but not by 3,5-diiodo-D-tyrosine; 3,5-dinitro-L-tyrosine; 3-nitro-L-tyrosine or 3-amino-L-tyrosine.

(c) 2-oxoglutarate (1.00) could be replaced by oxaloacetate (0.30) but not by pyruvate.

The aminotransferase has also been purified (1900 x) from rabbit kidney. This enzyme has different specificity properties from the rat kidney enzyme. Thus, it utilized 3,5-dinitro-L-tyrosine much more effectively than any other substrate tested. (1)

Inhibitors

The enzyme (rat kidney) was inhibited by 3,5-diiodo-4-hydroxyphenylacetate (mixed type; $Ki = 4.4 \times 10^{-5}$M) and by several other aromatic compounds. (2)

References

1. Soffer, R.L., Hechtman, P. & Savage, M. (1973) JBC, 248, 1224.
2. Nakano, M. (1967) JBC, 242, 73.

TRYPTOPHAN AMINOTRANSFERASE

(L-Tryptophan: 2-oxoglutarate aminotransferase)

L-Tryptophan + 2-oxoglutarate = indolepyruvate + L-glutamate

Ref.

Molecular weight

source	value	conditions	Ref.
Clostridium sporogenes	97,000	sucrose density gradient	(1)

Specificity and Michaelis constants

source	substrate	K_m (M)	Ref.
C. sporogenes (purified 200-fold)	tryptophan[a]	2.68×10^{-3}	(1)
	2-oxoglutarate	1.58×10^{-4}	(1)
	pyridoxal 5'phosphate[b]	2.18×10^{-6}	(1)

(a) tryptophan (1.00) could be replaced by phenylalanine (0.08) or tyrosine (0.07). (Conditions: pH 8.4, Tris, 37°)

(b) pyridoxal 5'-phosphate, which was required by the enzyme, could be replaced by pyridoxamine 5'-phosphate but not by pyridoxol, pyridoxal or pyridoxamine.

Light absorption data

The enol form of indolepyruvate has an absorption band of 305 nm (molar extinction coefficient = 3,900 $M^{-1}cm^{-1}$). (1)

References

1. O'Neil, S.R. & deMoss, R.D. (1968) ABB, 127, 361.

DIAMINE AMINOTRANSFERASE

(Diamine: 2-oxoglutarate aminotransferase)

An α,ω-diamine + 2-oxoglutarate = an ω-aminoaldehyde + L-glutamate

Ref.

Specificity and Michaelis constants

source	substrate	K_m (M)	conditions	
Escherichia coli (purified 80 x)	2-oxoglutarate[a]	8.8×10^{-4}	pH 9, Tris	(1)
	pyruvate	2.7×10^{-3}	pH 9, Tris	(1)

(a) oxaloacetate was inactive.

The enzyme (*E. coli*) could utilize the following as substrates in the presence of 2-oxoglutarate: 1,4-diaminobutane (putrescine, 1.00); 1,5-diaminopentane (cadaverine, 1.07); and 1,7-diaminoheptane (0.30). The following were inactive as substrates:- 1,3-diaminopropane; lysine; ornithine; spermidine; β-alanine; 4-aminobutanol; histamine; tyramine and 4-aminobutyrate. The enzyme requires pyridoxal-phosphate for activity. (1)

References

1. Kim, K. (1964) JBC, 239, 783.
2. Kim, K. & Tchen, T.T. (1971) Methods in Enzymology, 17B, 812.

GLYCINE-OXALOACETATE AMINOTRANSFERASE

(Glycine: oxaloacetate aminotransferase)

Glycine + oxaloacetate = glyoxylate + L-aspartate

Ref.

Equilibrium constant

$$\frac{[\text{glyoxylate}]\ [\text{L-aspartate}]}{[\text{glycine}]\ [\text{oxaloacetate}]} = 0.0164 \text{ (pH 7.1, Pi, 25°)}$$ (1)

Specificity and Michaelis constants

source	substrate	K_m (mM)	conditions	
Micrococcus denitrificans (purified 40 x)	glyoxylate[a]	0.43	pH 7.1, Pi, 25°	(1)
	L-aspartate[b]	1.9	pH 7.1, Pi, 25°	(1)

(a) no other compound tested was active.

(b) L-aspartate (1.00) could be replaced by L-serine (0.16), L-asparagine (0.07) or *threo*-L-3-hydroxyaspartate (0.02).

The enzyme requires pyridoxal 5'-phosphate for activity but there was no metal ion requirement. (1)

Inhibitors

The enzyme was inhibited by *erythro*-DL-2-hydroxyaspartate (C(L-aspartate), Ki = 0.65 mM) and by NH_4^+ and high concentrations of Tris. (1)

References

1. Gibbs, R.G. & Morris, J.G. (1970) *Methods in Enzymology*, *17A*, 982.

LYSINE 6-AMINOTRANSFERASE

(L-Lysine: 2-oxoglutarate 6-aminotransferase)

L-Lysine + 2-oxoglutarate = 2-aminoadipate 5-semialdehyde + L-glutamate

Ref.

Equilibrium constant

Only the forward reaction has been observed. 2-Aminoadipate 5-semialdehyde is converted nonenzymically into the intramolecularly dehydrated form, Δ'-piperideine 6-carboxylate. (1)

Molecular weight

source	value	conditions	
Achromobacter liquidum	116,000	pH 7.0	(1)

The enzyme contains 2 moles of pyridoxal phosphate per 116,000 daltons. Incubation of the enzyme with L-lysine in the presence of high Pi gave an inactive form of the enzyme (semiapoenzyme) which could be reactivated with pyridoxal phosphate. (1)

Specific activity

A. liquidum (180 x)	19	L-lysine (pH 8.0, Pi, 37°)	(1)

Specificity and Michaelis constants

source	substrate[(a)]	cosubstrate	relative velocity	K_m (M)	
A. liquidum	L-lysine	2-oxoglutarate	1.00	2.8×10^{-3}	(1)
	L-ornithine	2-oxoglutarate	0.55	2.0×10^{-3}	(1)
	2-oxoglutarate	L-lysine	-	5.0×10^{-4}	(1)
	2-oxoglutarate	L-ornithine	-	1.3×10^{-4}	(1)
	pyridoxal 5'-phosphate	-	-	3.6×10^{-7}	(1)
	pyridoxamine 5'-phosphate	-	-	7.1×10^{-6}	(1)

(a) conditions: pH 8.0, Pi, 37°.

The following compounds were inactive in the place of L-lysine: 5-hydroxylysine; N^2-acetyl-L-lysine; N^6-acetyl-L-lysine, N^6-methyl-L-lysine; L-2,4-diaminobutyrate; 6-aminocaproate; 5-aminovalerate; 4-aminobutyrate; DL-norleucine; DL-norvaline; L-leucine; L-valine, cadaverine; putrescine; L-arginine; L-phenylalanine and L-aspartate. The following compounds were inactive in the place of 2-oxoglutarate: glyoxylate, pyruvate, oxaloacetate, 2-oxobutyrate, 2-oxovalerate and 2-oxocaproate. (1)

Inhibitors

inhibitor[(a)]	type	K_i (M)	conditions	
5-aminovalerate	C(L-lysine)	7.7×10^{-5}	pH 8.0, Pi, 37°	(2)
5-hydroxylysine	C(L-lysine)	1.4×10^{-2}	pH 8.0, Pi, 37°	(2)

Ref.

Inhibitors continued.

(a) with the enzyme from A. liquidum. Other inhibitors were O-(2-aminoethyl)-DL-serine and Tris buffers.

References

1. Soda, K. & Misono, H. (1968) B, 7, 4110.
2. Soda, K. & Misono, H. (1971) Methods in Enzymology, 17B, 222.

2-AMINOADIPATE AMINOTRANSFERASE

(L-2-Aminoadipate: 2-oxoglutarate aminotransferase)

L-2-Aminoadipate + 2-oxoglutarate = 2-oxoadipate + L-glutamate

Ref.

Equilibrium constant

$$\frac{[\text{2-oxoadipate}]\ [\text{L-glutamate}]}{[\text{2-aminoadipate}]\ [\text{2-oxoglutarate}]} = 1.32 \text{ (pH 7.5, Pi, 37°)}$$ (1)

Molecular weight

source	value	conditions	
Bakers' yeast, enzyme I	100,000	Sephadex G 200, pH 7.0	(2)
enzyme II	140,000	Sephadex G 200, pH 7.0	(2)

Specificity and Michaelis constants

source	enzyme	substrate	K_m (M)	conditions	
Bakers' yeast	I	2-oxoadipate	1.4×10^{-2}		
(purified	II	2-oxoadipate	1.4×10^{-2}		
several fold)	I	glutamate[a]	1.0×10^{-2}	pH 7.0,	(2)
	II	glutamate[b]	(c)	Pi, 25°	
	I	pyridoxal phosphate	2.2×10^{-6}		
	II	pyridoxal phosphate	5.0×10^{-6}		
Rat liver		2-aminoadipate[d]	9.0×10^{-3}		
mitochondria		2-oxoglutarate	1.3×10^{-3}		
purified		pyridoxal phosphate	4.4×10^{-6}	pH 7.5,	(1)
100 x)		pyridoxamine phosphate	6.3×10^{-5}	Pi, 37°	
		2-oxoadipate	5×10^{-4}		
		glutamate	5×10^{-3}		

(a) glutamate (1.00) could be replaced by aspartate (0.30) but not by alanine.
(b) aspartate and alanine were inactive.
(c) the glutamate dependence curves were sigmoidal in shape.
(d) L-2-aminoadipate (1.00) could be replaced by DL-2-aminopimelate (0.14) or L-norleucine (0.15) but not by any of the protein amino acids or by the following L-amino acids: 2-aminobutyrate; norvaline; 6-aminocaproate; homoserine; citrulline or ornithine.

With enzyme I from bakers' yeast the ratio forward rate : reverse rate was 1 : 11. With enzyme II the ratio was 1 : 1. Neither enzyme could use 2-oxoadipate or pyruvate in the place of 2-oxoglutarate. (2)

Inhibitors

Enzyme I (bakers' yeast) was inhibited by 2-oxoglutarate (NC(glutamate or 2-oxoadipate)) but not by lysine. (2)

References

1. Nakatani, Y., Fujioka, M. & Higashino, K. (1970) BBA, 198, 219.
2. Matsuda, M. & Ogur, M. (1969) JBC, 244, 3352; 5153.

PHOSPHOSERINE AMINOTRANSFERASE

(O-Phospho-L-serine: 2-oxoglutarate aminotransferase)

O-Phospho-L-serine + 2-oxoglutarate =

3-O-phosphohydroxypyruvate + L-glutamate

Ref.

Equilibrium constant

$$\frac{[\text{phosphohydroxypyruvate}]\ [\text{L-glutamate}]}{[\text{phosphoserine}]\ [\text{2-oxoglutarate}]} = 8.25 \text{ (pH 8.2, Tris, 25°)}$$ (1)

Molecular weight

source	value	conditions	
Sheep brain	96,000	pH 7.2	(1)

The enzyme contains tightly bound pyridoxal phosphate which is required for activity. (1)

Specific activity

Sheep brain (465 x)	1.3	phosphohydroxypyruvate (pH 8.2, Tris, 25°)	(1)

Specificity and Michaelis constants

source	substrate	K_m (mM)	conditions	
Sheep brain	phosphohydroxypyruvate[a]	0.25	pH 8.2, Tris, 25°	(1)
	glutamate[b]	0.70	pH 8.2, Tris, 25°	(1)

(a) hydroxypyruvate was inactive.

(b) glutamate (1.0) could be replaced by alanine (0.1); low activities were obtained with glutamine and aspartate.

References

1. Hirsch, H. & Greenberg, D.M. (1967) JBC, 242, 2283.

GLUTAMATE SYNTHASE

(L-Glutamine : 2-oxoglutarate aminotransferase (reduced NADP oxidizing))

L-Glutamine + 2-oxoglutarate + reduced NADP = 2 L-glutamate + NADP

Molecular properties

source	value	conditions	Ref.
Escherichia coli	800,000 [~8]	pH 7.0; sucrose density gradient; gel filtration; gel electrophoresis (SDS); amino acid analysis.	(1)

The enzyme contains 8 moles of flavin (about equal amounts of FAD and FMN), 32 moles of iron and 32 moles of labile sulphide per 800,000 daltons. Molybdenum was not detected. The enzyme might be composed of 4 of each of two types of dissimilar subunits of molecular weights 135,000 and 53,000 daltons. (1)

Specific activity

E. coli (480 x) 26.2 glutamine (pH 7.5, HEPES, 37°) (1)

Specificity and catalytic properties

source	substrate	K_m(μM)	conditions	
E. coli	L-glutamine[(a)]	250	pH 7.5, HEPES, 30°	(1)
	2-oxoglutarate[(b)]	7.3	pH 7.5, HEPES, 30°	(1)
	reduced NADP[(c)]	7.7	pH 7.5, HEPES, 30°	(1)

(a) D-glutamine; NH_4Cl; L-asparagine and alkylated glutamine analogues were inactive.

(b) pyruvate and oxaloacetate were inactive.

(c) reduced NAD was inactive.

The enzyme does not require added metal ions for activity. It was not inhibited by EDTA; instead it was inhibited by several divalent metal ions such as Mg^{2+}, Mn^{2+} or Ca^{2+}. (1)

Of more than 50 naturally occurring compounds tested, NADP; D- and L-aspartate; D-glutamate and L-methionine produced 50% inhibition at concentrations of less than 7mM. Oxaloacetate and pyruvate caused irreversible inactivation. Nucleoside triphosphates were poor inhibitors and cyclic 3',5'-monophosphate, AMP, ADP and also S-adenosyl-L-methionine, folate and a number of D- and L-amino acids did not inhibit at all. (1)

References

1. Miller, R.E. & Stadtman, E.R. (1972) JBC, 247, 7407.

PYRIDOXAMINE PHOSPHATE AMINOTRANSFERASE

(Pyridoxamine 5'-phosphate:
2-oxoglutarate aminotransferase (D-glutamate forming))

Pyridoxamine 5'-phosphate + 2-oxoglutarate =
pyridoxal 5'-phosphate + D-glutamate

Ref.

Equilibrium constant

The reaction is reversible. (1)

Specific activity

Clostridium kainantoi (2700 x) 0.28 pyridoxamine 5'-phosphate (pH 8.0, Tris, 37°) (1)

Specificity and Michaelis constants

source	substrate	K_m (M)	conditions	
C. kainantoi	pyridoxamine 5'-phosphate[(a)]	4.0×10^{-6}	pH 8.0, Tris, 37°	(1)
	2-oxoglutarate	1.5×10^{-5}	pH 8.0, Tris, 37°	(1)
	pyridoxal 5'-phosphate	1.0×10^{-5}	pH 8.0, Tris, 37°	(1)
	D-glutamate[(b)]	9.0×10^{-3}	pH 8.0, Tris, 37°	(1)

(a) pyridoxamine 5'-phosphate (1.00) could be replaced by pyridoxamine (0.20) but not by the protein amino acids or by 4-aminobutyrate; D-cysteine; taurine; β-alanine; DL-ethionine, D-asparagine or DL-ornithine.

(b) L-glutamate was inactive.

The aminotransferase does not contain bound pyridoxal 5'-phosphate. It does not require metal ions for activity. (1)

References

1. Tani, Y., Ukita, M. & Ogata, K. (1972) Agr. Biol. Chem. (Tokyo), 36, 181.

TAURINE: 2-OXOGLUTARATE AMINOTRANSFERASE

(Taurine: 2-oxoglutarate aminotransferase)

Taurine + 2-oxoglutarate = sulphoacetaldehyde + glutamate

Ref.

Molecular weight

source	value	
Achromobacter superficialis	156,000	(1)

The enzyme contains 4 moles of a vitamin B_6 compound per 156,000 daltons. (1)

Specific activity

A. superficialis	(65 x)	0.52 taurine (with 2-oxoglutarate as the amino group acceptor; pH 8.0, Pi, 30°)	(1)

Specificity

The enzyme (*A. superficialis*) could use DL-3-aminoisobutyrate, β-alanine or 3-aminopropanesulphonate in the place of taurine (2-aminoethane sulphonate). It was inactive with pyruvate, phenylpyruvate or oxaloacetate in the place of 2-oxoglutarate. (1)

References

1. Toyama, S., Misono, H. & Soda, K. (1972) BBRC, 46, 1374.

PHOSPHORIBOKINASE

(ATP: D-ribose-5-phosphate 1-phosphotransferase)

ATP + D-ribose 5-phosphate = ADP + D-ribose 1,5-diphosphate

Ref.

Specificity and Michaelis constants

source	substrate	K_m(M)	conditions	
Pseudomonas saccharophilia (purified 40 x)	ATP	1×10^{-3}	pH 9.1, Tris, 37°	(1)
	ribose 5-phosphate	1×10^{-3}	pH 9.1, Tris, 37°	(1)

ATP (1.00) could be replaced by adenoside tetraphosphate (0.25) but not by ITP, cyclic 3',5'-AMP, UTP or ADP. Ribose 5-phosphate could not be replaced by ribose 1-phosphate; glucose 6-phosphate; glucose 1-phosphate or deoxyribose 5-phosphate. The enzyme requires Mg^{2+} for activity. (1)

Inhibitors

The enzyme from P. saccharophilia was inhibited by high concentrations of ATP or ribose 5-phosphate. (1)

References

1. DeChatelet, L.R. & Alpers, J.B. (1970) JBC, 245, 3161.

DEPHOSPHO-CoA KINASE

(ATP: dephospho-CoA 3'-phosphotransferase)

ATP + dephospho-CoA = ADP + CoA

Ref.

Specificity and catalytic properties

Dephospho-CoA kinase has been purified 262 x from rat liver but it has not been separated from the preceding enzyme involved in the biosynthesis of CoA, namely, dephospho-CoA pyrophosphorylase (EC 2.7.7.3). This enzyme catalyzes the following reaction:

ATP + pantetheine 4'-phosphate = PPi + dephospho-CoA

The 2 enzymes may occur as a bifunctional complex. The ratio of the specific activities (kinase to pyrophosphorylase) in crude extracts or in highly purified preparations is about 2. Both enzymes require Mg^{2+} and L-cysteine for activity. (1,2)

source	substrate	activity	K_m (M)	conditions	
Rat liver	ATP	kinase	3.6×10^{-4}	pH 8.0, Tris, 37°	(1,2)
	dephospho-CoA[a]	kinase	1.2×10^{-4}		
	ATP	pyrophosphorylase	1.0×10^{-3}		
	pantetheine 4'-phosphate[b]	pyrophosphorylase	1.4×10^{-4}		

(a) 3'-dephospho-α-carboxy CoA was inactive.

(b) pantetheine 4'-phosphate-L-cysteine was inactive.

References

1. Abiko, Y. (1970) Methods in Enzymology, 18A, 358.
2. Suzuki, T., Abiko, Y. & Shimizu, M. (1967) J. Biochem. (Tokyo), 62, 642.

TRIOKINASE

(ATP: D-glyceraldehyde 3-phosphotransferase)

ATP + D-glyceraldehyde = ADP + D-glyceraldehyde 3-phosphate

Ref.

Molecular weight

source	value	conditions	Ref.
Pig kidney cortex	90,000	pH 7.0	(1)

Specific activity

Triokinase from pig kidney cortex has been purified to homogeneity. (1)

Specificity and Michaelis constants

source	substrate	K_m (M)	conditions	Ref.
Bovine liver (purified 50 x)	D-glyceraldehyde	1.25×10^{-4}	pH 7.5, triethanolamine, 20°	(2)
	dihydroxyacetone[a]	2.0×10^{-5}	pH 7.5, triethanolamine, 20°	(2)
Rat liver	D-glyceraldehyde	1×10^{-5}	pH 7.0, imidazole, 22°	(3)

(a) in the place of D-glyceraldehyde. The following compounds were inactive: L-glyceraldehyde; glycerol; glucose; fructose; D-glycerate and glycolaldehyde.

Triokinase requires Mg^{2+} for activity. (2)

References

1. Tabor, M.W. & Prairie, R.L. (1971) ACS Sept 1971, 188 BIOL.
2. Heinz, F. & Lamprecht, W. (1961) Hoppe-Seylers Z. Phys. Chem., 324, 88.
3. Sillero, M.A.G., Sillero, A. & Sols, A. (1969) EJB, 10, 345.

PANTOTHENATE KINASE

(ATP: D-pantothenate 4'-phosphotransferase)

ATP + D-pantothenate = ADP + D-4'-phosphopantothenate

Ref.

Specificity and catalytic properties

source	substrate	K_m (M)	conditions	
Rat liver	D-pantothenate[a]	1.1×10^{-5}	pH 6.1, Tris, 37°	(1)
(purified 700 x)	ATP-Mg	1.0×10^{-3}	pH 6.1, Tris, 37°	(1)

(a) L-pantothenate and 2'-oxopantetheine were inactive as substrates but they were both competitive inhibitors of D-pantothenate phosphorylization. (L-pantothenate Ki = 0.6 mM; 2'-oxopantetheine, Ki = 0.83 mM). The following were also inhibitory: CoA (end product); 4'-phosphopantothenate (product) and pantothenoylcysteine 4'-phosphate. Dephospho-CoA and 4'-phosphopantetheine were less effective inhibitors.

The enzyme reaction requires Mg^{2+}. (1)

References

1. Abiko, Y., Ashida, S-I. & Shimizu, M. (1972) BBA, 268, 364.

FUCOKINASE

(ATP: 6-deoxy-L-galactose 1-phosphotransferase)

ATP + L-fucose = ADP + β-L-fucose 1-phosphate

Ref.

Specificity and Michaelis constants

source	substrate	K_m(M)	conditions	
Pig liver	L-fucose	1.2×10^{-4}	pH 8, Tris, 37°	(1)

The enzyme (purified 70-fold) phosphorylated L-fucose (1.00), D-glucose (1.07), D-ribose (0.87), L-rhamnose (0.41), D-arabinose (0.28) and L-arabinose (0.31). The following sugars were phosphorylated at rates of less than 5% of L-fucose: D-mannose, D-galactose, D- and L-xylose, D-fucose and L-fuculose. The following nucleotides acted as phosphoryl donors: ATP (1.00), CTP (0.60), UTP (0.59), GTP (0.56) and dTTP (0.03). Mg^{2+} (1.00), which was required by the enzyme, could be replaced by Mn^{2+} (0.87); Ca^{2+} (0.85); Fe^{2+} (0.29) or Co^{2+} (0.15). (1)

References

1. Ishihara, H., Massaro, D.J. & Heath, E.C. (1968) JBC, 243, 1103.

1-PHOSPHOFRUCTOKINASE

(ATP: D-fructose 1-phosphate 6-phosphotransferase)

ATP + D-fructose 1-phosphate = ADP + D-fructose 1,6-diphosphate

Ref.

Molecular weight

source	value	conditions	Ref.
Aerobacter aerogenes	75,000	Sephadex G 100, pH 7.5	(1)

Specificity and Michaelis constants

source	substrate	K_m (mM)	conditions	Ref.
A. aerogenes (purified 315 x)	D-fructose 1-phosphate[a]	0.7	pH 7.5, glygly, 25°	(1)
	ATP[b]	0.7	pH 7.5, glygly, 25°	(1)

(a) the following were inactive: L-fructose 1-phosphate; D-mannose 6-phosphate; D-fructose; D-fructose 6-phosphate; D-glucose 6-phosphate and D-glucose 1-phosphate.

(b) ATP (1.00) could be replaced by GTP (0.35) or ITP (0.43). Very low activities were obtained with CTP, dTTP or UTP.

The enzyme requires Mg^{2+} for activity. The effect of varying the Mg^{2+} concentration was complex and plots of Mg^{2+} concentration versus reaction rate gave S-shaped curves. (1)

1-Phosphofructokinase has also been purified (15 x) from Bacteroides symbiosus. This enzyme can utilize ATP, ITP, GTP or UTP as phosphoryl donor. Sorbose 1-phosphate and fructose 6-phosphate were neither substrates nor inhibitors. The enzyme requires Mg^{2+} or Mn^{2+} for activity. (2)

Inhibitors

source	inhibitor[a]	K_i (mM)	Ref.
A. aerogenes	D-fructose 6-phosphate	1	(1)
	D-fructose 1,6-diphosphate	4	(1)
	citrate	0.85	(1)

(a) C(D-fructose 1-phosphate). Conditions: pH 7.5, glygly, 25°. High concentrations of ATP were inhibitory but nucleoside mono- or diphosphates or Pi did not inhibit.

References

1. Sapico, V. & Anderson, R.L. (1969) JBC, 244, 6280.
2. Reeves, R.E., Warren, L.G. & Hsu, D.S. (1966) JBC, 241, 1257.

ISOPRENOID-ALCOHOL KINASE

(ATP: C_{55}-isoprenoid-alcohol phosphotransferase)

ATP + C_{55}-isoprenoid-alcohol = ADP + C_{55}-isoprenoid-alcohol phosphate

Ref.

Molecular properties

Isoprenoid-alcohol kinase (Staphylococcus aureus) is a lipoprotein. The enzyme activity is associated with a single polypeptide chain of molecular weight 17,000 (gel electrophoresis (SDS); amino acid composition). This is the most apolar protein known: it is soluble in n-butanol. (1,2,4)

Specific activity

S. aureus (6900 x) 0.54 C_{55}-isoprenoid alcohol from Streptococcus faecalis. (This value was obtained with the enzyme apoprotein which requires the presence of a phospholipid cofactor - or a non-ionic detergent - for activity. For other details see below). (1,2)

Specificity and Michaelis constants

source	substrate[a]	K_m (μM)	
S. aureus	C_{55}-isoprenoid alcohol[b]	57	(2)
	ATP	57	(2)

(a) the conditions were: pH 8.5, Tris, 25°. The reaction mixtures also contained 10% dimethyl sulphoxide and 0.3% Triton X-100. The enzyme preparation used had been purified 627 x; it contained bound phospholipid: the apoprotein was inactive. A neutral lipid fraction of S. aureus, added phospholipid or a neutral detergent (e.g. Span-20) restores the activity of the enzyme apoprotein.

(b) isolated from S. faecalis. A C_{55}-isoprenoid-alcohol from rubber plant (C_{55}-ficaprenol) had similar kinetic properties.

The enzyme requires Mg^{2+} for activity. The specificity of the enzyme towards its alcohol substrate has been determined, it is without activity on low molecular weight substrates such as choline, mevalonate, glucose or glycerol. (3)

References

1. Sandermann, H. & Strominger, J.L. (1971) PNAS, 68, 2441.
2. Sandermann, H. & Strominger, J.L. (1972) JBC, 247, 5123.
3. Higashi, Y., Siewert, G. & Strominger, J.L. (1970) JBC, 245, 3683.
4. Sandermann, H. (1973) FEBS lett, 29, 256.

PHOSPHOHISTIDINOPROTEIN-HEXOSE PHOSPHOTRANSFERASE

(Phosphohistidinoprotein: hexose phosphotransferase)

Phosphohistidinoprotein + hexose = protein + hexose 6-phosphate

Ref.

Escherichia coli has a complex consisting of 2 enzymes (I and II) and a histidine containing and heat-stable protein, HPr. The sequence of reactions catalysed by the complex is:

(a) phospho*enol*pyruvate + HPr = pyruvate + P-HPr
(b) P-HPr + hexose = HPr + hexose 6-phosphate

Reaction (a) is catalysed by enzyme I (phospho*enol*pyruvate-protein phosphotransferase; EC 2.7.3.9) and reaction (b) by enzyme II (phosphohistidinoprotein-hexose phosphotransferase, EC 2.7.1.69). Enzymes I and II have been separated and purified (enzyme I 400 x and enzyme II 330 x). The system requires Mg^{2+} for activity. (Mg^{2+} could be replaced by Mn^{2+}, Zn^{2+} or Co^{2+} but Ca^{2+} or Cu^{2+} were highly inhibitory). The pH optimum for the complete system was 7.2 - 7.4 and the Km values for phospho*enol*pyruvate and *N*-acetylmannosamine were both 0.6 mM. The following D-sugars were active in the place of *N*-acetylmannosamine: glucose (Km = 0.4 mM); mannose (2 mM); glucosamine (3 mM); mannosamine (1 mM); *N*-acetylglucosamine (0.9 mM) and *N*-glycolylmannosamine (2 mM). The following D-sugars were inactive:- galactose, galactosamine, *N*-acetylgalactosamine, fructose, xylose, arabinose, ribose, glucose 1-phosphate and glucose 6-phosphate. Phospho*enol*pyruvate could not be replaced by the mono-, di- or triphosphates of the common nucleosides; by cyclic-AMP, creatine phosphate, PPi, Pi, phosphoramidate, phosphohistidine, *N*-phosphoglycine, thiaminepyrophosphate, phosphoglycerate, phosphoserine, or by coenzyme A + glutathione. (1)

Abbreviations

P-HPr   phosphohistidinoprotein

References

1. Kundig, W., Ghosh, S. & Roseman, S. (1964) PNAS, 52, 1067.

PROTAMINE KINASE

(ATP: protamine _O_-phosphotransferase)

ATP + protamine = ADP + _O_-phosphoprotamine

Ref.

Protamine kinase, an enzyme which phosphorylates serine residues in protamine, has been purified 37 x from rainbow trout testis. At high ionic strengths (e.g. 0.3 M NaCl) the enzyme was specific for protamine; other basic proteins (e.g. the histones) were poor substrates and casein and serine were inactive. At low ionic strengths (e.g. 0.05 M Tris, pH 7.4) the enzyme was at once less active and less specific and it phosphorylated protamine (1.00); histone I (0.14); histone II (0.74) and histone III + IV (0.15). Casein and serine were phosphorylated at very low levels. The kinase was highly specific for ATP which could not be replaced by CTP, GTP or UTP. (1)

The enzyme requires Mg^{2+} specifically (Mn^{2+}, Ca^{2+}, Co^{2+} and Zn^{2+} were inactive) and also a thiol compound for activity. It was stimulated by cyclic AMP (Km = 5×10^{-8}M).

Rainbow trout testis also has a phosphatase which removes the phosphate groups from phosphoprotamine. (1)

Abbreviations

Histone I contains 18-19 arginyl residues per molecule; histone II 21-22 and histone III 24-25.

References

1. Jergil, B. & Dixon, G.H. (1970) _JBC_, _245_, 425.

DEOXYCYTIDINE KINASE

(NTP: deoxycytidine 5'-phosphotransferase)

NTP + deoxycytidine = NDP + dCMP

Ref.

Molecular weight

source	value	conditions	Ref.
L 1210 murine leukaemia cells	60,000	Sephadex G 200, pH 7.0	(1)
Calf thymus	56,000	sucrose density gradient; Sephadex G 150, pH 8.0	(2)
Lactobacillus acidophilus	35,000	-	(2)

Specificity and Michaelis constants

source	substrate(cosubstrate)	K_m(M)	conditions	Ref.
L 1210 cells	deoxycytidine (ATP)	1.1×10^{-5}	pH 7.0, TAES, 37°	(1)
(purified	ATP (deoxycytidine)	2.0×10^{-4}	pH 7.0, TAES, 37°	(1)
150 x)	UTP (deoxycytidine)	5.8×10^{-4}	pH 7.0, TAES, 37°	(1)

Deoxycytidine kinase is more specific for its phosphate acceptor substrate than for its phosphate donor substrate. Thus, the enzyme from calf thymus could utilize deoxycytidine (1.20); cytosine arabinoside (1.00); deoxyguanosine (0.30) or deoxyadenosine (0.20) but not adenosine, guanosine, uridine, cytidine or deoxythymidine. With deoxycytidine as the acceptor substrate, the following donors were about equally active: ATP, GTP, UTP, CTP, dATP, dGTP and dTTP. dCTP was inactive. (3)

The enzyme has been purified (500 x) from calf thymus. With deoxycytidine as the variable substrate and ATP as the phosphate donor several abrupt transitions were observed in slopes of double reciprocal plots, suggesting negative homotropic cooperativity in the binding of this substrate (but see also Ref. 3). Normal kinetics were observed with deoxyguanosine or with cytosine arabinoside as the variable phosphate acceptor substrate. (2)

Deoxycytidine kinase requires a divalent metal ion for activity. (1,2)

Inhibitors

source	inhibitor	type	K_i(M)	conditions	Ref.
L 1210	dCMP	C(deoxycytidine)	9.0×10^{-6}	pH 7.0, TAES, 37°	(1)
cells	dCDP	C(deoxycytidine)	2.2×10^{-6}	pH 7.0, TAES, 37°	(1)
	dCTP	C(deoxycytidine)	6.7×10^{-6}	pH 7.0, TAES, 37°	(1)
	cytosine arabinoside	C(deoxycytidine)	6.0×10^{-4}	pH 7.0, TAES, 37°	(1)

Deoxycytidine kinase from calf thymus is also inhibited by dCTP ("end product inhibition"). The inhibition is complex. (2)

References

1. Kessel, D. (1968) JBC, 243, 4739.
2. Durham, J.P. & Ives, D.H. (1970) JBC, 245, 2276; 2285.
3. Momparler, R.L. & Fischer, G.A. (1968) JBC, 243, 4298.

DEOXYADENOSINE KINASE

(ATP : deoxyadenosine 5'-phosphotransferase)

ATP + deoxyadenosine = ADP + dAMP

Ref.

Molecular weight

source	value	conditions	
Calf thymus	63,000	Sephadex G 150, pH 8	(1)

The enzyme did not aggregate in the presence of the inhibitor dCTP. (1)

Specificity and catalytic properties

source	substrate	cosubstrate	V relative	relative velocity	K_m (mM)	
Calf thymus (purified 140 x)	deoxyadenosine[a]	ATP	1.0	1.00	0.7	(1)
	deoxyguanosine	ATP	1.2	0.96	1.1	(1)
	cytidine	ATP	-	0.18	0.6	(1)
	deoxycytidine	ATP	-	0.13	-	(1)
	ATP[b]	deoxyadenosine	-	-	0.2	(1)

(a) deoxyadenosine could not be replaced by adenosine, guanosine, uridine or deoxythymidine (pH 8.0, Tris, 37°).

(b) ATP (1.00) could be replaced by GTP (0.98); UTP (0.90); dTTP (0.76); dGTP (0.29); CTP (0.19) or dATP (0.10) but not by dCTP or ADP.

Deoxyadenosine kinase requires Mg^{2+}; Mn^{2+} and Ca^{2+} were less effective activators. (1)

The kinase is inhibited by nucleotides containing adenine, guanine or cytosine. The potency of the inhibition by cytosine derivatives is dependent on the nature of the pentose moiety: deoxyribose > arabinose > ribose. (1)

inhibitor	type	cosubstrate	K_i (μM)	
deoxyguanosine	C(deoxyadenosine)	ATP	1100	(1)
deoxyadenosine	C(deoxyguanosine)	ATP	780	(1)
deoxyadenosine	C(cytidine)	ATP	350	(1)
dATP	C(ATP)	deoxyadenosine	27	(1)
CTP	C(ATP)	deoxyadenosine	80	(1)
dCTP	C(ATP)	deoxyguanosine	0.16	(1)

References

1. Krygier, V. & Momparler, R.L. (1971) JBC, 246, 2745; 2752.

NUCLEOSIDE PHOSPHOTRANSFERASE

(Nucleotide: 3'-deoxynucleoside 5'-phosphotransferase)

A nucleotide + 3'-deoxynucleoside =

a nucleoside + 3'-deoxynucleoside 5'-monophosphate

Ref.

Molecular weight

source	value	conditions	Ref.
Carrot	44,000 [2][a]	Sephadex G 100, pH 5.0; sucrose density gradient, amino acid composition.	(1)

(a) the 2 subunits have different molecular weights and isoelectric points. On its own each subunit is enzymically inactive.

Specific activity

			Ref.
Carrot (6000 x)	35	phenylphosphate (acceptor substrate = uridine; pH 5.0, acetate, 37°)	(1)
	12	phenylphosphate (acceptor substrate = H_2O; pH 5.0, acetate, 37°)	(1)

Both transfer and hydrolytic functions are carried out by the same protein molecule (also see EC 3.1.3.2) (1,2)

Specificity and catalytic properties

Nucleoside phosphotransferase (carrot) has 2 hydrolytic sites (which bind the phosphate donor) and 1 transfer site (which binds the nucleoside phosphate acceptor). The enzyme can utilize a number of acceptor substrates; thus with phenylphosphate as the donor substrate the following were phosphate acceptors: cytidine (1.4); uridine (1.2); deoxycytidine (0.8); ribothymidine (0.9); thymidine (0.6); deoxyuridine (0.4); deoxyguanosine (0.9); guanosine (0.9); deoxyadenosine (0.1) and adenosine (0.1). 3'-Deoxyadenosine was a better substrate than 2'-deoxyadenosine or adenosine. The following compounds were active donors: phenylphosphate (phosphotransferase = 1.0; phosphohydrolase = 1.0); 3'-adenylate (0.8, 0.3); 5'-adenylate (0.7, 0.3); ribose 5-phosphate (0.3, 0.9); β-glycerophosphate (0.2, 0.5); (2' + 3')-uridylate (0.1, 0.1); glucose 6-phosphate (0.1, 0.2) and glucose 1-phosphate (0, 0.1). (2,3)

Ref.

Specificity and catalytic properties continued

Details of the kinetic constants with various substrates or inhibitors are given below:

substrate or inhibitor	K_m (mM) phospho transferase	K_m (mM) phospho hydrolase	K_i (mM) phospho transferase	K_i (mM) phospho hydrolase	Ref.
phenylphosphate	3.5	0.7 and 3	-	-	(2)
uridine	3.5	-	-	4.3[a]	(2)
ribose 5-phosphate	-	2 and 60	-	-	(2)

(a) NC(phenylphosphate). Other acceptor substrates also inhibited the phenylphosphatase activity (conditions: pH 5.0, acetate, 37°)

The effect of metal ions on the 2 reactions is complex. Mg^{2+}, Cu^{2+} and Co^{2+} enhance the phosphatase function considerably but the transferase activity was only increased to a small extent. Dialysis against EDTA reduced the phosphotransferase to 36% and the phosphohydrolase activity to 65% of the respective initial activities. Co^{2+} and Cu^{2+} reactivated the enzyme but Mg^{2+} had only a limited effect. (3)

References

1. Rodgers, R. & Chargaff, E. (1972) JBC, 247, 5448.
2. Brunngraber, E.F. & Chargaff, E. (1970) JBC, 245, 4825.
3. Brunngraber, E.F. & Chargaff, E. (1967) JBC, 242, 4834.

POLYNUCLEOTIDE 5'-HYDROXYL-KINASE

(ATP: 5'-dephosphopolynucleotide 5'-phosphotransferase)

ATP + 5'-dephospho-DNA = ADP + 5'-phospho-DNA

Ref.

Specificity and Michaelis constants

source	substrate	V relative	K_m (μM)	conditions	Ref.
Bacteriophage T2; host = *Escherichia coli* (8650 x)	ATP [a]	1.00	14.3	pH 6.0, succinate, 37°	(1)
	UTP	0.83	15.0		
	GTP	1.65	33.3		
	CTP	1.30	25.0		
	RNA (*E.coli*) [b]	1	200	pH 6.0, succinate, 37°	(1)
	DNA (calf thymus)	1	33		
	dCMP	1	15		
	2'(3') CMP	1	13		

(a) phosphate acceptor = ribosomal RNA from *E. coli*.

(b) phosphate donor = ATP. In addition to polynucleotides, a variety of small, acid-soluble oligonucleotides act as phosphate acceptors. Thus, adenine oligonucleotides of different chain lengths and 3'-AMP were phosphorylated at equal rates. Adenosine; 2'-AMP and all 5'-mononucleotides tested were inactive.

Polynucleotide 5'-hydroxyl-kinase only phosphorylates certain hydroxyl groupsin the acceptor substrate. For instance, when DNA (from *Micrococcus lysodeikticus*) was phosphorylated, the phosphate appeared almost exclusively in the dAMP and dTMP regions: very little was found in the dCMP or dGMP regions. (1)

The enzyme requires Mg^{2+} (Km = 5 x 10^{-4}M) for activity; Mn^{2+} (Km = 1 x 10^{-4}M), Zn^{2+} and Co^{2+} were equally effective as activators. (1)

A polynucleotide 5'-hydroxyl-kinase has also been isolated from T4 infected *E. coli* (purified 1400 x) but T1 or T5 bacteriophage infected *E. coli* does not produce the enzyme. (1,2)

References

1. Novogrodsky, A., Tal, M., Traub, A. & Hurwitz, J. (1966) JBC, 241, 2923; 2933.
2. Richardson, C.C. (1965) PNAS, 54, 158.

PYROPHOSPHATE-SERINE PHOSPHOTRANSFERASE

(Pyrophosphate: L-serine O-phosphotransferase)

PPi + L-serine = Pi + O-phospho-L-serine

Ref.

Equilibrium constant

$\frac{[Pi]\ [O\text{-phospho-L-serine}]}{[PPi]\ [L\text{-serine}]}$ = 950 (pH 7.7, Tris, 37° + 1mM $MgSO_4$) (1)

Molecular weight

source	value	conditions	
Propionibacterium shermanii	65,000	Sephadex G 200	(1)

Specificity and Michaelis constants

source	substrate	K_m (M)	conditions	
P. shermanii (purified 102 x)	PPi [a]	1.0×10^{-4}	pH 7.8, Tris, 37°	(1)
	L-serine [b]	1.9×10^{-3}	pH 7.8, Tris, 37°	(1)
	Mg^{2+}	2×10^{-4}	pH 7.8, Tris, 37°	(1)

(a) PPi (1.00) could be replaced by tripolyphosphate (0.32) but not by the 5'-di- or triphosphates of adenosine, cytosine, uridine or guanosine or by 4-carboxyphenylphosphate or 4-nitrophenylphosphate.
(b) L-serine (1.00) could be replaced by N-chloroacetyl-L-serine (0.19); glycyl-L-serine (0.16); L-threonine (0.08); α-methylserine (0.07) and DL-homoserine (0.07) but not by N-acetylserinamide; isoserine; serinol; ethanolamine; L-serylglycine; glycine; L-cysteine; hydroxypyruvate; D-serine; glycerol; glucose or Tris.

Mg^{2+} (1.00), which is required by the enzyme, could be replaced by Mn^{2+} (0.34) or Co^{2+} (0.42). Ca^{2+} was a strong inhibitor. The enzyme does not require pyridoxal phosphate. (1)

Inhibitors

source	inhibitor	type	K_i (M)	conditions	
P. shermanii	glycine [a]	C(L-serine)	8×10^{-4}	pH 7.8, Tris, 37°	(1)
	Pi	C(PPi)	-	-	(1)

(a) D-serine and L-alanine were also inhibitory but isoserine; hydroxypyruvate; β-alanine; ethanolamine and acetate were not.

References

1. Cagen, L.M. & Friedmann, H.C. (1972) JBC, 247, 3382.

HYDROXYLYSINE KINASE

(GTP: 5-hydroxy-L-lysine O-phosphotransferase)

GTP + 5-hydroxy-L-lysine = GDP + 5-O-phosphohydroxy-L-lysine

Ref.

Specificity and Michaelis constants

source	substrate	V relative	K_m (μM)	conditions	
Rat liver (purified 1200 x)	hydroxy-L-lysine[a]	1	5.6	pH 8.0, HEPES, 37°	(1)
	allohydroxy-L-lysine	about 2	6.2	pH 8.0, HEPES, 37°	(1)

(a) hydroxy-D-lysine and allohydroxy-D-lysine were inactive.

The kinase (rat liver) could utilize the following phosphoryl donors: GTP (1.00); ITP (0.90); CTP (0.06); UTP (0.06) or ATP (0.02). It requires Mg^{2+} for activity; other metal ions tested were poor activators (Mn^{2+}, Co^{2+}, Ca^{2+}). The enzyme was inhibited by excess GTP or Mg^{2+}, by L-lysine, L-ornithine, L-2,4-diamino-n-butyrate, 2-amino-5-hydroxyadipate and by hydroxy-L-proline. (1)

References

1. Hiles, R.A. & Henderson, L.M. (1972) JBC, 247, 646.

β-GLUCOSIDE KINASE

(ATP: β-D-glucoside O-phosphotransferase)

ATP + cellobiose = ADP + cellobiose monophosphate

Ref.

Molecular weight

source	value	conditions	Ref.
Aerobacter aerogenes	150,000	sucrose density gradient	(1)

Specificity and Michaelis constants

source	substrate	V relative	K_m (mM)
A. aerogenes (purified 91 x)	methyl β-D-glucoside[a]	1.10	13.3
	cellobiose	1.00	1.0
	salicin	0.87	2.5
	gentiobiose	0.70	1.5
	phenyl β-D-glucoside	0.70	2.5
	cellotriose	0.47	6.7
	arbutin	0.36	0.6
	cellotetraose	0.33	6.3
	cellobiitol	0.25	4.0
	amygdalin	0.21	3.1
	sophorose	0.19	3.1
	ATP[b]	-	1.7

(a) the following were inactive as substrates or inhibitors: 4-nitrophenyl β-D-glucoside; methyl α-D-glucoside; sucrose; maltose; melibiose; trehalose; lactose; turanose; melezitose; raffinose; inulin; D-glucose; D-fructose; D-galactose; D-mannose; D-fucose; L-sorbose; D-ribose; L-arabinose; D-mannitol and sorbitol. (Conditions: pH 7.5, glycylglycine, Ref. 1).

(b) ATP (1.00) could be replaced by GTP (0.15), CTP (0.12), ITP (0.07) or UTP (0.03) but the following were inactive as phosphoryl donors: ADP; acetyl phosphate; α-glycerophosphate; D-fructose 1,6-diphosphate; phosphoramidate; choline phosphate; creatine phosphate, 3-phosphoglycerate, phenyl phosphate and phosphoenolpyruvate.

Abbreviations

cellobiose monophosphate — 6-O-phosphoryl-β-D-glucopyranosyl-(1→4)-D-glucose

References

1. Palmer, R.E. & Anderson, R.L. (1972) JBC, 247, 3415.

PSEUDOURIDINE KINASE

(ATP: pseudouridine 5'-phosphokinase)

ATP + pseudouridine = ADP + pseudouridylate

Ref.

Specificity and Michaelis constants

source	substrate	K_m (M)	conditions	
Escherichia coli (cell extract)	ATP	2.9×10^{-3}	pH 7.65, Pi, 37°	(1)
	Mg^{2+}	3.3×10^{-3}	pH 7.65, Pi, 37°	(1)
	K^+	4×10^{-2}	pH 7.65, Pi, 37°	(1)
	pseudouridine	1.8×10^{-4}	pH 7.65, Pi, 37°	(1)

The enzyme requires Mg^{2+} and K^+ for activity. K^+ could be replaced by NH_4^+ but not by Na^+ or Li^+. The reaction was not inhibited by uridine or deoxythymidine.

References

1. Solomon, L.R. & Breitman, T.R. (1971) BBRC, 44, 299.

ETHANOLAMINE KINASE

(ATP: ethanolamine O-phosphotransferase)

ATP + ethanolamine = ADP + O-phosphoethanolamine

Ref.

Specificity and kinetic properties

source	substrate	K_m (M)	conditions	
Rat liver (soluble fraction)	ethanolamine	1×10^{-4}	pH 8.5, glygly, 37°	(1)

Ethanolamine kinase (rat liver) was inhibited by choline (C(ATP); NC (ethanolamine); Ki = 2×10^{-5}M) and by dimethylethanolamine; monomethylethanolamine; phosphorylcholine; 2-amino-2-methylpropanol; phosphoethanolamine and high concentrations of ATP or Mg-ATP. Betaine, CDPcholine and ADP did not inhibit. (1)

Ehrlich ascites-carcinoma cells also have ethanolamine kinase activity. This enzyme requires Mg^{2+} (not replaced by Mn^{2+} or Ca^{2+}) and a monovalent ion (e.g. Na^{+}). The Km for ethanolamine was 0.12 mM. The kinase was inhibited by choline and a number of other structural analogues tested. (2)

References

1. Weinhold, P.A. & Rethy, V.B. (1972) BBA, 276, 143.
2. Sung, C-P. & Johnstone, R.M. (1967) BJ, 105, 497.

PHOSPHOENOLPYRUVATE SYNTHASE

(ATP: pyruvate 2-*O*-phosphotransferase (AMP))

ATP + pyruvate = AMP + Pi + phospho*enol*pyruvate

Ref.

Equilibrium constant

The reaction is reversible. (1,2)

Molecular properties

source	value	conditions	
Escherichia coli	250,000	Sephadex G 200	(1)

The enzyme bound 3-4 moles of Mn^{2+} per 250,000 daltons. A phospho-enzyme can be prepared with ATP or PEP; this material contains 1-2 moles of Pi per 250,000 daltons. (1,2)

Specificity and kinetic properties

substrate	K_m (M) E. coli[a]	K_m (M) E. coli B[b]
ATP-Mg	2.8×10^{-5}	-
ATP	2.4×10^{-5}	1.9×10^{-4}
pyruvate	2.0×10^{-5}	2.8×10^{-4}
Mg^{2+}	4.2×10^{-4}	-
AMP	-	1.1×10^{-4}
Pi	1.04×10^{-2}	3.8×10^{-2}
PEP	-	3.7×10^{-5}

(a) purified 60 x; specific activity = 8.9 pyruvate. Conditions: forward reaction, pH 8.0, Tris, 30°; reverse reaction, pH 6.8, Pi, 30°. From Ref. (2).
(b) purified 57 x; specific activity = 9.7 pyruvate. Conditions: forward reaction: pH 8.4, Tris, 30°; reverse reaction, pH 6.8, Pi, 30°. From Ref. (3).

The synthase could not utilize CTP, GTP or UTP in the place of ATP nor could pyruvate be replaced by phenylpyruvate, hydroxypyruvate, 2-oxo-butyrate or glyoxylate. It requires free divalent metal ions (Mg^{2} or Mn^{2+}) in addition to that bound to ATP. The free ions are involved in the binding of pyruvate to the enzyme. High concentrations of Mg^{2+} or Mn^{2+} were, however, inhibitory as was Ca^{2+}. The enzyme was inhibited by AMP and PEP in the forward reaction and by ATP in the reverse reaction. A very potent inhibitor of phospho*enol*pyruvate formation is 5'-adenylyl-methylene diphosphonate (C(ATP); K_i = 2.1 μM). (2,3)

The mechanism of the reaction has been investigated and compared with the pyruvate-phosphate dikinase reaction (EC 2.7.9.1) (2)

Ref.

Abbreviations

PEP phospho*enol*pyruvate

References

1. Cooper, R.A. & Kornberg, H.L. (1967) BJ, 105, 49C.
2. Berman, K.M. & Cohn, M. (1970) JBC, 245, 5309, 5319.
3. Cooper, R.A. & Kornberg, H.L. (1969) Methods in Enzymology, 13, 309.

ACETYLGLUTAMATE KINASE

(ATP: N-acetyl-L-glutamate 5-phosphotransferase)

ATP + N-acetyl-L-glutamate = ADP + N-acetyl-L-glutamate 5-phosphate

Ref.

Equilibrium constant

The reaction is reversible. The reverse reaction is favoured. (1)

Molecular weight

source	value	conditions	
Escherichia coli	$S_{20,w}$ = 4.2 S	sucrose density gradient	(2)

Specificity and Michaelis constants

source	substrate	K_m (mM)	conditions	
E. coli (purified 68 x)	N-acetylglutamate[a]	6.0	pH 7.4, Pi, 37°	(2)
	ATP[b]	1.0	pH 7.4, Pi, 37°	(2)
Chlamydomonas reinhardti (purified 120 x)	N-acetylglutamate	15	pH 5.5, succinate, 37°	(1)
	ATP	1.6	pH 5.5, succinate, 37°	(1)

(a) L-glutamate, D-glutamate and N-benzoylglutamate were inactive.

(b) GTP was inactive.

The enzyme requires Mg^{2+} for activity which in the case of the enzyme from C. reinhardti can be replaced by Co^{2+}. Mn^{2+} was less effective. (1)

Inhibitors

L-Arginine is an allosteric (end product) inhibitor of the enzyme (C. reinhardti). The inhibition was C(N-acetylglutamate), Ki = 0.35 mM and NC(ATP). There are 2 binding sites for arginine on each molecule of the enzyme. L-Canavanine and L-citrulline (but not L-homoarginine) were also inhibitory. The effect of L-arginine was suppressed by urea in a competitive manner. On its own, urea had no effect on the activity of the enzyme. (1)

References

1. Faragó, A. & Dénes, G. (1967) BBA, 136, 6.
2. Vogel, H.J. & McLellan, W.L. (1970) Methods in Enzymology, 17A, 251.

AMMONIA KINASE

(ATP: ammonia phosphotransferase)

ATP + NH_3 = ADP + phosphoramide

Ref.

Specificity and kinetic properties

source	substrate	K_m (mM)	conditions	
Bakers' yeast (30 x)	ADP[a]	0.115	pH 7.0, maleate	(1)
	phosphoramide[b]	0.139	pH 7.0, maleate	(1)

(a) ADP (1.00) could be replaced by dADP (0.91), GDP (0.74), CDP (0.67), dTDP(0.62), dCDP (0.54), IDP (0.52) or UDP (0.35). AMP was inactive.

(b) phosphoramide (1.00) could be replaced by L-phosphoiodohistidine (2.07) and also by N-phosphorylhistidine and N-phosphorylglycine. D-Phosphoiodohistidine; Pi and creatine phosphate were inactive.

Mg^{2+} ions were not required and were found to be inhibitory with either phosphoramide or phosphoiodohistidine as substrate. Inhibitors of oxidative phosphorylation (e.g. oligomycin and sodium azide but not 2,4-dinitrophenol) were inhibitors of ammonia kinase. (1)

A phosphoenzyme intermediate is formed in the presence of phosphoramide. (1)

References

1. Dowler, M.J. & Nakada, H.I. (1968) JBC, 243, 1434.

PHOSPHOMEVALONATE KINASE

(ATP: 5-phosphomevalonate phosphotransferase)

ATP + 5-phosphomevalonate = ADP + 5-diphosphomevalonate

Ref.

Equilibrium constant

$$\frac{[\text{ADP}]\ [\text{5-diphosphomevalonate}]}{[\text{ATP}]\ [\text{5-phosphomevalonate}]} = \text{approx. 1 (pH 7.3, Tris, 37°)}$$ (1,2)

Specificity and Michaelis constants

source	substrate	K_m (M)	conditions	
Pig liver (purified 16 x)	5-phosphomevalonate	3×10^{-4}	pH 7.3, Tris, 37°	(1,2)

Phosphomevalonate kinase (pig liver) is closely associated with mevalonate kinase (EC 2.7.1.36). It requires ATP specifically: all other nucleotide triphosphates tested were inactive. It also requires a divalent metal ion of which Mg^{2+} was the most active; Mn^{2+} was less effective and Zn^{2+} was inhibitory. (1,2)

The enzyme from yeast could utilize Mg^{2+} and Zn^{2+} with equal facility. (1)

References

1. Popják, G. (1969) Methods in Enzymology, 15, 413.
2. Hellig, H. & Popják, G. (1961) J. Lipid Res, 2, 235.

dTMP KINASE

(ATP: deoxythymidinemonophosphate phosphotransferase)

ATP + dTMP = ADP + dTDP

Ref.

Equilibrium constant

The reaction is reversible. (1)

Molecular weight

source	value	conditions	Ref.
Mouse ascites hepatoma	35,000	Sephadex G 75, pH 7.4	(1)
Escherichia coli B	65,000	Sephadex G 100, pH 7.8; sucrose density gradient	(2)

Specific activity

Mouse hepatoma	(7500 x)	13	dTMP (pH 7.4, Tris, 37°)	(1)
E. coli B	(4800 x)	20	dTMP (pH 7.8, Tris, 37°)	(2)

Specificity and Michaelis constants

source	substrate	K_m(M)	conditions	Ref.
Mouse hepatoma	dTMP[a]	1.9×10^{-4}	pH 7.4, Tris, 37°	(1)
	ATP-2Mg [b]	1.4×10^{-3}	pH 7.4, Tris, 37°	(1)
E. coli B	dTMP[c]	2.4×10^{-4}	pH 7.8, Tris, 37°	(2)
	ATP-Mg	1.2×10^{-3}	pH 7.8, Tris, 37°	(2)

(a) dUMP had 23% the activity of dTMP but the following were inactive:- dAMP, AMP, dCMP, CMP, dGMP, GMP, dUMP, UMP, dCDP, CDP, UDP and thymidine.

(b) dATP had 70% the activity of ATP but the following were inactive: GTP, CTP, UTP, dCTP, dGTP and dTTP. Mg^{2+} could to a limited extent be replaced by Mn^{2+} but not by Ca^{2+}.

Ref.

Specificity and Michaelis constants continued

(c) 5-iodo-2'-dUMP had 70% and dUMP 15% the activity of dTMP but the following were inactive as substrates or inhibitors: dCMP, dAMP, dGMP, AMP, GMP, UMP, CMP, ribo-TMP, thymidine, 4-N-hydroxy-dCMP, 5-methyl-4-N-hydroxy-dCMP and cyclic 3',5'-AMP.

The enzyme requires Mg^{2+} for activity. (1,2)

Inhibitors

source	inhibitor	type	K_i (mM)	conditions	
Mouse hepatoma	ADP[a]	mixed (ATP)	0.25	pH 7.4, Tris, 37°	(1)
E. coli B	5-iodo-2'-dUMP[b]	C(dTMP)	0.4	pH 7.8, Tris, 37°	(2)
	dUMP	C(dTMP)	3.0	pH 7.8, Tris, 37°	(2)

(a) dADP, and to a lesser extent dTDP, were also inhibitory. A large number of other nucleotides tested had little or no effect on the reaction.

(b) ATP, dCTP or dTTP in excess of the stoicheiometrically equivalent concentration of Mg^{2+} was strongly inhibitory.

References

1. Kielley, R.K. (1970) JBC, 245, 4204.
2. Nelson, D.J. & Carter, C.E. (1969) JBC, 244, 5254.
   Cheng, Y-C. & Prusoff, W.H. (1973) B, 12, 2612.

NUCLEOSIDETRIPHOSPHATE-ADENYLATE KINASE

(Nucleosidetriphosphate: AMP phosphotransferase)

NTP + AMP = NDP + ADP

Ref.

Equilibrium constant

$\frac{[GDP]\ [ADP]}{[GTP]\ [AMP]} = 0.82$ (pH 8.5, triethanolamine, 25°; in the presence of 5 mM Mg^{2+}) (1)

Molecular weight

source	value	conditions	
Beef heart (mitochondria)	52,000	Sephadex G 200, pH 7.5	(1)

Specificity and kinetic constants

source	substrate (cosubstrate) or inhibitor	V relative	K_m (M)	K_i (M)	type of inhibition
Beef heart[a] (purified 41 x)	ITP(AMP)	1.00	6.3×10^{-4}	-	-
	GTP(AMP)	0.82	5.6×10^{-5}	8.0×10^{-4}	C(AMP)
	UTP(AMP)	0.19	7.4×10^{-3}	-	-
	CTP(AMP)	0.13	9.1×10^{-3}	-	-
	ATP(AMP)	0.07	1.0×10^{-3}	7.4×10^{-4}	C(GTP)
	ATP	-	-	7.3×10^{-4}	C(AMP)
	AMP(GTP)[b]	-	3.3×10^{-5}	1.6×10^{-3}	C(GTP)
	GDP(ADP)	-	1.2×10^{-6}	7.7×10^{-4}	NC(ADP)
	ADP(GDP)[c]	-	2.9×10^{-4}	9.0×10^{-4}	NC(GDP)

(a) the conditions were: pH 8.5, triethanolamine, 25° (Ref. 1).

(b) AMP could not be replaced by 3'-GMP, IMP, CMP or UMP.

(c) ADP could not be replaced by any other substance tested.

The enzyme required Mg^{2+} for full activity; Mn^{2+} or Ca^{2+} were also activators. (1)

The mechanism of the reaction has been studied. In the presence of GTP, the enzyme forms a GTP-enzyme complex. There was no evidence for a phosphoryl-enzyme intermediate. (1)

References

1. Albrecht, G.J. (1970) B, 9, 2462.

DEOXYNUCLEOSIDEMONOPHOSPHATE KINASE

(ATP: deoxynucleosidemonophosphate phosphotransferase)

ATP + deoxynucleoside monophosphate =

ADP + deoxynucleoside diphosphate

Ref.

Equilibrium constant

$\frac{[ADP]\ [dTDP]}{[ATP]\ [dTMP]}$ = 1.4 (pH 7.4, 37°; in the presence of 8 mM Mg^{2+}) (1)

Specificity and kinetic properties

source	substrate or inhibitor	relative velocity	K_m (mM)	K_i (mM)	type of inhibition	Ref.
Bacteriophage T5. Host = *Escherichia coli* (purified 500 x)	dAMP [(a)]	1.00	0.22	0.15	C(dTMP)	(1)
	dTMP	0.71	0.17	-	-	(1)
	dGMP	0.64	0.22	0.15	C(dTMP)	(1)
	dCMP	0.30	0.03	0.03	C(dTMP)	(1)
	dUMP	0.06	2.80	-	-	(1)
	ATP [(b)]	-	0.36	-	-	(1)

(a) dAMP (1.00) could also be replaced by 5-bromodeoxyuridine 5'-phosphate (0.44); AMP (0.05) or CMP (0.04) but not by dIMP, dXMP, dHMP, UMP, GMP, 5-methyl dCMP or thymidine (conditions: pH 7.5, Tris, 37°; donor = ATP).

(b) with dTMP as phosphoryl acceptor. dATP was as effective as ATP as phosphoryl donor but UTP, GTP, CTP, dTTP, dGTP and dCTP were inactive.

The enzyme requires Mg^{2+} for activity; Mn^{2+} and Ca^{2+} were also good activators but Fe^{2+} was less effective. (1)

References

1. Bessman, M.J., Herriott, S.T. & van Bibber Orr, M.J. (1965) JBC, 240, 439.

HYDROXYMETHYLDIHYDROPTERIDINE PYROPHOSPHOKINASE

(ATP: 2-amino-4-hydroxy-6-hydroxymethyl-7,8-dihydropteridine 6'-pyrophosphotransferase)

ATP + 2-amino-4-hydroxy-6-hydroxymethyl-7,8-dihydropteridine =
AMP + 2-amino-4-hydroxy-6-hydroxymethyl-7,8-dihydropteridine 6'-diphosphate

Ref.

Molecular weight

source	value	conditions	Ref.
Escherichia coli	15,000	Sephadex G 100, pH 7.0	(1,2)

Specificity and Michaelis constants

source	substrate	K_m(M)	conditions	Ref.
E. coli (purified 413 x)	ATP[a]	1.5×10^{-5}	pH 8.6, Tris, 37°	(1)
	hydroxymethyl-dihydro-pteridine	1.5×10^{-5}	pH 8.6, Tris, 37°	(1)

(a) dATP was less active than ATP and GTP, CTP and UTP were inactive.

The enzyme requires Mg^{2+} or Mn^{2+}; Co^{2+} was less effective and Ca^{2+}, Cu^{2+}, Zn^{2+} and monovalent ions were ineffective. (1)

The pyrophosphokinase has also been purified (50 x) from Lactobacillus plantarum. (3)

Abbreviations

hydroxymethyldihydropteridine — 2-amino-4-hydroxy-6-hydroxymethyl-7,8-dihydropteridine

References

1. Richey, D.P. & Brown, G.M. (1971) Methods in Enzymology, 18B, 765.
2. Richey, D.P. & Brown, G.M. (1969) JBC, 244, 1582.
3. Shiota, T., Baugh, C.M., Jackson, R. & Dillard, R. (1969) B, 8, 5022.

SULPHATE ADENYLYLTRANSFERASE (ADP)

(ADP: sulphate adenylyltransferase)

ADP + sulphate = Pi + adenylylsulphate

Ref.

Equilibrium constant

The reaction is reversible. (1)

Molecular weight

source	value	conditions	
Yeast	200,000	sucrose density gradient	(2)

Specificity and Michaelis constants

source	substrate	K_m (M)	conditions	
Saccharomyces cerevisiae (purified 319 x)	adenylyl-sulphate(a)	3.6×10^{-4} 6.2×10^{-5}	Pi = 2 mM Pi = 20 mM	(1)

(a) at pH 7.5, Tris and 30°. Adenylylsulphate (1.00) could be replaced by ADP (0.50) but not by AMP, ATP or PPi.

The kinetics with Pi as the variable substrate are complex. The enzyme catalyzes an exchange of Pi into ADP, UDP, IDP, GDP or CDP. It is not activated by metals (e.g. Mg^{2+}, Mn^{2+} or K^+). The mechanism of the reaction has been investigated. (1,2)

References

1. Adams, C.A. & Nicholas, D.J.D. (1972) BJ, 128, 647.
2. Grunberg-Manago, M., del Campillo-Campbell, A., London, L. & Michelson, A.M. (1966) BBA, 123, 1.

tRNA ADENYLYLTRANSFERASE

(ATP: tRNA adenylyltransferase)

$ATP + tRNA_n = PPi + tRNA_{n+1}$

Ref.

Molecular weight

source	value	conditions	Ref.
Escherichia coli	50,000 [1]	gel electrophoresis (SDS); Sephadex G 100; Sepharose 4B in 6M guanidine HCl	(1)

Specific activity

E. coli (4000 x)	0.63	CTP(acceptor = tRNA-X)	
	0.61	CTP(acceptor = tRNA-X-C)	
	0.48	ATP(acceptor = tRNA-X-C-C)	
	(conditions: pH 9.2, glycine, 37°)		(1)

Specificity and catalytic properties

tRNA adenylyltransferase catalyzes the addition of 2CMP residues and then 1AMP residue to the 3'-OH end of tRNA from which the terminal -pC-C-A has been removed. Undigested tRNA or tRNA with more than pC-C-A removed was inactive. All suitably digested aminoacyl tRNA species from E. coli and also heterologous tRNA species (including eukaryocytes) were acceptor substrates. The following Michaelis constants have been obtained (at pH 9.2, glycine, 37° and with the enzyme from E. coli):

substrate	cosubstrate	approx K_m (μM)
CTP	tRNA-X or tRNA-X-C [a]	10
ATP	tRNA-X-C-C [a]	100
tRNA-X [a]	CTP	0.2
tRNA-X-C [a]	CTP	0.2
tRNA-X-C-C [a]	ATP	0.2

(a) unfractionated tRNA from E. coli. Purified tRNA had similar kinetic properties. tRNA-X and tRNA-X-C (but not tRNA-X-C-C) concentration dependence curves were sigmoidal with 1.5 binding sites for tRNA per mole of enzyme.

The Ki values for ATP inhibiting CMP incorporation into tRNA-X and tRNA-X-C and for CTP inhibiting AMP incorporation into tRNA-X-C-C were 2.6 mM, 3.6 mM and 1.1 mM, respectively. In each case the inhibition was competitive (nucleotide substrate). Furthermore, tRNA-X-C-C inhibits CMP incorporation into tRNA-X or tRNA-X-C and the latter 2 substances inhibit AMP incorporation into tRNA-X-C-C. All the 3 reactions were inhibited by intact tRNA. (1)

Ref.

Abbreviations

tRNA-X-C-C; tRNA-X-C, tRNA-X: tRNA in which one, two or three nucleotides are missing from the 3'-OH end. Thus, tRNA-X-C-C-A possesses the complete 3'-terminal sequence where C = CMP and A = AMP.

EC 2.7.7.20, EC 2.7.7.21 and EC 2.7.7.25 have overlapping specificities. For details of the specificity properties of EC 2.7.7.20 and EC 2.7.7.21 see Enzyme Handbook (1969), Vol 1, p 479-480.

References

1. Miller, J.P. & Philipps, G.R. (1971) JBC, 246, 1274.

HEXOSE 1-PHOSPHATE NUCLEOTIDYLTRANSFERASE

(Nucleoside triphosphate: hexose-1-phosphate nucleotidyltransferase)

NTP + hexose 1-phosphate = PPi + NDPhexose

Ref.

Equilibrium constant

$$\frac{[PPi]\ [GDPglucose]}{[GTP]\ [glucose\ 1\text{-}phosphate]} = 0.25$$ (pH 7.8, 30°. Mg^{2+} was present) (1)

Specificity and Michaelis constants

source	substrate	V relative	K_m (M)	Ref.
Calf liver (purified 530 x)	GTP	1.0	2.6×10^{-3}	(1)
	glucose 1-phosphate	1.0	2.6×10^{-3}	(1)
	PPi	0.3	8.2×10^{-4}	(1)
	GDPglucose[a]	0.3	6.2×10^{-4}	(1)

(a) GDPglucose (1.00) could be replaced by GDPmannose (0.49; Km = 1×10^{-7}M); IDPglucose (0.07) or IDPmannose (0.21). The conditions were: pH 7.8, Tris, 25°.

The enzyme has also been purified from rat liver and rat mammary gland; it requires Mg^{2+} for full activity. (1)

Inhibitors

source	inhibitor	type	K_i (M)	conditions	Ref.
Calf liver	GDPmannose	complex	1.7×10^{-7}	pH 7.8, Tris, 25°	(1)
	mannose 1-phosphate[a]	complex	4.7×10^{-5}	pH 7.8, Tris, 25°	(1)

(a) the following were inactive as inhibitors: mannose 6-phosphate; glucose 6-phosphate; fructose 6-phosphate; ribose 5-phosphate; galactose 6-phosphate; fructose 1-phosphate; galactose 1-phosphate; ribose 1-phosphate and lactose 1-phosphate.

References

1. Verachtert, H., Rodriguez, P., Bass, S.T. & Hansen, R.G. (1966) JBC, 241, 2007.

FUCOSE 1-PHOSPHATE GUANYLYLTRANSFERASE

(GTP: β-6-deoxy-L-galactose-1-phosphate guanylyltransferase)

GTP + L-fucose 1-phosphate = PPi + GDP-L-fucose

Ref.

Specificity and Michaelis constants

source	substrate	K_m (mM)	conditions	
Pig liver	GDP-L-fucose	0.12	pH 8, Tris, 37°	(1)

In the presence of GTP, the enzyme (purified 46-fold) could utilize β-L-fucose 1-phosphate (1.00); α-D-mannose 1-phosphate (0.18) and α-D-glucose 1-phosphate (0.01) but not α-L-fucose 1-phosphate or α-D-galactose 1-phosphate. In the presence of β-L-fucose 1-phosphate, GTP (1.00) and UTP (0.19) were active but the following nucleotides were inactive: CTP, dTTP and ATP. The enzyme requires a divalent metal ion for activity: Mg^{2+} (1.00) could be replaced by Co^{2+} (0.24) and Mn^{2+} (0.08) but not by Ca^{2+} or Fe^{2+}. (1)

References

1. Ishihara, H. & Heath, E.C. (1968) JBC, 243, 1110.

RIBOSE 5-PHOSPHATE ADENYLYLTRANSFERASE

(ADP: D-ribose 5-phosphate adenylyltransferase)

ADP + D-ribose 5-phosphate = Pi + ADPribose

Ref.

Equilibrium constant

The reaction is irreversible [R] with the enzyme from *Euglena gracilis* as catalyst. This enzyme also catalyzes a reversible exchange reaction between ADP and Pi and between ADPribose and ribose 5-phosphate. (1,2,3)

Specificity and Michaelis constants

substrate	reaction [a]	K_m (μM)	conditions	
ADPribose [b]	P	40-50	pH 7.8, Tris, 30°	(1)
Pi [c]	P	400-500	pH 7.8, Tris, 30°	(1)
Pi	E	400-500	pH 7.8, Tris, 30°	(1)
ADP [d]	E	600	pH 7.8, Tris, 30°	(1)

(a) P = phosphorolysis reaction; E = ADP-Pi exchange reaction. With the enzyme from *E. gracilis* (purified 65 x).
(b) ADPribose could be replaced by IDPribose but not by ADPglucose, UDPglucose or adenosinetriphosphoribose.
(c) when Pi was replaced by arsenate, ADPribose yielded 5'-AMP and ribose 5-phosphate.
(d) ADP (1.00) could be replaced by IDP (0.16) but not by CDP, UDP or GDP.

The enzyme did not require added metal ions for activity. (1,2,3)

Inhibitors

inhibitor [a]	type	K_i (mM) P	K_i (mM) E	
ribose 5-phosphate	NC(ADPribose)	4.0	-	(1,3)
	C(Pi)	-	0.11	(1,3)
5'-AMP	C(ADPribose)	1.0	-	(1,3)
	C(ADP)	-	1.5	(1,3)
ATP	C(ADPribose)	2.9	-	(1,3)
	C(ADP)	-	1.3	(1,3)

(a) with the enzyme from *E. gracilis* (conditions = pH 7.8, Tris, 30°). P = phosphorolysis reaction; E = ADP-Pi exchange reaction. The following did not inhibit either reaction: ribose; deoxyadenosine; 3'-AMP; CMP; CDP; UDP; reduced NAD and NADP. The phosphorolysis of ADP-ribose was not inhibited by ADPglucose, UDPglucose or NAD.

References

1. Evans, W.R. (1971) Methods in Enzymology, 23A, 566.
2. Evans, W.R. & san Pietro, A. (1966) ABB, 113, 236.
3. Stern, A.I. & Avron, M. (1966) BBA, 118, 577.

3-DEOXY-manno-OCTULOSONATE CYTIDYLYLTRANSFERASE

(CTP: 3-deoxy-D-*manno*-octulosonate cytidylyltransferase)

CTP + 3-deoxy-D-*manno*-octulosonate =

PPi + CMP-3-deoxy-D-*manno*-octulosonate

Ref.

Equilibrium constant

The reaction is reversible, the forward direction being strongly favoured. CMP-3-deoxy-D-*manno*-octulosonate is unstable. (1)

Specificity and Michaelis constants

source	substrate	K_m (M)	conditions	
Escherichia coli (purified 170 x)	CTP[a]	2.2×10^{-4}	pH 8, Tris, 37°	(1)
	3-deoxy-D-*manno*-octulosonate[b]	8×10^{-4}	pH 8, Tris, 37°	(1)

(a) ITP and dTTP exhibited slight activities but ATP, CDP, GTP and UTP were inactive.

(b) the following were inactive: 3-deoxy-D-*manno*-octulosonate-8-phosphate; 3-deoxy-D-*arabino*-heptulosonate; 3-deoxy-D-*erythro*-hexulosonate and *N*-acetylneuraminate.

The transferase requires Mg^{2+}. Mn^{2+} had 60% and Ca^{2+} 80% the activity of Mg^{2+}. No other divalent ion tested was active. The enzyme also requires reduced glutathione for full activity. (1)

References

1. Ghalambor, M.A. & Heath, E.C. (1966) JBC, 241, 3216.

PHOSPHATIDATE CYTIDYLYLTRANSFERASE

(CTP: phosphatidate cytidylyltransferase)

CTP + phosphatidate = PPi + CDPdiglyceride

Ref.

Specificity and kinetic properties

source	substrate	K_m (mM)	conditions	
Micrococcus cerificans (particulate)	CTP[(a)]	1.2	pH 8.1, 38°	(1)
	phosphatidate[(b)]	0.3	pH 8.1, 38°	(1)

(a) CTP (1.00) could be replaced by dCTP (0.49). ATP, GTP, UTP, ITP and dTTP were neither substrates nor inhibitors.

(b) prepared from lecithin isolated from yeast or soybean.

The enzyme requires Mg^{2+} (which could be replaced by Mn^{2+}) and K^+ (which could be replaced by Rb^+ or NH_4^+ but not by Na^+, Cs^+ or Li^+). High concentrations of CTP, phosphatidate or Mg^{2+} were inhibitory. Long chain CoA esters were also inhibitory. CDP, CMP and PPi were non-inhibitory. (1)

The cytidylyltransferase is also present in chick brain mitochondrial fraction and guinea pig liver microsomes. The microsomal enzyme was inhibited by PPi. (2,3)

References

1. McCaman, R.E. & Finnerty, W.R. (1968) JBC, 243, 5074.
2. Petzold, G.L. & Agranoff, B.W. (1967) JBC, 242, 1187.
3. Carter, J.R. & Kennedy, E.P. (1966) J. Lipid Res., 7, 678.

GLUTAMINE-SYNTHETASE ADENYLYLTRANSFERASE

(ATP: [L-glutamate: ammonia ligase (ADP)] adenylyltransferase)

ATP + [L-glutamine: ammonia ligase (ADP)] =

PPi + adenylyl-[L-glutamate: ammonia ligase (ADP)]

Ref.

Equilibrium constant

$$\frac{\text{[glutamine synthetase-su-AMP] [PPi]}}{\text{[glutamine synthetase-su] [ATP]}} = 8.5$$ (pH 7.36, 30°, in the presence of 10 mM Mg^{2+}) (2)

Molecular weight

source	value	conditions	Ref.
Escherichia coli B	115,000	pH 7.6	(3)
	145,000	Sephadex G 150, pH 7.0	(4)
E. coli	70,000	sucrose density gradient	(6)

Specificity and catalytic properties

Adenylyltransferase (purified 300 x) catalyzes the transfer of the adenyl moiety of ATP to a specific tyrosyl residue present in each of the 12 subunits of glutamine synthetase (EC 6.3.1.2). This process leads to the inactivation of the synthetase. Deadenylylation (reactivation) is catalyzed not by the adenylyltransferase but by a separate complex enzyme system. The adenylylation and deadenylylation of glutamine synthetase are phenomena of central regulatory importance for *E. coli*. Each subunit of glutamine synthetase reacts independently with ATP (or with PPi in the reverse reaction). The adenylyltransferase requires glutamine (glutamate was inhibitory), Mg^{2+} and a sulphydryl compound for activity. It is highly specific for intact glutamine synthetase (the individual subunits were inactive) and the following proteins were inactive: calf thymus histone, casein, protamine, bovine haemoglobin, egg-white lysozyme, bovine α-chymotrypsinogen, bovine ribonuclease, acid phosphatase (potato), pyruvate kinase (rabbit muscle), glutamate dehydrogenase (bovine liver), ovalbumin and bovine serum albumin. The enzyme is also highly specific for ATP which could not be replaced by any other nucleoside triphosphate. (1,2,3,4,7)

The mechanism of the reaction has been studied. The purified enzyme may not catalyze an ATP-Pi exchange reaction. With ATP-Mg as the variable substrate, negative cooperativity was observed. (5)

The adenylyltransferase was inhibited by glutamate, 2-oxoglutarate, RNA, tRNA, CTP, UTP, GTP, GDP, UDP, ADP, PPi, Pi and Mn^{2+}. (1)

Specificity and catalytic properties continued

The deadenylation of adenylyl-glutamine synthetase is catalyzed by an enzyme consisting of two protein components of different molecular weights (EC 3.1.4.15). Both components are required for enzyme activity. The enzyme, which has been purified 20-fold from E. coli, requires Mg^{2+} (or Mn^{2+}) for activity and it was stimulated by 2-oxoglutarate, Pi and by UTP and ATP. PPi and glutamine were inhibitory and AMP inhibited the activation by UTP and ATP. (7,8)

Abbreviations

glutamine synthetase-su	one of the 12 identical subunits of glutamine synthetase (EC 6.3.1.2) each of molecular weight 50,000.

References

1. Stadtman, E.R., Shapiro, B.M., Ginsburg, A., Kingdon, H.S. & Denton, M.D. (1968) Brookhaven Symp. Biol, 21, 378.
2. Mantel, M. & Holzer, H. (1970) PNAS, 65, 660.
3. Wolf, D., Ebner, E. & Hinze, H. (1972) EJB, 25, 239.
4. Ebner, E., Wolf, D., Gancedo, C., Elsässer, S. & Holzer, H. (1970) EJB, 14, 535.
5. Wohlhueter, R.M., Ebner, E. & Wolf, D.H. (1972) JBC, 247, 4213.
6. Hennig, S.B. & Ginsburg, A. (1971) ABB, 144, 611.
7. Shapiro, B.M. (1969) B, 8, 659.
8. Heilmeyer, L., Battig, F. & Holzer, H. (1969) EJB, 9, 259.
   Stadtman, E.R. (1973) The Enzymes, 8A, 40.

GLUCURONATE 1-PHOSPHATE URIDYLYLTRANSFERASE

(UTP: 1-phospho-α-D-glucuronate uridylyltransferase)

UTP + 1-phospho-D-glucuronate = PPi + UDP-D-glucuronate

Ref.

Equilibrium constant

$$\frac{[PPi]\ [UDP\text{-}D\text{-}glucuronate]}{[UTP]\ [1\text{-}phospho\text{-}D\text{-}glucuronate]} = 0.34 \quad (pH\ 8.2,\ Tris,\ 30°)$$ (1)

Molecular weight

source	value	conditions	Ref.
Hordeum vulgaris (Barley seedlings)	35,000	Sephadex G 100, pH 8.2. The buffer contained mercaptoethanol and EDTA.	(1)

Specificity and catalytic properties

source	substrate	V relative	K_m (mM)	conditions	Ref.
H. vulgaris (purified 80 x)	1-phospho-D-glucuronate	1.0	0.33	pH 8.0, Tris, 30°	(1)
	UDP-D-glucuronate	0.8	0.5	pH 8.2, Tris, 30°	(1)

The enzyme could utilize CTP (0.16) in the place of UTP (1.00) but GTP, ATP, ITP and dTTP were inactive. It requires a divalent cation for activity in both the forward and reverse reactions. Thus, in the forward reaction Mg^{2+} (1.00); Mn^{2+} (0.53 $^{+}$); Co^{2+} (0.40); Zn^{2+} (0.37); Ca^{2+} (0.36) and Cu^{2+} (0.06) were all active. Excess Mg^{2+} was inhibitory. Mg^{2+} (or Mn^{2+}) forms a complex with UDP-D-glucuronate which is stable to ion exchange chromatography and electrophoresis. (1)

References

1. Roberts, R.M. (1971) JBC, 246, 4995.

SERINE ETHANOLAMINE PHOSPHATE SYNTHASE

(CDPethanolamine: L-serine ethanolaminephosphotransferase)

CDPethanolamine + L-serine = CMP + L-serine ethanolamine phosphate

Ref.

Specificity and catalytic properties

source	substrate	K_m(M)	conditions	
Chicken	L-serine[a]	1×10^{-3}	pH 7.0, imidazole, 37°	(1)
intestinal	CDPethanolamine[b]	8.5×10^{-5}	pH 7.0, imidazole, 37°	(1)
mucosa (microsomes)	CMPaminoethyl-phosphonate	1.1×10^{-5}	pH 7.0, imidazole, 37°	(1)

(a) L-serine could be replaced by D-serine or DL-α-methylserine but not by N-acetyl-L-serine; L-homoserine; 4-hydroxy-L-proline; 5-hydroxy-DL-lysine; 3-hydroxy-DL-glutamate; L-threonine; ethanolamine or 3-hydroxypropionate.

(b) CDP-2-amino-2-methylpropanol or CMPaminoethylphosphonate could replace CDPethanolamine but CDPserine + ethanolamine or CDPcholine could not.

The enzyme requires Mg^{2+} for activity; Mn^{2+} and Co^{2+} were less effective as activators. (1)

The mechanism of the reaction has been investigated. (1)

The enzyme was inhibited by L-threonine, L-alanine and L-homoserine. CMP (C(CDPethanolamine), NC(L-serine)); serine ethanolamine phosphate (NC(L-serine)) and CTP were also inhibitory. (1)

References

1. Allen, A.K. & Rosenberg, H. (1968) BBA, 151, 504.

HOLO ACYL CARRIER PROTEIN SYNTHASE

(CoA: apo-acyl carrier protein pantetheinephosphotransferase)

CoA + apo-ACP = 3',5'-ADP + holo-ACP

Ref.

Equilibrium constant

The reaction is reversible. (2)

Molecular weight

source	value	conditions	Ref.
Escherichia coli B	50,000	gel filtration; sucrose density gradient	(2)

Specificity and Michaelis constants

source	substrate	K_m (M)	conditions	Ref.
E. coli B (purified 780 x)	CoA[a]	1.5×10^{-4}	pH 8, Tris, 33°	(1,2)
	apo-ACP[b]	4×10^{-7}	pH 8, Tris, 33°	(1,2)

(a) CoA could not be replaced by dephospho-CoA or oxidized CoA.

(b) low concentrations of apo-ACP were inhibitory. Native apo-ACP (*E. coli*) has 77 amino acid residues of known sequence. The synthase can utilize a peptide composed of residues 1 to 74 but not the following fragments: 7 to 77 or 19 to 61.

The enzyme requires Mg^{2+} or Mn^{2+} for activity. Both cations gave complex saturation curves. (2)

Abbreviations

ACP acyl carrier protein (see EC 2.3.1.39)

References

1. Prescott, D.J. & Vagelos, P.R. (1972) *Advances in Enzymology*, *36*, 284.
2. Elovson, J. & Vagelos, P.R. (1968) *JBC*, *243*, 3603.

PYRUVATE-PHOSPHATE DIKINASE

(ATP: pyruvate, orthophosphate phosphotransferase)

ATP + pyruvate + Pi = AMP + phospho*enol*pyruvate + PPi

Ref.

Equilibrium constant

$$\frac{[AMP]\ [PEP]\ [PPi]\ [H^+]^2}{[ATP]\ [pyruvate]\ [Pi]} = 10^{-17}\ M^2 \qquad (1)$$

Specificity and catalytic properties

source	substrate	K_m (M)	conditions	Ref.
Bacteroides symbiosus (purified 19 x)	ATP[a]	1×10^{-4}	-	(2)
	pyruvate	8×10^{-5}	-	(2)
	Pi	6×10^{-4}	-	(2)
	AMP[b]	3.5×10^{-6}	pH 6.8, imidazole, 25°	(2)
	PEP	6×10^{-5}	pH 6.8, imidazole, 25°	(2)
	PPi	1×10^{-4}	pH 6.8, imidazole, 25°	(2)
	NH_4^+[c]	2.5×10^{-3}	pH 6.8, imidazole, 25°	(2)

(a) at pH 7.4, ATP could not be replaced by dATP, GTP, UTP, ITP or CTP. Of these, all except GTP and UTP were inhibitory.
(b) at pH 6.4, AMP (1.00) could be replaced by dAMP (0.27). IMP, UMP, GMP and CMP were inactive as substrates or inhibitors.
(c) in the reverse reaction. NH_4 could be replaced by K^+ but not Na^+.

The dikinase (*B. symbiosus*) requires a divalent metal ion for activity. In the reverse reaction, Ni^{2+}, Mn^{2+}, Mg^{2+} and Co^{2+} were increasingly effective whereas in the forward reaction only Mg^{2+} was effective. (2)

The mechanism of the reaction catalyzed by the dikinase present in *Propionibacterium shermanii* has been investigated and compared with the mechanism of phospho*enol*pyruvate synthase (EC 2.7.1.dd). A phosphoenzyme is formed in the presence of PEP and Mg^{2+}. (3)

Abbreviations

PEP phospho*enol*pyruvate

References

1. Reeves, R.E., Menzies, R.A. & Hsu, D.S, (1968) JBC, 243, 5486.
2. Reeves, R.E. (1971) BJ, 125, 531.
3. Evans, H.J. & Wood, H.G. (1968) PNAS, 61, 1448.
   Sugiyama, T. (1973) B, 12, 2862.

3β-HYDROXYSTEROID SULPHOTRANSFERASE

(3'-Phosphoadenylylsulphate: 3β-hydroxysteroid sulphotransferase)

3'-Phosphoadenylylsulphate + a 3β-hydroxysteroid =
adenosine 3',5'-disphosphate + a steroid 3β-sulphate

Ref.

Molecular properties

source	value	conditions	Ref.
Human adrenal gland	65,000[a]	Sephadex G 200, pH 7.5; sucrose gradient	(1)
Human liver	50,000	Sephadex G 200, pH 7.4	(2)

(a) this is the molecular weight of the monomer. Dimer, trimer and possibly higher associated states exist in a slowly reversible association-dissociation equilibrium. PAPS causes complete dissociation to monomer whereas DHEA or Mg^{2+} and cysteine favour association. The different forms of the enzyme have different catalytic properties and this results in the enzyme having complex kinetics (see below).

Specificity and kinetic properties

source	substrate	K_m (μM)	conditions	Ref.
Human liver	DHEA[a]	6.7	pH 7.4, Tris, 37°	(2)
(purified 22 x)	PAPS	4.0	pH 7.4, Tris, 37°	(2)

(a) DHEA (1.00) could be replaced by 3α-hydroxy-5β-androstan-17-one (0.45), oestriol (0.31), oestradiol-17β (0.15) or oestrone (0.08) but testosterone and cholesterol were inactive.

The enzyme (human adrenal gland, high speed supernatant) could utilize DHEA (1.00), pregnenolone (0.46) and cholesterol (0.01) as acceptor substrates. 17α-hydroxypregnenolone was also active. The enzyme was activated by Mg^{2+} and cysteine and it exhibits complex kinetics with DHEA as the variable substrate, multipeaks being observed in velocity-substrate plots. (1)

The enzyme (human liver) was activated by divalent metal ions (e.g. Co^{2+}, Ni^{2+}. Ca^{2+} or Mn^{2+}) but it was not inhibited by EDTA. It was inhibited by high concentrations of DHEA. (2)

Abbreviations

PAPS	3'-phosphoadenylylsulphate
DHEA	dehydroepiandrosterone; Δ^5-androsten-3β-ol-17-one

References

1. Adams, J.B. & Edwards, A.M. (1968) BBA, 167, 122.
2. Gugler, R., Rao, G.S. & Breuer, H. (1970) BBA, 220, 69.

CHOLINE SULPHOTRANSFERASE

(3'-Phosphoadenylylsulphate: choline sulphotransferase)

3'-Phosphoadenylylsulphate + choline =

adenosine 3',5'-diphosphate + choline sulphate

Ref.

Equilibrium constant

The reaction is irreversible [F] with the enzyme from *Aspergillus nidulans* or *A. sydowi* as catalyst. (1,2)

Specificity and kinetic properties

source	substrate	K_m (M)	conditions	
A. nidulans (purified 3 x)	choline	1.2×10^{-2}	pH 7.3, Pi, 37°	(1)
	dimethylethyl-aminoethanol	2.0×10^{-2}	pH 7.3, Pi, 37°	(1)
	dimethylamino-ethanol	2.5×10^{-2}	pH 7.3, Pi, 37°	(1)
	PAPS	2.2×10^{-5}	pH 7.3, Pi, 37°	(1)

Choline sulphotransferase is highly specific and is unable to transfer sulphate to phenols, γ-pyrones, phenolic steroids, 3β-hydroxysteroids, aliphatic alcohols and arylamines. A number of choline analogues tested were inhibitory but only 2 were substrates (see above). 4-Nitrophenol (Ki = 9.8 mM), carnitine (Ki = 36.5 mM) and dimethylaminopropan-1-ol (Ki = 30.1 mM) were competitive (choline) inhibitors. (1)

The enzyme from *A. sydowi* required Mg^{2+} for activity but that from *A. nidulans* did not. (1,2)

Abbreviations

PAPS 3'-phosphoadenylylsulphate

References

1. Orsi, B.A. & Spencer, B. (1964) J. Biochem (Tokyo), 56, 81.
2. Kaji, A. & Gregory, J.D. (1959) JBC, 234, 3007.

3-OXOADIPATE ENOL-LACTONASE

(4-Carboxymethylbut-3-enolide (1,4) enol-lactone hydrolase)

4-Carboxymethylbut-3-enolide (1,4) + H_2O = 3-oxoadipate

Ref.

Molecular weight

source	value	conditions	
Pseudomonas putida	33,000	Sephadex G 200, pH 7.1	(1,2)

Specificity and Michaelis constants

source	substrate	K_m (μM)	conditions	
P. putida (purified 422 x)	4-carboxymethylbut-3-enolide (1,4)	12	pH 8.0, Tris, 25°	(1,2)

Light absorption data

4-Carboxymethylbut-3-enolide (1,4) has an absorption band at 230 nm (molar extinction coefficient about 2000 $M^{-1}cm^{-1}$). (2)

References

1. Ornston, L.N. (1966) JBC, 241, 3787.
2. Ornston, L.N. (1970) Methods in Enzymology, 17A, 546.

GALACTOLIPASE

(2,3-Di-O-acyl-1-O-(β-D-galactosyl)-D-glycerol acylhydrolase)

2,3-Di-O-acyl-1-O-(β-D-galactosyl)-D-glycerol + $2H_2O$ =

1-O-(β-D-galactosyl)-D-glycerol + 2 fatty acid anions

Ref.

Molecular weight

source	value	conditions	Ref.
Phaseolus multiflorus (runner bean)	110,000	Sephadex G 200, pH 7.0	(1)

Specificity and catalytic properties

source	substrate[a]	V relative	K_m (mM)	conditions	Ref.
P. multiflorus (purified 50 x)	GDG	1.38	0.65	pH 7.0, Pi, 30°	(1)
	GGDG	1.00	0.31	pH 5.6, Pi, 30°	(1)

(a) a less purified preparation (3 x) hydrolyzed the following: digalactosyldilinolenin (0.54); monogalactosyldilinolenin (0.42); runnerbean leaf-lecithin (0.03) and 2,3-di-glyceride derived from leaf-lecithin (0.03) but not monogalactosyldistearin; digalactosyldistearin or triolein (Ref. 2).

There are conflicting reports as to whether one single enzyme protein removes fatty acid residues from both GDG and GGDG (Ref. 1) or whether two proteins are responsible: one specific for GDG and the other for GGDG (Ref. 2).

The enzyme requires a reductant such as sodium dithionite for full activity but the naturally occurring reductant ascorbic acid was ineffective and cysteine was inhibitory. (1)

Abbreviations

GDG monogalactosyl diglyceride
GGDG digalactosyl diglyceride

Both were obtained from spinach leaves. (1)

References

1. Helmsing, P.J. (1969) BBA, 178, 519.
2. Sastry, P.S. & Kates, M. (1964) B, 3, 1280.

ACYLCARNITINE HYDROLASE

(O-Acylcarnitine hydrolase)

O-Acylcarnitine + H_2O = a fatty acid + L-carnitine

Ref.

Molecular weight

source	value	
Rat liver	$S_{20,w}$ = 4.5 S	(1)

Specificity and kinetic properties

source	substrate	K_m(mM)	conditions	
Rat liver (microsomal; purified 18 x)	palmitoyl-L-carnitine	5	pH 7.5, Pi, 37°	(1)
	decanoyl-L-carnitine	3.2	pH 7.5, Pi, 37°	(1)

The enzyme is highly specific for higher fatty acid (C_6 to C_{18}) esters of L-carnitine. The following were hydrolyzed: decanoyl-L-carnitine (1.00); lauroyl-L-carnitine (0.72); octanoyl-L-carnitine (0.42); myristoyl-L-carnitine (0.35); hexanoyl-L-carnitine (0.32); palmitoyl-L-carnitine (0.29) and stearoyl-L-carnitine (0.10). Decanoyl-D-carnitine; acetyl-L-carnitine; propionyl-L-carnitine; cholesterol palmitate; palmitoyl-CoA; 4-nitrophenyl palmitate; glycerol tripalmitate and glycerol monopalmitate were not hydrolyzed. (1)

The enzyme did not require any added cofactor or metal ions for activity. It was not inhibited by EDTA. (1)

Decanoyl-D-carnitine inhibited the hydrolysis of both decanoyl-L-carnitine and palmitoyl-L-carnitine. The inhibition was not overcome by excess decanoyl-L-carnitine. L-Carnitine and D-carnitine did not inhibit. (1)

References

1. Mahadevan, S. & Sauer, F. (1969) JBC, 244, 4448.

AMINOACYL-tRNA HYDROLASE

(Aminoacyl-tRNA hydrolase)

N-Substituted aminoacyl-tRNA + H_2O = N-substituted amino acid + tRNA

Ref.

Specificity and catalytic properties

The hydrolase cleaves the ester bond between the N-acetylamino acid and the 3'-hydroxyl group of the 3'- end of the terminal nucleotide in the tRNA. The tRNA, therefore, remains intact and retains its pCpCpA-terminal sequence. (1)

The enzyme requires Mg^{2+} for activity and EDTA was inhibitory. (1,2)

source	substrate	K_m (μM)	conditions	
Yeast (purified 230 x)	N-acetylphenyl-alanyl-tRNA[a]	1	pH 7.0, Tris, 30°	(1)
Escherichia coli (purified 160 x)	N-acetylleucyl-tRNA[b]	~0.1	pH 7.3, Tris, 37°	(2)

(a) free aminoacyl-tRNAs are resistant or very slowly hydrolyzed but the following are good substrates: N-benzoyl-Gly-Gly-Phe-tRNA and N-benzoyl-Gly-Gly-Gly-Phe-tRNA. N-Acetylvalyladenosine is not hydrolyzed and N-acetylvalyloligonucleotide and yeast oligolysyl-tRNA are very poor substrates. The enzyme is competitively inhibited by uncharged yeast tRNA but not by purified ribosomal RNA.

(b) all the N-acetyl-L-aminoacyl-tRNAs tested so far are cleaved by the enzyme except for N-acetylmethionyl-$tRNA_F$. N-Acetyl-D-aminoacyl-tRNAs are not hydrolyzed. Further, since yeast and E. coli N-acetyl-phenylalanyl-tRNAsare hydrolyzed at about the same rate, species specificity of the enzyme is probably not rigorous. N-Acetylamino-acyloligonucleotides are cleaved extremely slowly, if at all.

References

1. Jost, J-P. & Bock, R.M. (1969) JBC, 244, 5866.
2. Kössel, H. & Rajbhandary, U.L. (1968) J. Mol. Biol., 35, 539.

ACTINOMYCIN LACTONASE

(Actinomycin lactone-hydrolase)

Actinomycin + H_2O = actinomycinic monolactone

Ref.

Molecular properties

source	value	conditions	
Actinoplanes missouriensis	60,000	ultracentrifugation	(1)

The light absorption spectrum of the enzyme revealed the absence of a prosthetic chromophore. (1)

Specific activity

A. missouriensis	(80 x)	0.004 actinomycin[a] (pH 7.8, Tris, 38°)	(1)

(a) this is a maximum value.

Specificity and catalytic properties

source	substrate	K_m (µM)	conditions	
A. missouriensis	actinomycin[a]	11.4	pH 7.8, Tris, 38°	(1)

(a) actinomycin could not be replaced by echinomycin; etamycin; staphylomycin S; stendomycin; thiostrepton or vernamycin Bα. The product of the reaction, actinomycinic monolactone, was inhibitory. Actinomycinic acid did not inhibit.

The enzyme does not require metal ions for activity.

References

1. Hou, C.T. & Perlman, D. (1970) JBC, 245, 1289.

ORSELLINATE DEPSIDE ESTERASE

(Orsellinate depside hydrolase)

Lecanorate + H_2O = 2 orsellinate

Ref.

Molecular weight

source	value	conditions	Ref.
Lasallia pustulata (a lichen)	42,000 [1]	gel electrophoresis (SDS)	(1)

Specific activity

L. pustulata (135 x)	1300	lecanorate (pH 6.8, Pi, 25°)	(1)

Specificity and catalytic properties

source	substrate	K_m (μM)	conditions	Ref.
L. pustulata	lecanorate	56	pH 6.8, Pi, 25°	(1)

The enzyme is highly specific for depsides based on orsellinate:

lecanorate = 2 orsellinate
evernate = orsellinate + everninate
gyrophorate = 3 orsellinate

Methyl lecanorate and erythrin were also hydrolyzed but the following were not: iso-evernate; phenyl benzoate and 3-digallate. The enzyme was inhibited by diisopropylfluorophosphate (1)

Light absorption data

The transformation of 1 mole of lecanorate to 2 moles of orsellinate is accompanied by an increase in absorption at 250 nm (molar extinction coefficient = 7610 $M^{-1}cm^{-1}$). (1)

Abbreviations

orsellinate	2,4-dihydroxy-6-methylbenzoate
everninate	2-hydroxy-4-methoxy-6-methylbenzoate
depsides	substances composed of two or more phenolic acids linked by ester bonds.

References

1. Schultz, J. & Mosbach, K. (1971) EJB, 22, 153.

1L-myo-INOSITOL 1-PHOSPHATASE

(1L-*myo*-Inositol-1-phosphate phosphohydrolase)

1L-*myo*-Inositol 1-phosphate + H_2O = *myo*-Inositol + Pi

Ref.

Specificity and Michaelis constants

source	substrate	K_m (mM)	conditions	
Yeast (purified 110 x)[a]	inositol 1-phosphate	1.67	pH 7.7, Tris, 29°	(1)

(a) the preparation obtained also contained glucose 6-phosphate cyclo-isomerase activity (EC 5.5.1.4). The two activities can be separated (Ref. 2) and the preparation thus obtained will hydrolyse inositol 1-phosphate (1.00); inositol 3-phosphate (0.77); 2'-AMP (0.72); β-glycerophosphate (0.70); fructose 6-phosphate (0.20); glucose 1-phosphate (0.19); ribose 5-phosphate (0.17); mannose 6-phosphate (0.14); α-glycerophosphate (0.15); 2' and 3'-CMP mixture (0.17); 3'-AMP (0.14); 4-nitrophenyl phosphate (0.13) and ATP (0.10) but not inositol 2-phosphate; glucose 6-phosphate; 6-phosphogluconate; 5'-AMP; 3',5'-cyclic AMP; 5'-GMP; 5'-IMP; 5'-UMP; 2',3'-cyclic CMP or NAD.

The phosphatase has an absolute requirement for Mg^{2+}. (1,3)

Inositol 1-phosphatase is present in a number of mammalian tissues and its specificity is similar to that of the yeast enzyme. Thus, the testicular enzyme (rat) has a preference for phosphates in the equatorial conformation and it does not distinguish between optical isomers. (3)

Inhibitors

The enzyme (yeast) is inhibited by a number of sugar phosphates. (1)

References

1. Charalampous, F. & Chen, I-W. (1966) Methods in Enzymology, 9, 698.
2. Chen, I-W. & Charalampous, F. (1966) ABB, 117, 154.
3. Eisenberg, F. (1967) JBC, 242, 1375.

NUCLEOTIDASE

(Nucleotide phosphohydrolase)

Nucleotide + H_2O = nucleoside + Pi

Ref.

Molecular weight

source	value	conditions	Ref.
Rat liver (lysosomes)	79,500	sucrose density gradient	(1)

Specific activity

Rat liver	25	5'-dAMP (pH 4.8, malonate, 37°)	(1)

Specificity and catalytic properties

source	substrate	relative velocity	K_m(mM)	conditions	Ref.
Rat liver	5'-dAMP[a]	1.00	0.125	pH 4.8, malonate, 37°	(1)
	3'-AMP	0.75	0.33	pH 4.8, malonate, 37°	(1)
	2'-AMP	0.50	0.58	pH 4.8, malonate, 37°	(1)
	5'-AMP	0.55	-	pH 4.8, malonate, 37°	(1)

(a) a number of other nucleotides and also 4-nitrophenyl phosphate; α-napthyl phosphate and β-glycerophosphate were active. The following were inactive: sugar phosphates; PPi; 3',5'-cyclic AMP; UDPglucose; phosphorylcholine; bis-(4-nitrophenyl) phosphate and the pyridine nucleotides. 5'-IMP and 5'-XMP were poor substrates.

The enzyme does not require the addition of any divalent metal ion for activity but it is inhibited by EDTA. Tartrate is a competitive inhibitor and Pi a non-competitive inhibitor. (1)

References

1. Arsenis, C. & Touster, O. (1968) JBC, 243, 5702.

THYMIDYLATE 5'-PHOSPHATASE

(Thymidylate 5'-phosphohydrolase)

Thymidylate + H_2O = thymidine + Pi

Molecular weight

source	value	conditions	Ref.
Rat liver	70,000	Sephadex G 100, pH 8.0	(1)
Bacteriophage PBS 2[a]	40,000	Sephadex G 100, pH 7.5	(2)

(a) host = Bacillus subtilis

Specificity and Michaelis constants

source	substrate	K_m(M)	conditions	Ref.
Rat liver (purified 100 x)	dTMP[a]	5.7×10^{-3}	pH 6.0, Tris, 37°	(1)
Bacteriophage PBS 2 (purified 350 x)	dTMP[b]	1.1×10^{-5}	pH 6.2, MES, 37°	(2)
	dUMP	8×10^{-4}	pH 6.2, MES, 37°	(2)
	dGMP	7×10^{-4}	pH 6.2, MES, 37°	(2)
	Mg^{2+}	6×10^{-5}	pH 6.2, MES, 37°	(2)

(a) dTMP (1.00) could be replaced by dUMP (1.00); dGMP (0.80); dAMP (0.26); UMP (0.21); 4-nitrophenyl phosphate (0.20); dCMP (0.12); GMP (0.12); β-glycerophosphate (0.09); AMP (0.07) or CMP (0.07). dTTP was inactive.

(b) dTMP (1.00) could be replaced by ribo-TMP (1.10) and derivatives; by dUMP (1.10) and derivatives; dGMP (0.59); dIMP (0.39); dXMP (0.69); dAMP (0.03); UMP (0.03). Deoxynucleoside di- and triphosphates; ribose 5-phosphate; deoxyribose 5-phosphate; CMP; AMP and GMP were inactive.

The nucleotidase requires Mg^{2+} for activity. (1,2)

Inhibitors

dTMP hydrolysis was competitively inhibited by several substrate deoxyribonucleotides tested (e.g. dUMP, dGMP). The Ki values obtained with these substrates as inhibitors were similar to their Km values. (1,2)

References

1. Magnusson, G. (1971) EJB, 20, 225.
2. Price, A.R. & Fogt, S.M. (1973) JBC, 248, 1372.

3-PHOSPHOGLYCERATE PHOSPHATASE

(D-Glycerate-3-phosphate phosphohydrolase)

3-Phosphoglycerate + H_2O = glycerate + Pi

Ref.

Molecular weight

source	value	conditions	Ref.
Sugar cane (*Saccharum* sp.)	160,000 [3-4]	sucrose gradient; Sephadex G 200, pH 6.3; gel electrophoresis (SDS)	(1)

Specific activity

Sugar cane (2530 x)	740	3-phosphoglycerate (pH 5.9 cacodylate, 30°)	(1)

Specificity and kinetic properties

source	substrate	K_m(M)	conditions	Ref.
Sugar cane	3-phosphoglycerate	2.85×10^{-4}	pH 5.9, cacodylate, 30°	(1)

The enzyme from sugar cane was active with the following phosphates:- 3-phosphoglycerate (1.00); 4-nitrophenylphosphate (0.66); phosphoenolpyruvate (0.64); phenolphthalein diphosphate (0.48); α-phenylphosphate (0.48); DL-glyceraldehyde 3-phosphate (0.41); propanediolphosphate (0.37); ATP (0.37); fructose 1,6-diphosphate (0.31); CTP (0.28); GTP (0.26); 2-phosphoglycerate (0.24); dihydroxyacetone phosphate (0.24); ADP (0.18); 6-phosphogluconate (0.17); phosphohydroxypyruvate (0.17); 3-glycerolphosphate (0.16); 2-glycerolphosphate (0.14); UMP (0.14); phosphoglycoaldehyde (0.13); phosphoglycollate (0.11); phospholactate (0.11); reduced NADP (0.11); pyridoxal 5'-phosphate (0.11); ribulose 1,5-diphosphate (0.11); phosphoserine (0.10); 3'-AMP (0.09); PPi (0.09); NADP (0.07); ribose 5-phosphate (0.07); CMP (0.05); fructose 6-phosphate (0.06); phytic acid (0.05); AMP (0.02); phosphoethanolamine (0.02), and phosphocholine (0.01). The following were inactive: 3'-deoxyadenosine monophosphate; bis-4-nitrophenylphosphate; glucose 1-phosphate; carbamoylphosphate, creatinephosphate, pentaphosphate and decaphosphate. The enzyme did not require any added cofactor (e.g. isocitrate) or metal ions for activity. It was not inhibited by EDTA. (1)

Glycidol phosphate (1,2-epoxypropanediol phosphate) is an irreversible inhibitor (C(3-phosphoglycerate)). The enzyme was also inhibited by tartrate. (1)

References

1. Randall, D.D. & Tolbert, N.E. (1971) JBC, 246, 5510.

SEDOHEPTULOSE 1,7-DIPHOSPHATASE

(D-Sedoheptulose-1,7-diphosphate 1-phosphohydrolase)

D-Sedoheptulose 1,7-diphosphate + H_2O = D-sedoheptulose 7-phosphate + Pi

Ref.

Molecular weight

source	value	conditions	Ref.
Candida utilis	75,000 [2]	sucrose density gradient; gel electrophoresis (SDS); amino acid composition.	(1)

Specific activity

C. utilis (2200 x)	11.5	sedoheptulose 1,7-diphosphate (pH 6.0, maleate, 37°)	(1)

Specificity and catalytic properties

The enzyme (C. utilis) is highly specific for sedoheptulose 1,7-diphosphate (Km = 1 mM) which could not be replaced by sedoheptulose 1-phosphate; sedoheptulose 7-phosphate; fructose 1,6-diphosphate; fructose 1-phosphate; fructose 6-phosphate; glucose 1-phosphate; glucose 6-phosphate; ribose 5-phosphate; ribulose 5-phosphate or erythrose 4-phosphate. The enzyme does not require divalent metal ions such as Mg^{2+} or Mn^{2+}. It was not inhibited by EDTA. (1,2)

Sedoheptulosediphosphatase was not activated by CoA or cysteamine nor was it inhibited by AMP, fructose 1,6-diphosphate or high concentrations of sedoheptulose 1,7-diphosphate. (1)

References

1. Traniello, S., Calcagno, M. & Pontremoli, S. (1971) ABB, 146, 603.
2. Racker, E. (1962) Methods in Enzymology, 5, 270.

SPHINGOMYELIN PHOSPHODIESTERASE

(Sphingomyelin cholinephosphohydrolase)

Sphingomyelin + H_2O = N-acylsphingosine + choline phosphate

Ref.

Equilibrium constant

The reaction is essentially irreversible [F] with the enzyme from rat brain as the catalyst. (1)

Specificity and catalytic properties

source	substrate	K_m (mM)	conditions	
Rat brain (purified 19 x)	sphingomyelin[a]	0.13	pH 5.0, acetate, 37°	(1,2)
Rat liver (purified 64 x)	sphingomyelin[b]	0.18	pH 4.5, acetate, 37°	(3,4)

(a) from bovine brain (1.00). The following were also active: sphingomyelin from the spleen of humans with Niemann-Picks' disease (0.94); N-palmitoyl-D-*erythro*-sphingosylphosphorylcholine (0.96); N-palmitoyl-DL-*erythro*-dihydrosphingosylphosphorylcholine (0.91); N-palmitoyl-L-*erythro*-sphingosylphosphorylcholine (0.20) and N-palmitoyl-DL-*threo*-dihydrosphingosylphosphorylcholine (0.29). Phosphorylcholine and lecithin were not hydrolyzed. The enzyme required detergents for full activity but there was no divalent metal ion requirement.

(b) lecithin; phosphatidylethanolamine; glucosylceramide; galactosylceramide; glucosylsphingosine; galactosylsphingosine and galactosylglucosylsphingosine were not hydrolyzed. The enzyme was activated by sodium cholate; it was not inhibited by EDTA.

The enzyme from rat brain was inhibited by ceramide but not by cholinephosphate. Other inhibitors are sphingosine, fatty acids, lecithin, phosphatidylserine and inositolphosphatides; cetavlon and sodium lauryl sulphate. (1,2)

The enzyme from rat liver was inhibited by lecithin (Ki = 0.2 mM) but not by galactosylceramide; glucosylsphingosine or lysolecithin. (3,4)

Abbreviations

sphingomyelin	1-O-phosphorylcholine-2-N-acylsphingosine
ceramide	2-N-acylsphingosine

References

1. Gatt, S. & Barenholz, Y. (1969) Methods in Enzymology, 14, 144.
2. Barenholz, Y., Roitman, A. & Gatt, S. (1966) JBC, 241, 3731.
3. Kanfer, J.N. & Brady, R.O. (1969) Methods in Enzymology, 14, 131.
4. Kanfer, J.N., Young, O.M., Shapiro, D. & Brady, R.O. (1966) JBC, 241, 1081.

ACYL CARRIER PROTEIN PHOSPHODIESTERASE

(Acyl carrier protein 4'-pantetheine phosphohydrolase)

ACP + H_2O = 4'-phosphopantetheine + apo-ACP

Ref.

Specificity and Michaelis constants

source	substrate	K_m (M)	conditions	
Escherichia coli (purified 30 x)	*E. coli* ACP[a]	1.7×10^{-6}	pH 8.6, Tris, 33°	(2,3)

(a) *E. coli* ACP (1.0) could be replaced by *Clostridium butyricum* ACP (0.3) or *E. coli* acetyl-ACP. 4'-Phosphopantetheine was not removed from large peptides of *E. coli* ACP or from animal fatty acid synthetase complex (see EC 2.3.1.38). CoA and glycerolphosphoryl serine were inactive. The enzyme is, therefore, highly specific for ACP.

The enzyme requires Mn^{2+} (partially replaced by Mg^{2+}, Co^{2+}, Fe^{2+} or Zn^{2+} but not by Cu^{2+}, Cd^{2+}, Ca^{2+}, Fe^{3+} or Al^{3+}) and dithiothreitol or mercaptoethanol for full activity. (2,3)

Abbreviations

ACP acyl carrier protein (see EC 2.3.1.39)

References

1. Prescott, D.J. & Vagelos, P.R. (1972) *Advances in Enzymology*, 36, 283.
2. Vagelos, P.R. & Larrabee, A.R. (1969) *Methods in Enzymology*, 14, 81.
3. Vagelos, P.R. & Larrabee, A.R. (1967) *JBC*, 242, 1776.

RIBONUCLEASE II

(Ribonucleate 3'-oligonucleotidohydrolase)

Endonucleolytic cleavage of RNA similar to EC 3.1.4.22 (formerly EC 2.7.7.16) but cleaves at the 3'- position of purine nucleotide residues as well as at pyrimidine nucleotide residues

Molecular properties

source	value	conditions	Ref.
Aspergillus saitoi or A. oryzae	36,000	pH 4.5	(1)
Klebsiella sp.[a]	24,000	Sephadex G 100, pH 7.5	(3)

(a) the crude enzyme is bound to nucleic acid.

Specificity and kinetic properties

source	substrate	V(10^3mole substrate min^{-1} mole enzyme^{-1})	Km (10^{-4}M)
A.[a] saitoi	UpU; UpG; UpC; UpA	7.7; 13.7; 19.1; 6	2.1; 1.2; 0.8; 0.5
	GpU; GpG; GpC; GpA	3.8; 8.1; 8.4; 3.4	1.6; 0.8; 0.7; 0.7
	CpU; CpG; CpC; CpA	3.6; 6; 5; 1.9	1.2; 0.6; 0.4; 0.1
	ApU; ApG; ApC; ApA	2.3; 2.4; 5.1; 1.6	0.4; 0.4; 0.5; 0.3
A.[b] oryzae	UpU; UpA	26; 23	1.2; 0.5
	GpU; GpA	12; 12.7	1.1; 0.6
	CpU; CpA	20; 11.3	0.7; 0.4
	ApU; ApG; ApC; ApA	10.3; 7.2; 15.3; 3.6	0.8; 0.7; 0.7; 0.4

(a) RNase M, purified 125 x. Conditions = pH 5.5, acetate, 25° (2)
(b) RNase T_2, purified 1000 x. Conditions = pH 5.5, acetate, 25° (2)

Ribonuclease II from Klebsiella sp. (purified 320 x) was active with the following substrates: yeast RNA (1.00); poly (A) (3.35); poly (C) (2.16) and poly (U) (7.16) but not poly (G); poly (I); poly (A) . poly (U); poly (I) . poly (C); poly (A) . poly (U) . poly (U); poly (X) or poly (hU). Hydrolysis was inhibited by poly (G) and to a lesser extent by poly (I). A number of mononucleotides tested did not inhibit. (4)

Ribonuclease II does not require any added cofactor or metal ion for activity. The enzyme from A. saitoi, for instance, was not inhibited by EDTA. (1,4)

Light absorption data

The change of the molar absorption coefficient resulting when each of the 16 ribonucleotidyl-(3',5')-ribonucleosides is hydrolyzed has been determined. (2)

Ref.

Abbreviations

UpU, UpG etc. refer to ribonucleotidyl-(3',5')-ribonucleosides where U = uridine; G = guanosine; C = cytidine and A = adenosine.

References

1. Irie, M. (1967) J. Biochem. (Tokyo), 62, 509.
2. Imazawa, M., Irie, M. & Ukita, T. (1968) J. Biochem. (Tokyo), 64, 595.
3. Friedling, S.P., Schmukler, M. & Levy, C.C. (1972) BBA, 268, 391.
4. Schmukler, M., Friedling, S.P. & Levy, C.C. (1972) BBA, 268, 403.

UV-ENDONUCLEASE

Ref.

Molecular weight

source	value	conditions	
Micrococcus luteus	14,500	Sephadex G 75, pH 7.5, Pi	(1)

Specificity and catalytic properties

The enzyme from M. luteus (purified 5100 x) catalyzes the formation of single-strand incisions in UV-irradiated double-stranded DNA. The incisions occur either adjacent to or one base removed from thymine-containing dimers and leaves a 3'-phosphomonoester and a 5'-hydroxyl with the photoproduct lying 5' to the break. DNA treated with the cross-linking reagent mitomycin C is also attacked by the enzyme. The enzyme is without activity on native DNA, denatured DNA, irradiated denatured DNA or DNA treated with the intercalating agent 3',4'-benzpyrene. (1,2)

The endonuclease requires Mg^{2+} (Mn^{2+} and Ca^{2+} did not activate) for full activity. At high concentrations, Mg^{2+}, Mn^{2+} and Ca^{2+} were inhibitory. EDTA did not cause any reduction of the enzymic activity observed in the absence of Mg^{2+}. (1,2)

An UV-exonuclease, also from M. luteus (purified 1000 x), has been described. This enzyme, which requires Mg^{2+} for activity, is required for the excision of photoproducts from UV-irradiated DNA (see above). The combined action of the endo- and exonucleases results in the quantitative removal of photoproduct regions from the UV-irradiated DNA. About 5 nucleotides are released for every single-strand incision. (2,3)

References

1. Kushner, S.R. & Grossman, L. (1971) Methods in Enzymology, 21, 244.
2. Kaplan, J.C., Kushner, S.R. & Grossman, L. (1969) PNAS, 63, 144.
3. Kaplan, J.C. & Grossman, L. (1971) Methods in Enzymology, 21, 249.

CHOLINESULPHATASE

(Choline-O-sulphate sulphohydrolase)

Choline sulphate + H_2O = choline + sulphate

Ref.

Equilibrium constant

The reaction is essentially irreversible [F] with the enzyme from *Pseudomonas aeruginosa* as the catalyst. (1)

Molecular weight

source	value	conditions	
P. aeruginosa	175,000	Sephadex G 200, pH 7.5	(1)
Penicillium chrysogenum	345,000	Sephadex G 200, pH 7.5	(1)

Specificity and kinetic properties

source	substrate	K_m (mM)	conditions	
P. aeruginosa (purified 12 x)	choline sulphate[a]	2	pH 7.5, Tris, 30°	(1)
P. chrysogenum	choline sulphate	8.5	pH 7.5, Tris, 30°	(1)

(a) the enzyme is highly specific for choline sulphate.

Cholinesulphatase from *P. aeruginosa* was inhibited by the following compounds: choline (UC; Ki = 0.2 mM); tetramethylamine (UC; Ki = 0.25 mM); trimethylamine (UC; Ki = 0.19 mM); dimethylaminoethanol (UC; Ki = 0.35 mM); dimethylamine (UC; Ki = 5 mM); choline-O-phosphate (C; Ki = 1.5 mM) and betaine (C; Ki = 2.3 mM). Inorganic sulphate did not inhibit. (1)

The enzyme did not require any added cofactor or metal ion for activity. It was not inhibited by diisopropylfluorophosphate. The mechanism of the reaction has been investigated. (1)

References

1. Lucas, J.J., Burchiel, S.W. & Segel, I.H. (1972) ABB, 153, 664.

OLIGO-1,6-GLUCOSIDASE

(Dextrin 6-α-glucanohydrolase)

Endohydrolysis of 1,6-α-glucosidic linkages in isomaltose and dextrins produced from starch and glycogen by α-amylase (EC 3.2.1.1).

Ref.

Molecular properties

In rabbit small intestine, oligo-1,6-glucosidase occurs as a complex with sucrase (β-fructofuranosidase; EC 3.2.1.26). The complex, which has been highly purified, is a membrane-bound glycoprotein of molecular weight 220,000 daltons. It consists of 2 subunits each of molecular weight about 115,000: one subunit is oligo-1,6-glucosidase (EC 3.2.1.10) and the other β-fructofuranosidase (EC 3.2.1.26). The amino-acid and sugar compositions of the two subunits have been determined. (1,2)

Oligo-1,6-glucosidase is a single polypeptide chain of molecular weight 113,000 daltons. (3)

A human oligo-1,6-glucosidase - sucrase complex has been described. (4)

Catalytic properties

Oligo-1,6-glucosidase (rabbit small intestine) is active on its own as well as when it is part of the complex with sucrase. For instance, with palatinose as the substrate (in sodium maleate buffer, pH 6.8 and 37°) the following kinetic constants were obtained: in native complex, relative velocity = 1.0, Km = 4.3 mM and as isolated subunit, relative velocity = 2.0, Km = 4.3 mM. (2)

The enzyme requires a monovalent metal ion for full activity (e.g. Na^+, Li^+). NH_4^+ and Tris were inhibitory. (2,5,6)

The glucosidase was not inhibited by diisopropylfluorophosphate. (6)

References

1. Cogoli, A., Mosimann, H., Vock, C., von Balthazar, A-K. & Semenza, G. (1972) EJB, 30, 7.
2. Cogoli, A., Eberle, A., Sigrist, H., Joss, C., Robinson, E., Mosimann, H. & Semenza, G. (1973) EJB, 33, 40.
3. Mosimann, H., Semenza, G. & Sund, H. (1973) EJB, 36, 489.
4. Yamashiro, K.M. & Gray, G.M. (1970) Gastroenterology, 58, 1056.
5. Kolinska, J. & Semenza, G. (1967) BBA, 146, 181.
6. Larner, J. & Gillespie, R.E. (1956) JBC, 223, 709.

β-MANNOSIDASE

(β-D-Mannoside mannohydrolase)

Hydrolysis of terminal, non-reducing β-D-mannose residues in β-mannosides

Ref.

Molecular weight

source	value	conditions	
Hen oviduct	100,000	Sephadex G 200, pH 4.6	(1)

Multiple forms have been observed. (1)

Specificity and catalytic properties

source	substrate	V relative	K_m (mM)	
Hen oviduct (purified 10,000 x)	4-nitrophenyl β-D-mannopyranoside[a]	1	2.9	(1)
	Asn-$(GlcNAc)_2(Man)_1$	1	16.9	(1)
Achatina fulica (snails; purified 100 x)	phenyl-β-D mannoside[b]	-	6.5	(2)

(a) 4-nitrophenyl β-D-mannopyranoside (1.00) could also be replaced by $(GlcNAc)_2(Man)_1$ (0.33) and possibly by methyl β-D-mannopyranoside (0.007) but not by 4-nitrophenyl α-D-mannopyranoside; methyl α-D-mannopyranoside; α-D-Man (1→4) GlcNAc or α-D-Man (1→6) GlcNAc. (Conditions = pH 4.6, citrate, 37°).
(b) at pH 4.5, citrate-Pi.

The enzyme does not need any added cofactor or metal ion for activity. The enzyme from hen oviduct, for instance, was not inhibited by EDTA. (1)

β-Mannosidase is inhibited by mannono- (1→4) and -(1→5) lactones. With the enzyme from hen oviduct and D-mannono-(1→5) lactone, the inhibition was competitive (4-nitrophenyl β-D-mannopyranoside; Ki = 17 μM). (1,3)

The seeds of *Leucaena glauca* contain a β-mannosidase which has been purified about 1000-fold. This enzyme hydrolyses galactomannan mucilage. (4)

Abbreviations

Asn-$(GlcNAc)_2$ $(Man)_1$ — 2-acetamido-4-*O*-[*O*-mannopyranosyl(1→4)-2-acetamido-2-deoxy-β-D-glucopyranosyl]-*N*-(4-L-aspartyl)-2-deoxy-β-D-glucopyranosylamine.

References

1. Sukeno, T., Tarentino, A.L., Plummer, T.H. & Maley, F. (1972) B, 11, 1493.
2. Sagahara, K., Okumura, T. & Yamashina, I. (1972) BBA, 268, 488.
3. Levvy, G.A., Hay, A.J. & Conchie, J. (1964) BJ, 91, 378.
4. Hylin, J.W. & Sawai, K. (1964) JBC, 239, 990.

EXO-1,4-β-XYLOSIDASE

(1,4-β-D-Xylan xylohydrolase)

Hydrolysis of 1,4-β-D-xylans so as to remove successive D-xylose residues from the non-reducing termini

Ref.

Specificity and catalytic properties

source	substrate	V relative	K_m (mM)	Ref.
Bacillus pumilus (purified 34 x)	4-nitrophenyl-β-D-xylopyranoside[a]	1.00	1.43	(1)
	2-nitrophenyl-β-D-xylopyranoside	1.64	1.06	(1)

(a) 4-nitrophenyl-β-D-xylopyranoside could be replaced by phenyl β-D-xylopyranoside (0.07) but not by the following: phenyl-α-D-xylopyranoside; phenyl-β-D-xylofuranoside; 4-nitrophenyl β-D-glucopyranoside; 2-nitrophenyl α-L-arabinopyranoside; 2-nitrophenyl β-D-galactopyranoside; phenyl β-D-ribopyranoside; benzyl β-D-xylopyranoside; methyl β-D-xylopyranoside and butyl β-D-xylopyranoside. (pH 7.2, Pi, 25°).

β-Xylosidase removes xylose residues from β-1,4-xylan and it hydrolyses 4-*O*-β-D-xylopyranosyl-D-xylose. The enzyme was activated by divalent or trivalent anions; high concentrations were, however, inhibitory. Divalent metal ions such as Mg^{2+} or Mn^{2+} were inhibitory. (1)

β-Xylosidase from mould is less specific and it hydrolyses alkyl as well as aryl derivatives of D-xylopyranoside. This enzyme has transferase activity. (1)

inhibitor[a]	K_i (mM)
phenyl 1-thio-β-D-xylopyranoside	3.88
methyl β-D-xylopyranoside	47.6
butyl β-D-xylopyranoside	1.42
methyl 1-thio-β-D-xylopyranoside	8.95
butyl 1-thio-β-D-xylopyranoside	2.77
phenyl β-D-glucopyranoside	75.0
D-xylose	26.2
Tris	6.77

(a) with the enzyme from *B. pumilus* (conditions: pH 7.2, Pi, 25°). A number of other sugars tested were also inhibitory. The inhibition was competitive (Ref. 1).

References

1. Kersters-Hilderson, H., Loontiens, F.G., Claeyssens, M. & de Bruyne, C.K. (1969) *EJB*, *7*, 434.

ENDO-1,3-β-GLUCANASE

(1,3-β-D-Glucan glucanohydrolase)

Hydrolysis of 1,3-β-glucosidic linkages in 1,3-β-D-glucans.

Ref.

Molecular weight

source	value	conditions	Ref.
Nicotiana glutinosa (leaves)	45,000	Sephadex G 75, pH 6.5	(1)
Phaseolus vulgaris (red kidney bean)	34,000	0.15 M NaCl	(2)
	12,000	Sephadex G 200, pH 7.0; amino acid analysis	(2)

Specificity and Michaelis constants

source	substrate	K_m (mM)	conditions	Ref.
N. glutinosa (purified 280 fold)	O-carboxymethyl-pachyman[(a)]	0.13	pH 5.0, acetate, 40°	(1)

(a) molecular weight = 46,000. Pachyman is an homogeneous, linear β-1,3-glucan obtained from Poria cocos (a fungus).

β-1,3-Glucan hydrolase (N. glutinosa) has an endo-action pattern and is highly specific for linear β-1,3-glucan substrates which are de-polymerized to yield a mixture of β-1,3-oligoglucosides of degree of polymerization 2-7 and a trace of glucose. Laminaribiose and -triose are not hydrolyzed but laminaritetraose and -pentaose are slowly hydrolyzed. The enzyme does not catalyze the transfer of glycosyl groups. It was not inhibited by glucono-(1,5)-lactone. (1)

β-1,3-Glucan hydrolase (P. vulgaris, purified 15 x) only attacked polysaccharides having β-1,3-linkages. The enzyme had no effect on β-1,2; β-1,4; β-1,6; α-1,4; or α-1,6 linked glucans. It was not inhibited by EDTA. (2)

The chemistry and biochemistry of β-1,3-glucans have been reviewed. (3)

References

1. Moore, A.E. & Stone, B.A. (1972) BBA, 258, 238; 248.
2. Abeles, F.B., Bosshart, R.P., Forrence, L.E. & Habig, W.H. (1970) Plant Physiol, 47, 129.
3. Bull, A.T. & Chester, C.G.C. (1966) Advances in Enzymology, 28, 325.

GDP GLUCOSE GLUCOHYDROLASE

(GDPglucose glucohydrolase)

GDPglucose + H_2O = GDP + D-glucose

Ref.

Specificity and catalytic properties

source	substrate	K_m(mM)	conditions	
Bakers' yeast (purified 135 x)	GDPglucose[(a)]	0.23	pH 7.2, glycerophosphate, 30°	(1)

(a) ADPglucose; UDPglucose; dTDPglucose; GDPmannose; GDPgalactose; GDPglucose and glucose 1-phosphate were inactive.

The enzyme was inhibited by GDP (C(GDPglucose), Ki = 80 µM). The inhibition was relieved by Mg^{2+} as the GDP-Mg complex has little affinity for the enzyme. The enzyme did not require the addition of metal ions for activity and there was no inhibition by EDTA. (1)

References

1. Sonnino, S., Carminatti, H. & Cabib, E. (1966) ABB, 116, 26.

α-N-ACETYLGALACTOSAMINIDASE

(2-Acetamido-2-deoxy-α-D-galactoside acetamidodeoxygalactohydrolase)

R-2-acetamido-2-deoxy-α-D-galactoside + H_2O = ROH +
2-acetamido-2-deoxy-D-galactose

Ref.

Molecular properties

Bovine liver N-acetylgalactosaminidase exists in dilute solutions as a mixture of active oligomers (highest molecular weight = 155,000 daltons) and the inactive monomer (molecular weight = about 36,000 daltons). Association (i.e. activation) is enhanced by increased enzyme concentration, increased substrate concentration or by decreased temperature. This effect gives rise to complex kinetics; it has also been observed with the corresponding enzymes from rabbit, turtle and frog but not from pig, rat, guinea-pig, chicken, fish or earthworm. (1)

Multiple forms have been observed with the enzyme from pig liver. (2)

Specificity and catalytic properties

source	substrate	K_m (mM)	conditions	
Beef liver (purified 700 x)[a]	phenyl N-acetyl-α-galactosaminide	10.8	pH 4.7, citrate, 33°	(1,2)
Ox spleen	phenyl N-acetyl-α-galactosaminide	50	pH 4.4, 37°	(3)

(a) the enzyme was tetrameric under the assay conditions.

The action of bovine liver N-acetylgalactosaminidase on glycoproteins isolated from the submaxillary glands of cattle and sheep has been investigated. (2)

The action of N-acetylgalactosaminidase from Helix pomatia (purified 300 x) on soluble blood group substance A has been investigated. (4)

The enzyme (beef liver) does not require any added cofactor or metal ions. It was not inhibited by EDTA. (2)

The enzyme from beef liver was inhibited by N-acetylgalactosamine (C(phenyl N-acetylgalactosaminide; Ki = 8.7 mM). The pig liver enzyme was also inhibited by N-acetylgalactosamine but not by N-acetylglucosamine; N-acetylmannosamine; N-acetylglucosaminolactone; acetamide; sodium acetate or methyl N-acetyl-α-D-glucosaminide. (2)

References

1. Wang, C-T. & Weissmann, B. (1971) B, 10, 1067.
2. Weissmann, B. & Hinrichsen, D.F. (1969) B, 8, 2034.
3. Werriers, E., Wollek, E., Gottschalk, A. & Buddecke, E. (1969) EJB, 10, 445.
4. Tuppy, H. & Staudenbauer, W.L. (1966) B, 5, 1742.

α-L-FUCOSIDASE

(α-L-Fucoside fucohydrolase)

An α-L-fucoside + H_2O = an alcohol + L-fucose

Ref.

Molecular weight

source	value	conditions	
Rat epididymis	215,000 [4]	pH 6.0; gel electrophoresis (SDS); amino acid composition	(1)

The enzyme has 2 pairs of dissimilar subunits (47,000 and 60,000 daltons). It is a glycoprotein (0.6% carbohydrate). (1)

Specific activity

Rat epididymis (180 x)	12.8	4-nitrophenyl-α-L-fucopyranoside (pH 6.5, citrate, 37°)	(1)

Specificity and catalytic properties

α-L-Fucosidase (rat epididymis) removes L-fucose specifically from native glycoproteins and peptides with smaller substrates being strongly preferred. For instance, with the hormone-specific chain of luteinizing hormone (LH-β) as substrate, only L-fucose was released. (1)

2 α-L-Fucosidases have been obtained from the liver of *Heliotis gigantea* (abalone). One was optimally active at about pH 5 on 4-nitrophenyl α-L-fucopyranoside but not on the fucosidic linkages of porcine submaxillary mucin and the other was optimally active at about pH 2 on both substrates. Abalone liver contains a specific inhibitor of the pH 2 enzyme. (2)

α-L-Fucosidase does not require any added cofactor or divalent metal ion for activity. It was not inhibited by EDTA. (1)

References

1. Carlsen, R.B. & Pierce, J.G. (1972) JBC, 247, 23.
2. Tanaka, K., Nakano, T., Noguchi, S. & Pigman, W. (1968) ABB, 126, 624

α-L-ARABINOFURANOSIDASE

(α-L-Arabinofuranoside arabinohydrolase)

An α-L-arabinofuranoside + H_2O = an alcohol + L-arabinose

Ref.

Molecular weight

source	value	conditions	
Aspergillus niger	53,000	Sephadex G 100, pH 6.8; amino acid composition	(1)

The enzyme is a glycoprotein. (1)

Specific activity

A. niger (108 x)	397	phenyl α-L-arabinofuranoside (pH 4.0, 30°)	(1)
Corticium rolfsii (purified 67 x)	44	phenyl-α-L-arabinofuranoside (pH 2.5, citrate, 30°)	(3)

Specificity and catalytic properties

The enzyme is highly specific for α-L-arabinofuranosides. The enzyme from A. niger, for example, released only L-arabinose from beet araban (Km = 0.26 g per l) or from phenyl α-L-arabinofuranoside (Km = 1.1 g per l). It was inactive with nitrophenyl α-L-arabinopyranoside. (2)

The enzyme from C. rolfsii removed L-arabinose from the following: phenyl α-L-arabinofuranoside (V = 1.00, Km = 2.86 mM); beet arabinan (V = 0.43, Km = 8.47 g per l) and 1,5-arabinan (V = 0.14, Km = 28.6 g per l). It was inactive with 4-nitrophenyl-β-D-galactopyranoside. (3)

α-L-Arabinofuranosidase did not require any added metal ion for activity. (2)

References

1. Kaji, A. & Tagawa, K. (1970) BBA, 207, 456.
2. Kaji, A., Tagawa, K. & Ichimi, T. (1969) BBA, 171, 186.
3. Kaji, A. & Yoshihara, O. (1971) BBA, 250, 367.

ENDO-1,3-α-GLUCANASE

(1,3-(1,3; 1,4)-α-D-Glucan 3-glucanohydrolase)

Endohydrolysis of 1,3-α-glucosidic linkages in isolichenin, pseudonigeran and nigeran.

Ref.

Molecular properties

source	value	conditions	
Trichoderma viride	47,000	Sephadex G 100, pH 8.0	(1)

Specificity

The enzyme (T. viride; purified 63 x) is highly specific for the α-(1,3)-glucosidic linkage and the following linkages are not cleaved: β-(1,3); α-(1,4); α-(1,6) and β-(1,2). The hydrolysis occurs with the retention of the configuration of the anomeric carbon atom involved in the cleavage. The following carbohydrates were cleaved: pseudonigeran; nigeran; isolichenin; Polyporus tumulosus polysaccharide; nigerotriose; nigerotetraose and nigeropentaose. The following carbohydrates were not cleaved: soluble starch; waxy starch; dextran; Sclerotium rolfsii gum; pachyman; maltose; sucrose; gentiobiose; isomaltose and nigerose. (1)

The Km with pseudonigeran as the substrate was 0.41 mM (expressed as glucose equivalents; pH 4.5, citrate, 40°). (1)

The enzyme did not require any added metal ion for activity but some stimulation occurred with Co^{2+}. (1)

References

1. Hasegawa, S., Nordin, J.H. & Kirkwood, S. (1969) JBC, 244, 5460.

EXO-MALTOTETRAOHYDROLASE

(1,4-α-D-Glucan maltotetraohydrolase)

Hydrolysis of 1,4-α-glucosidic linkages in amylaceous polysaccharides so as to remove successive maltotetraose residues from the non-reducing chain ends

Ref.

Molecular properties

source	value	conditions	
Pseudomonas stutzeri	12,500	Bio-Gel P 150, pH 7.0	(1)

The enzyme exhibits concentration dependent association to give dimers, tetramers, hexamers, octamers and decamers. All the molecular forms were enzymically active. (1)

Specific activity

P. stutzeri (1000 x) 2500 μmoles glycosidic bonds hydrolyzed $min^{-1}mg^{-1}$ with soluble starch as substrate (pH 7.0, 40°) (1)

Specificity and catalytic properties

The hydrolase forms predominantly G_4 from G_n (e.g. soluble starch; the Schardinger dextrins were inactive). Thus, G_8 gave G_4; G_7 gave $G_4 + G_3$ and G_6 gave $G_4 + G_2$. At very high enzyme concentrations, G_4 was slowly hydrolyzed to give $G_1 + G_3$ and 2 G_2. (1)

The enzyme may require a divalent metal ion for activity. (1)

Abbreviations

G_n a linear molecule of n glucosyl units linked through α-1,4 glycosidic bonds.

References

1. Robyt, J.F. & Ackerman, R.J. (1971) ABB, 145, 105.

MYCODEXTRANASE

(1,3-1,4-α-D-Glucan 4-glucanohydrolase)

Endohydrolysis of 1,4-α-glucosidic linkages in α-glucans containing both 1,3- and 1,4-bonds.

Ref.

Molecular weight

source	value	conditions	
Penicillium melinii	40,000 [1]	pH 4.5 with or without guanidine-HCl; amino acid composition	(1)

P. melinii contains two enzymes with mycodextranase activity. The enzymes have identical molecular weights and similar catalytic properties but different amino acid compositions and electrophoretic mobilities. Each enzyme (purified 70-fold) contains about 15% carbohydrate. (1)

Specific activity

P. melinii (70 x)	27	μmoles of glucose-reducing equivalents released from nigeran (pH 4.5, acetate, 50°)	(1)

Specificity and catalytic properties

Mycodextranase operates via an endo-multichain mechanism. It hydrolyzes the glucan nigeran (mycodextran) and certain oligosaccharides of nigeran to yield nigarose and the tetrasaccharide O-α-D-glucopyranosyl-(1,3) -O-α-D-glucopyranosyl-(1,4) -O-α-D-glucopyranosyl-(1,3) -D-glucose as sole end products. Glucose is not produced. It is thought that the enzyme has a substrate-binding site for 8 glucopyranose units and that a catalytic site is located midway along the binding site. (1)

The following compounds were inactive: methyl α-D-glucopyranoside; nigerose; isomaltose; cellobiose; trehalose; melibiose; lactose; nigerotriose; amylose; amylopectin; cellulose; laminarin and pseudonigeran. Maltose and maltotriose were hydrolyzed very slowly. (1)

The enzyme was inhibited by the following compounds (C(nigeran)): nigerose (Ki = 60 mM); maltose (Ki = 70 mM); maltotriose (Ki = 152 mM); nigeran trisaccharide (Ki = 30 mM) and nigeran tetrasaccharide (Ki = 4.8 mM). (1)

Mycodextranase did not require metal ions for activity. (1)

Abbreviations

nigeran — an unbranched polyglucan with alternating α-(1,4)- and α-(1,3)-linkages. Nigeran is a constituent of the hyphal cell wall matrix of various Aspergillus species.

References

1. Tung, K.K., Rosenthal, A. & Nordin, J.H. (1971) JBC, 246, 6722.

1,2-α-L-FUCOSIDASE

(2-O-α-L-Fucopyranosyl-β-D-galactoside fucohydrolase)

2-O-α-L-Fucopyranosyl-β-D-galactoside + H_2O = L-fucose + D-galactose

Ref.

Molecular weight

source	value	conditions	
Clostridium perfringens	>200,000	Sephadex G 200, pH 7.0	(1)

Multiple forms have been observed. (1)

Specificity and catalytic properties

source	substrate	K_m (mM)	conditions	
Aspergillus niger (purified 590 x)	methyl-2-O-α-L-fucopyranosyl-β-D-galactoside	8.3	pH 4.0, acetate, 37°	(2)

1,2-α-L-Fucosidase is highly specific for the 1,2-α-L-fucosidic linkage to D-galactose. Thus, the enzyme from A. niger removed fucose from the following compounds:- 2-O-α-L-fucopyranosyl-D-galactose; methyl 2-O-α-L-fucopyranosyl-β-D-galactoside; 2-O-α-L-fucopyranosyllactose; lacto-N-fucopentaose I and porcine and canine submaxillary mucin. A large number of other substances tested were inactive; these included 4-nitrophenylfucopyranosides; fucopyranosyl-L-fucoses and L-fucopyranosyl-D-galactoses. The 1,3-α-L-fucosidic linkage to D-glucose or to D-N-acetylglucosamine was also resistant. (2)

The enzyme from C. perfringens (purified 600 x) is also highly specific and it acts on oligosaccharides and glycoproteins but not on simple methyl and nitrophenyl fucosides. The enzyme was activated by Na^+, Mg^{2+}, Ca^{2+} or NH_4^+. (1)

References

1. Bahl, O.P. (1970) JBC, 245, 299.
2. Aminoff, D. & Furukawa, K. (1970) JBC, 245, 1659.

ISOAMYLASE

(Glycogen 6-glucanohydrolase)

Hydrolysis of 1,6-α-glucosidic branch linkages in glycogen, amylopectin and their β-limit dextrins

Ref.

Molecular weight

source	value	conditions	
Pseudomonas sp. (strain SB-15)	95,000	pH 4.0	(1)

Specificity and catalytic properties

Isoamylase (*Pseudomonas* sp.) has been purified (720 x) to homogeneity: the branching linkages in 1g of waxy maize amylopectin were completely hydrolyzed by 0.01 mg of the enzyme in 20 h (pH 3.5, acetate, 40°). The enzyme hydrolyzed all α-1,6-glucosidic inter-chains in glycogen, amylopectin and their phosphorylase (EC 2.4.1.1) limit dextrins but the branching points of β-amylase (EC 3.2.1.2) limit dextrins were not hydrolyzed completely. It released maltotriose and higher malto-saccharides (but no maltose) from its substrates. The specificity of isoamylase and pullulanase (EC 3.2.1.41) have been compared. (1,2)

The enzyme does not require any added cofactor. (1)

References

1. Yokobayashi, K., Misaki, A. & Harada, T. (1970) BBA, 212, 458.
2. Harada, T., Misaki, A., Akai, H., Yokobayashi, K. & Sugimoto, K. (1972) BBA, 268, 497.

CAPSULAR POLYSACCHARIDE DEPOLYMERASE

(Poly-α-1,3-galactosyl glycanohydrolase)

Hydrolysis of 1,3-α-galactosidic linkages in *Aerobacter aerogenes* capsular polysaccharide

Ref.

Molecular properties

source	value	conditions	
Phage K-2 (host = *A. aerogenes*)	379,000	sucrose density gradient; Sephadex G 200; gel-electrophoresis (SDS); amino acid composition	(1)

The enzyme is composed of 2 nonidentical subunits of molecular weights 63,200 and 36,400 daltons. The number of each type of subunit per molecule of native enzyme is not known. The enzyme exists in 2 forms: soluble and phage-bound. The 2 forms have identical catalytic properties. (1)

Specific activity

Phage K-2 235 μmoles reducing sugar (as galactose) formed min^{-1} mg^{-1} with *A. aerogenes* capsular polysaccharide as the substrate (pH 5.2, acetate, 42°) (1)

Specificity

The enzyme is a highly specific glycanohydrolase which cleaves the galactosyl-α-1,3-galactose linkages in *A. aerogenes* capsular polysaccharide. It was completely inactive with 4-nitrophenyl-α-D-galactopyranoside or the capsular polysaccharide of *A. aerogenes* strain 243 or strain A 3S1. The attack occurs at random along the capsular polysaccharide molecule and is most rapid with high molecular weight capsular polysaccharide. The smallest oligosaccharide that the depolymerase will degrade is a dodecasaccharide composed of three tetrasaccharide repeating units (see below). The only bond cleaved in this compound is the galactosyl-galactose linkage immediately adjacent to the terminal, non-reducing tetrasaccharide repeating unit. The pH optimum of the enzyme is 5.2. (1)

The enzyme does not require added cofactors or activators (e.g. metal ions). (1)

Ref.

Abbreviations

A. aerogenes capsular polysaccharide is composed of a linear sequence of tetrasaccharide repeating units of the following structure (see Ref. 2):

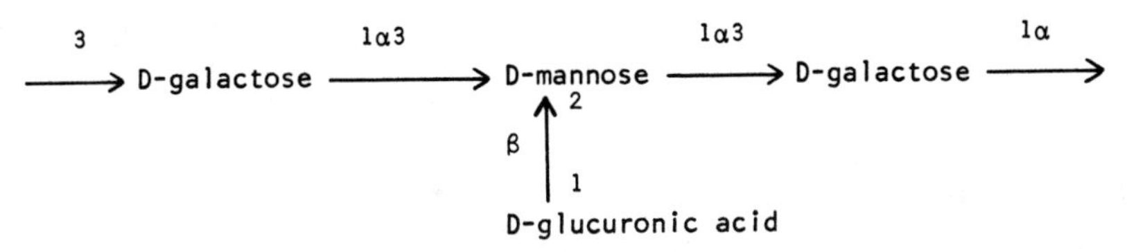

References

1. Yurewicz, E.C., Ghalambor, M.A., Duckworth, D.H. & Heath, E.C. (1971) JBC, 246, 5607.
2. Yurewicz, E.C., Ghalambor, M.A. & Heath, E.C. (1971) JBC, 246, 5596.

6-PHOSPHO-β-D-GALACTOSIDASE

(6-Phospho-β-D-galactoside 6-phosphogalactohydrolase)

A 6-phospho-β-D-galactoside + H_2O = an alcohol + 6-phospho-D-galactose

Ref.

Molecular weight

source	value	conditions	Ref.
Staphylococcus aureus	50,000 [1]	Sephadex; gel electrophoresis (SDS); end group analysis	(1)

Specific activity

S. aureus (30 x)	81[a]	2-nitrophenyl-β-D-galactoside 6-phosphate (pH 7.6, Tris, 30°)	(1)

(a) this is a maximum value

Michaelis constants

source	substrate	K_m (mM)	conditions	Ref.
S. aureus	2-nitrophenyl-β-D-galactoside 6-phosphate	3	pH 7.6, Tris, 30°	(1)

The enzyme does not require added metal ions for activity. (1)

References

1. Hengstenberg, W., Penberthy, W.K. & Morse, M.L. (1970) EJB, 14, 27.

6-PHOSPHO-β-D-GLUCOSIDASE

(6-O-Phosphoryl-β-D-glucopyranosyl-(1,4)-D-glucose glucohydrolase)

6-O-Phosphoryl-β-D-glucopyranosyl-(1,4)-D-glucose + H_2O =

D-glucose 6-phosphate + D-glucose

Ref.

Molecular weight

source	value	conditions	Ref.
Aerobacter aerogenes	52,000	sucrose density gradient	(1)

Specificity and Michaelis constants

source	substrate[a]	V relative	K_m (mM)	Ref.
A. aerogenes (purified 14 x)	phenyl β-D-glucose phosphate	1.00	0.50	(1)
	salicin phosphate	0.78	0.50	(1)
	gentiobiose phosphate	0.54	0.24	(1)
	arbutin phosphate	0.44	0.31	(1)
	cellobiose phosphate	0.41	0.23	(1)
	cellobiitol phosphate	0.39	0.25	(1)
	methyl β-D-glucoside phosphate	0.20	0.27	(1)

(a) the enzyme was inactive with the non-phosphorylated β-glucosides. (Conditions: pH 7.5, glygly, 25°).

Divalent metal ions (Mg^{2+}, Mn^{2+} or Ca^{2+}) are required for full activity. (1)

Abbreviations

cellobiose phosphate 6-O-phosphoryl-β-D-glucopyranosyl-(1,4)-D-glucose

References

1. Palmer, R.E. & Anderson, R.L. (1972) JBC, 247, 3420.

N-RIBOSYLPYRIMIDINE NUCLEOSIDASE

(Nucleoside ribohydrolase)

An N-ribosylpyrimidine + H_2O = a pyrimidine + D-ribose

Ref.

Specific activity

Pseudomonas fluorescens (purified 1600 x) 272 uridine (pH 8.5, Tris, 37°) (1)

Specificity and catalytic properties

source	substrate	V relative	K_m (mM)	pH[b]	K_i (mM)[c]
P. fluorescens (purified 300 x)	uridine[a]	1.00	0.8	8.5	-
	cytidine	1.15	1.0	8.5	-
	adenosine	0.28	0.5	8.5	0.5
	inosine	0.14	0.5	6.0	0.07
	5-bromouridine	0.90	2.5	8.5	-
	1-β-D-ribofuranosylthymidine	0.90	7.5	8.0	-
	guanosine	0.18	0.6	6.5	-
	xanthosine	0.20	6.5	7.0	-

(a) the enzyme is highly specific for nucleosides containing the β-D-ribofuranosyl carbon-nitrogen linkage. Purine ribonucleosides were hydrolyzed more slowly than pyrimidine ribonucleosides. The following compounds were inactive: deoxyuridine; deoxycytidine; thymidine; deoxyadenosine; deoxyguanosine; cordycepin; 5'-UMP; 5'-CMP; β-NAD; pseudouridine and orotidine. The data are from Ref. 1.

(b) pH optimum. The buffer used was Tris at 37°.

(c) C(uridine).

Although the enzyme was inhibited by EDTA, it did not require the addition of metal ions for activity. (1)

References

1. Terada, M., Tatibana, M. & Hayaishi, O. (1967) JBC, 242, 5578.

THIOGLUCOSIDASE

(Thioglucoside glucohydrolase)

A thioglucoside + H_2O = an isothiocyanate + glucose + bisulphate

Ref.

Molecular weight

source	value	conditions	Ref.
Sinapis alba[a] (mustard seed)	151,000 [2]	0.2M NaCl; Sephadex G 200; chromatography of the reduced and alkylated protein on Sepharose 4B in guanidinehydrochloride; amino acid composition.	(1)

(a) 3 thioglucosidase isoenzymes have been separated. The data given are for the major component which is a glycoprotein containing 18% carbohydrate.

Specific activity

S. alba (14 x)	60	sinigrin[a] (pH 5.5, citrate, 40°)	(1)

(a) sinigrin is hydrolyzed to allyl isothiocyanate, glucose and $KHSO_4$.

Specificity and catalytic properties

source	substrate	K_m (mM)	conditions	Ref.
Aspergillus sydowi (purified 150 x)	sinigrin[a]	3.6	pH 7.0, Pi, 37°	(2,3)

(a) sinigrin (1.0) could be replaced by sinalbin (0.8) or glucotropaeolin (0.7) but not by phenyl 1-thio-β-D-glucoside or 4-methoxyphenyl 1-thio-β-D-glucoside. The enzyme from S. alba has similar properties.

Thioglucosidase hydrolyzes a variety of thioglucosides (glucosinolates) to the corresponding goitrogenic isothiocyanate, glucose and bisulphate. The enzyme does not require any added cofactor or metal ions. It was not inhibited by EDTA. (1,2,3)

References

1. Björkman, R. & Janson, J-C. (1972) BBA, 276, 508.
2. Ohtsuru, M., Tsuruo, I. & Hata, T. (1969) Agr. Biol. Chem, (Tokyo), 33, 1309.
3. Reese, E.T., Clapp, R.C. & Mandels, M. (1958) ABB, 75, 228.

AMINOPEPTIDASE B

(Arginyl(lysyl)-peptide hydrolase)

Arginyl-peptide + H_2O = arginine + peptide

Ref.

Molecular weight

source	value	conditions	Ref.
Rat liver	95,000	Sephadex G 100, pH 7.0	(1)

Specificity and catalytic properties

The enzyme has been purified 1000 x from rat liver. It is a non-stereospecific and chloride-activated peptidase and it selectively catalyzes the hydrolysis of N-terminal arginine and lysine residues of peptides and naphthylamides. Thus, it cleaves L-Arg-β-naphthylamide (1.00, Km = 0.1 mM), and L-Lys-β-naphthylamide (0.50). L-His- and L-ornithine-β-naphthylamides were not cleaved. It hydrolyses the following peptides: L-Arg-L-Val; D-Arg-D-Val; L-Arg-L-Leu; D-Arg-L-Phe; L-Lys-L-Leu; L-Lys-L-Phe; L-Lys-L-Ala; L-Lys-L-Val; L-Lys-L-Lys; poly-L-Lys and poly-DL-Lys. The following were not hydrolyzed: compounds with an acylated α-amino group; compounds with amino acids other than arginine or lysine in the N-terminal position (of these only Gly-L-Lys was hydrolyzed); amino acid amides; dipeptide amides and naphtholic or other esters. (1)

The enzyme was inactivated by EDTA but activity was not restored by the addition of divalent metal ions. This enzyme may have a reactive thiol group at its active site and there was no evidence for the presence of a reactive serine residue. (1)

References

1. Hopsu, V.K., Mäkinen, K.K. & Glenner, G.G. (1966) ABB, 114, 557; 567.

PYROGLUTAMATE AMINOPEPTIDASE

(L-Pyroglutamyl-peptide hydrolase)

Pyroglutamyl-peptide + H_2O = pyroglutamate + a peptide

Ref.

Specificity and catalytic properties

source	substrate	K_m (mM)	conditions	Ref.
Pseudomonas fluorescens (purified 200 x)	pyr-ala[a]	2	pH 7.3, 30°	(2,3,5)
Rat liver (purified 31 x)	pyr-ala[b]	1.5	pH 7.3, 30°	(3)
Bacillus subtilis (purified 530 x)	pyr-β-naphthylamide[c]	1.7	pH 8.0, Pi, 37°	(4)

(a) pyr-ala (1.00) could be replaced by pyr-ile (0.50); pyr-val (0.22); pyr-leu (0.19); pyr-phe (0.14) or pyr-tyr (0.09) but not pyr-pro or D-pyr-ala. Pyr-D-ala was hydrolysed very slowly.

(b) pyr-ala (1.00) could be replaced by pyr-ile (0.35) or pyr-phe (0.29).

(c) pyr-β-naphthylamide (1.0) could be replaced by pyr-glutamic diethyl ester (3.0) or pyr-anilide (0.1); low activities were obtained with pyr-α-naphthylamide or α-glu-β-naphthylamide and the following were inactive: pro-; γ-glu-; α-asp-; gly-; ala-; leu-; chloroacetyl- and N-α-benzoyl-DL-arg-β-naphthylamides; and pro-gly; leu-gly; gly-phe; gly-leu and gly-gly.

Pyroglutamate amino peptidase removes selectively amino-terminal pyroglutamate residues from polypeptide chains. The enzyme requires EDTA and a sulphydryl compound for activity. It does not require divalent metal ions. (1,2,4)

Abbreviations

pyr L-pyroglutamyl (L-pyrrolidonyl) residue.

References

1. Orlowski, M. & Meister, A. (1971) The Enzymes, 4, 147.
2. Armentrout, R.W. & Doolittle, R.F. (1969) ABB, 132, 80.
3. Armentrout, R.W. (1969) BBA, 191, 756.
4. Szewczuk, A. & Mulczyk, M. (1969) EJB, 8, 63.
5. Uliana, J.A. & Doolittle, R.F. (1969) ABB, 131, 561.

AMINOACYLPROLINE AMINOPEPTIDASE

(Aminoacylprolyl-peptide hydrolase)

Aminoacylprolyl-peptide + H_2O = amino acid + prolyl-peptide

Ref.

Molecular weight

source	value	conditions	
Escherichia coli B	230,000	pH 5.6	(1)

Specific activity

E. coli (860 x)	55 μmoles L-proline formed per min per mg with poly-L-proline as substrate (pH 8.6, veronal, 40°).	(1)

Specificity and catalytic properties

The amino peptidase cleaves the bond between an *N*-terminal amino acid residue followed by a proline residue:

A ↓ - PRO - B - C ---

A free terminal amino group is essential; thus DNP-polyproline was resistant to hydrolysis. Dipeptides such as Gly.Pro, Val.Pro and Ala.Pro were cleaved at low rates. Pro.Pro.Ala; Pro.Pro.Ala OMe and Gly.Pro.Gly were better substrates and the (NH_2) Arg-Pro bond in bradykinin was cleaved very rapidly but the Pro-Pro bond more slowly. No other bond was cleaved in bradykinin. With reduced and carboxymethylated papain as substrate the only bond cleaved was the following: (NH_2) Ile-Pro. The enzyme cleaved poly-L-proline but not poly-L-hydroxyproline. It requires Mn^{2+} and it was inhibited by EDTA. (1)

Abbreviations

bradykinin	Arg-Pro-Pro-Gly-Phe-Ser-Pro-Phe-Arg.
A,B,C	amino acid residues

References

1. Yaron, A. & Mlynar, D. (1968) BBRC, 32, 658.

AEROMONAS PROTEOLYTICA AMINOPEPTIDASE

(α-Aminoacyl-peptide hydrolase (Aeromonas proteolytica)

An aminoacyl-peptide + H_2O = an amino acid + a peptide

Ref.

Molecular properties

source	value	conditions	
A. proteolytica	29,000 [1]	pH 7.0; amino acid composition	(1,2)

The enzyme, which has been purified to homogeneity (170 x), contains 2 gm atoms of Zn^{2+} per 29,000 daltons. The apoenzyme was inactive; activation occurred by the addition of Zn^{2+} or Co^{2+}; Mn^{2+} was less effective and Mg^{2+}, Ca^{2+} or Ni^{2+} were ineffective. (1,2)

Specificity and kinetic properties

substrate[(a)]	k_o (sec^{-1})	K_m (mM)
leucinamide[(b)]	220.0	5.1
norleucinamide	60.0	6.3
norvalinamide	66.7	10.8
isoleucinamide	6.2	1.0
valinamide	6.5	2.4
methioninamide	32.0	14.6
phenylalaninamide	6.2	1.8
leu-met; leu-arg	53.7; 39.3	0.35; 0.39
leu-phe; leu-trp	72.0; 58.0	0.86; 0.96
leu-tyr; leu-leu	85.4; 8.3	1.50; 0.18
leu-ile; leu-val	16.5; 15.2	0.38; 0.64
leu-ala; leu-gly	15.4; 31.6	1.0 ; 7.3
leu-OMe; phe-OMe	20.4; 41.6	12.2 ; 3.9

(a) conditions: pH 8.0, Tris, 25°. Lysinamide; argininamide; tyrosinamide and tryptophanamide were also hydrolyzed. The following were not hydrolyzed:- alaninamide; serinamide; threoninamide; prolinamide; histidinamide, isoglutamine; isoasparagine; 2-aminobutyramide; glycinamide; leu-glu; leu-pro; D-leu-leu and leu-D-leu. The data are from Ref. 3.

(b) hydrolyzed to leucine and NH_3

References

1. Prescott, J.M. & Wilkes, S.H. (1966) ABB, 117, 328.
2. Prescott, J.M., Wilkes, S.H., Wagner, F.W. & Wilson, K.J. (1971) JBC, 246, 1756.
3. Wagner, F.W., Wilkes, S.H. & Prescott, J.M. (1972) JBC, 247, 1208.

CARBOXYPEPTIDASE C

(Peptidyl-L-amino-acid (L-proline) hydrolase)

Peptidyl-L-aminoacid + H_2O = peptide + L-aminoacid

Molecular properties

source	value	conditions	Ref.
Gossypium hirsutum (cotton seed)	85,000 [3] (a)	pH 6.7; Sephadex G 150, pH 6.7; gel electrophoresis(SDS); aminoacid composition	(1)
Barley	90,000	Sephadex G 200, pH 5.2	(2)
Lemon (Citrus carboxy peptidase)	149,000	ultracentrifugation	(3)
	126,000	gel filtration	(3)
	176,000	ultracentrifugation in 8 M urea	(3)

(a) the individual chains, of molecular weights 33,000; 31,000 and 24,000 daltons, are linked by disulphide bridges. The enzyme has 2 active sites both of which are inactivated by diisopropylfluorophosphate.

Specificity and catalytic properties

source	substrate	specific activity	K_m (mM)	conditions	Ref.
G. hirsutum (purified 2000 x)	BAEE (a)	25.5	0.42	pH 6.6, 25°	(1)
	ATEE	23.6	13.4	pH 6.6, 25°	(1)
	CbZ-ala-4-nitrophenyl ester	23.2	0.24	pH 6.6, 25°	(1)
Barley (purified 430 x)	CbZ-phe-ala (b)	1286	6.7	pH 5.2, acetate, 30°	(2)

(a) BAEE (1.00) could be replaced by CbZ-arg-methyl ester (0.32); tosyl-arg-methyl ester (0.12) or CbZ-leu-gly-gly-methyl ester (0.08) but not by triacetin; 4-nitrophenyl acetate; 4-nitrophenyl laurate or cotton seed oil. The enzyme released amino acids from glucagon; A-chain of insulin; angiotensin II; denatured sea pansy luciferase, denatured yeast enolase and leu-gly-gly but not ala-ala; val-glu; trp-ala; his-ala; gly-asp; glu-val; benzoyl-arg-4-nitroanilide; hippuryl-phe; hippuryl-ser; casein or bovine serum albumin.

(b) a number of other CbZ-peptides tested were also hydrolyzed. CbZ-Peptides having carboxyl-terminal or penultimate proline and dipeptides; trigly; tetragly; pentagly; leu-4-nitroanilide and glutathione were not hydrolyzed.

The kinetic properties of carboxypeptidase C from French bean leaves have been investigated. (4)

Ref.

Specificity and catalytic properties continued

The enzyme did not require any added cofactor or metal ion for activity. The enzyme from barley, for instance, was not inhibited by EDTA. (2)

Carboxypeptidase C was inhibited by diisopropylfluorophosphate. (1,2,3)

The enzyme from *G. hirsutum* was also inhibited by phenylmethylsulphonylfluoride but not by tosyllysine chloromethylketone. The enzyme from barley was inhibited by 4-chloromercuribenzoate. (1,2)

Abbreviations

CbZ	benzyloxycarbonyl group
BAEE	*N*-benzoyl-arg-ethyl ester
ATEE	*N*-acetyl-tyr-ethyl ester

References

1. Ihle, J.N. & Dure, L.S. (1972) JBC, 247, 5034, 5041.
2. Visuri, K., Mikola, J. & Enari, T.M. (1969) EJB, 7, 193.
3. Zuber, V.H. (1968) Hoppe-Seylers Z. Physiol. Chem., 349, 1337.
4. Carey, W.F. & Wells, J.R.E. (1972) JBC, 247, 5573.

MURAMOYL-PENTAPEPTIDE CARBOXYPEPTIDASE

(UDP-N-acetylmuramoyl-tetrapeptidyl-D-Ala alanine-hydrolase)

UDP-N-acetylmuramoyl-L-Ala-D-Glu-2,6-diaminopimelyl-D-Ala-D-Ala + H_2O =
UDP-N-acetylmuramoyl-L-Ala-D-Glu-2,6-diaminopimelyl-D-Ala + D-alanine

Ref.

Specificity and catalytic properties

source	substrate or inhibitor	K_m (μM)	K_i (μM) [a]	conditions	Ref.
Escherichia coli B (purified 120 x)	UDP-MurNAc-pentapeptide[b]	600	-	pH 8.6, Tris, 38°	(1)
	penicillin G	-	0.016		
	ampicillin	-	0.006		
	cephalothin	-	3		

(a) C(UDP-MurNAc-pentapeptide). Several other penicillins and cephalosporins tested were also inhibitory.

(b) UDP-MurNAc-pentapeptide (1.00) could be replaced by P-MurNAc-pentapeptide (1.12); MurNAc-pentapeptide (0.18); GlcNAc-MurNAc-pentapeptide (0.11) or pentapeptide (0.51) but not by UDP-MurNAc-pentapeptide, P-MurNAc-pentapeptide or MurNAc-pentapeptide with 2,6-diaminopimelate replaced by L-lysine; or by D-Ala-D-Ala, N-phenacetyl-D-Ala-D-Ala or N-acetyl-D-Ala-D-Ala.

The enzyme requires Mg^{2+} for full activity. Several other divalent metal ions tested could replace Mg^{2+}. (1)

Muramoyl-pentapeptide carboxypeptidase releases only one D-alanine residue from UDP-MurNAc-pentapeptide and another carboxypeptidase (purified 25 x from *E. coli* B) releases D-alanine from UDP-MurNAc-tetrapeptide (Km = 100 μM). This enzyme was inactive with UDP-MurNAc-pentapeptide; it was not stimulated by Mg^{2+}, Ca^{2+} or Mn^{2+} nor was it inhibited by EDTA or by several penicillins or cephalosporins tested. (1)

Abbreviations

muramoyl pentapeptide or UDP-MurNAc-pentapeptide	UDP-N-acetylmuramoyl-L-alanyl-D-glutamyl-2,6-diaminopimelyl-D-alanyl-D-alanine.

References

1. Izaki, K. & Strominger, J.L. (1968) JBC, 243, 3193.

CARBOXYPEPTIDASE G_1

(N-Pteroyl-L-glutamate L-glutamate hydrolase)

Pteroylglutamate + H_2O = pteroate + L-glutamate

Ref.

Equilibrium constant

$\frac{\text{[pteroate] [L-glutamate]}}{\text{[pteroylglutamate]}}$ = 0.064 M (Tris, pH 7.3, 37°) (1)

Molecular weight

source	value	conditions	
Pseudomonas stutzeri	92,000 [2]	Sephadex G 100; gel electrophoresis (SDS)	(1,2)

The enzyme contains 4 gm atoms of Zn^{2+} (Cu^{2+}, Ni^{2+}, Co^{2+}, Hg^{2+} and Mn^{2+} were inactive) per 92,000 daltons. The Zn^{2+} is required for activity.

Specific activity

P. stutzeri (1050 x) 725 methotrexate (pH 7.3, Tris, 37°) (1)

Specificity and Michaelis constants

substrate[(a)]	relative velocity	K_m(μM)	
pteroylglutamate (folate)	1.00	1.1	(1)
5-formyltetrahydrofolate	0.90	18.1	(1)
5-methyltetrahydrofolate	-	12.9	(1)
methotrexate	0.69	3.9	(1)
aminopterin	0.92	8.3	(1)
4-amino-N^{10}-methylpteroylaspartate	0.94	580.0	(1)
4-aminopteroylaspartate	0.92	104.0	(1)

(a) with the enzyme from P. stutzeri. (pH 7.3, Tris, 37°)

Carboxypeptidase G_1 is an exopeptidase and removes the carboxyl-terminal L-glutamate from both reduced and non-reduced folate derivatives and also from oligopeptides and N-benzoyloxycarbonyl glutamates. It is less active towards aspartate carboxyl-terminal linkages and inactive towards D-glutamate carboxyl-terminal linkages. (1)

A carboxypeptidase (carboxypeptidase G) has been isolated from a soil pseudomonad. This enzyme did not hydrolyze aspartate -terminal peptides; it required Zn^{2+} for activity. (3)

Light absorption data

The hydrolysis of methotrexate is accompanied by a change in absorbance at 320 nm (molar extinction coefficient = 8,300 $M^{-1}cm^{-1}$). The light absorption data of other folate derivatives are known. (1)

Ref.

Abbreviations

Methotrexate 4-amino-N^{10}-methyl pteroylglutamate

References

1. McCullough, J.L., Chabner, B.A. & Bertino, J.R. (1971) JBC, 246, 7207.
2. Chabner, B.A. & Bertino, J.R. (1972) BBA, 276, 234.
3. Levy, C.C. & Goldman, P. (1967) JBC, 242, 2933.

γ-GLUTAMYLCARBOXYPEPTIDASE

(Pteroyloligoglutamate L-glutamate hydrolase)

Pteroyl (glutamate)$_n$ + H_2O = pteroyl (glutamate)$_{n-1}$ + L-glutamate

Ref.

Molecular weight

source	value	conditions	Ref.
Chicken pancreas	130,000 [4]	pH 7.5; gel electrophoresis (SDS)	(1)

γ-Glutamylcarboxypeptidase is a metalloprotein. (1)

Specific activity

Chicken pancreas	(1000 x)	9.2	N-formylpteroyl-γ-heptaglutamate (pH 7.8, Tris, 37°)	(1)

Specificity and catalytic properties

γ-Glutamylcarboxypeptidase attacks pteroyl oligoglutamates, at the free carboxyl terminus, in a stepwise manner the end product being pteroyl-γ-diglutamate. N-Formylpteroyl-γ-heptaglutamate, pteroyl-γ-triglutamate and 4-aminobenzoyl-γ-triglutamate are substrates. α-Glutamyl peptides are not cleaved. The Michaelis constant for N-formylpteroyl-γ-heptaglutamate is 3 μM (pH 7.8, Tris, 37°). (1)

The enzyme requires a high ionic strength for activity; it was not inhibited by diisopropylfluorophosphate or by thiol compounds. (1)

References

1. Kaferstein, H. & Jaenicke, L. (1972) Hoppe-Seyler's Z. Physiol. Chem., 353, 1153.

AMINOACYL-LYSINE DIPEPTIDASE

(Aminoacyl-L-lysine (-L-arginine) hydrolase)

Aminoacyl-L-lysine + H_2O = amino acid + L-lysine

Ref.

Equilibrium constant

The reaction is irreversible [F] with the enzyme from hog kidney as the catalyst. (1)

Specificity and catalytic properties

Aminoacyl-lysine dipeptidase hydrolyses dipeptides having lysine, arginine or ornithine as the C-terminal amino-acid. The specificity of this enzyme has been compared with that of carboxypeptidase B (EC 3.4.12.3; formerly EC 3.4.2.2). (1)

source	substrate	K_m (mM)	conditions	
Hog kidney (purified 62 x)	N^2-(4-aminobutyryl)-lysine[(a)]	9.8	pH 7.0, Tris, 37°	(1)

(a) N^2-(4-aminobutyryl) lysine (1.00) could be replaced by 4-amino-butyrylarginine (1.24); Ala.Arg (0.60); Gly.Arg (0.50); β-Ala.Lys (0.49); Phe.Lys (0.49); Leu.Lys (0.37); β-Ala.Arg (0.36); Arg.Ala (0.35); β-Alanyl-ornithine (0.27); Leu.Gly (0.26); 4-aminobutyryl-ornithine (0.18); Ile.Arg (0.10); Ala.Lys (0.09); Ser.Arg (0.08); N^2-(N-acetyl-4-aminobutyryl) lysine (0.08); Gly.Gly (0.04); Ile-Lys (0.03); Arg.Leu (0.03) or benzoyl-Gly-Lys (0.03) but not by carbo-benzoxy-Leu-Arg; 4-amino-butyrylhistidine; β-Ala.His; Glu.His; N^6-(4-aminobutyryl) lysine; N^2-(4-aminobutyryl) lysine amide; N^2-(4-aminobutyryl) lysine methylester; N^2-(carbobenzoxy-4-aminobutyryl)-(N^6-carbobenzoxy) lysine; N^2-(carbobenzoxy-β-alanyl) (N^6-carbobenzoxy) lysine; Asp.Glu; glutathione; glycyl-glycine methyl ester; Gly.Leu; glycine ethyl ester; histidine or leucine methyl ester; benzoyl-arginine amide; benzoyl-arginine-4-nitroanilide; tosyl-lysine methyl-ester or carbobenzoxy-Gly.Phe.

The enzyme was inhibited by o-phenanthroline and 4-chloromercuri-benzoate but not by EDTA, diisopropylfluorophosphate or KCN. Mn^{2+} (or Mg^{2+}) activated the reaction by about 50%. Pi, ornithine, lysine and arginine (but not histidine or the neutral and acidic amino acids) were inhibitory. (1)

References

1. Kumon, A., Matsuoka, Y., Kakimoto, Y., Nakajima, T. & Sano, I. (1970) BBA, 200, 466.

β-ASPARTYLDIPEPTIDASE

(β-L-Aspartyl-L-amino-acid hydrolase)

β-L-Aspartyl-L-leucine + H_2O = L-aspartate + L-leucine

Ref.

Molecular weight

source	value	conditions	
Escherichia coli	120,000	Sephadex G 200, pH 8.1	(1)

Multiple forms have been observed. (1)

Specificity and catalytic properties

β-Aspartyl dipeptidase (E. coli) is specific for β-aspartyl dipeptides. The enzyme does not require metal ions for activity and it is not a sulphydryl peptidase. (1)

A β-aspartyl peptidase has been purified (15 x) from rat liver. With this enzyme, β-aspartyl-Gly was the most active substrate and in addition to β-aspartyldipeptides, β-aspartyltripeptides were also hydrolyzed. The peptidase had no metal ion requirement. (2)

source	substrate	K_m (mM)	conditions	
E. coli	β-L-aspartyl-Leu[a]	0.81	pH 8.1, Tris, 37°	(1)

(a) β-L-aspartyl-Leu (1.00) could be replaced by β-L-aspartyl-Ser (0.82); -Met (0.68); -Val (0.56); -Gln (0.48); -Phe (0.38); -Ala (0.33); -Ile (0.19); -Thr (0.18) and -Asn (0.10) but not by the following:- β-L-aspartyl-Gly; -His; -Gly-Ala; -Gly-Gly; -Leu-Gly; by α-L-aspartyl-Leu; -Ser; -Phe; -Ala; -Ile; -Asn; -Gly-Ala; -Leu-Gly; or by N-acetyl-Met; asparagine; β-aspartyl-glucosylamine; γ-glutamyl-Leu; α-glutamyl-Leu; Leu-Gly; Asn-Leu or by carnosine.

References

1. Haley, E.E. (1968) JBC, 243, 5748.
2. Dorer, F.E., Haley, E.E. & Buchanan, D.L. (1968) ABB, 127, 490.

DIPEPTIDASE

(Dipeptide hydrolase)

Dipeptide + H_2O = 2 amino acid

Ref.

Molecular properties

source	value	conditions	Ref.
Mouse ascites tumour cells[(a)]	85,000	Sephadex G 150, pH 8.2	(1)
Hog kidney, cortex[(a)]	47,200	pH 8.0; amino acid composition	(2,3)
Mycobacterium phlei	88,000 [2]	gel filtration; gel electrophoresis (SDS); amino acid composition	(4)

(a) the enzyme contains 1 gm atom of Zn^{2+} per mole of protein.

Specific activity

Mouse ascites tumour cells	(800 x)	2600	Ala-Gly (pH 8.3, 40°)	(1)
Hog kidney	(300 x)	2.43	Gly-dehydrophenylalanine (pH 8.0, Tris, 35°)	(2)
		376	Gly-Gly (pH 8.0, Tris, 35°)	(2,3)
M. phlei	(1000 x)	28	Leu-Gly (pH 8.0, Tris, 37°)	(4)

Specificity and catalytic properties

Dipeptidase hydrolyses only L-α-dipeptides (the enzyme from M. phlei had low activities towards D-peptides) with a free amino and carboxyl group. For example, the enzyme from mouse ascites tumour hydrolyzed the following peptides: Ala-Gly (V relative = 1.00; Km = 2.5 mM); Ala-Ile (5.3; 1.2 mM); Gly-Ile (3.6; 7 mM); Gly-Val (2.9; 4 mM); Gly-Leu (2.3; 2.5 mM); Ala-Ala (2.2; 2 mM); Ala-Val (1.9; 1.1 mM); Leu-Ala (1.7; 1.2 mM); Gly-Met (1.6; 2.5 mM); Gly-Ala (1.1; 22 mM); and Val-Gly (0.92; 1.4 mM). Gly-Gly was hydrolyzed very slowly and Gly-Asp not at all. (1)

Dipeptidases from different sources vary in their metal ion requirements. The enzyme from mouse ascites tumour was activated by Mn^{2+} with the poor substrate Pro-Gly whereas the hydrolysis of the good substrate Ala-Gly was inhibited by Mn^{2+} and several other metal ions tested. With the enzyme from M. phlei, Co^{2+} activated the hydrolysis of Leu-Gly but inhibited the hydrolysis of Ala-Gly. The hydrolysis of both substrates was activated by Mn^{2+} or Mg^{2+}. (1,4)

Dipeptidase from hog kidney was inhibited by Pi (Ki = 2.8 mM); AMP (Ki = 1.5 mM); ADP (Ki = 1.2 mM); ATP (Ki = 0.9 mM); UTP (Ki = 0.7 mM); GTP (Ki = 0.5 mM) and CTP (Ki = 0.4 mM). The inhibition was competitive (Gly-dehydrophenylalanine). The enzyme was also inhibited by amino acids. (6)

Light absorption data

Gly-dehydrophenylalanine, which is hydrolyzed by the dipeptidase to glycine, phenylpyruvate and NH_3, has an absorption band at 275 nm (molar extinction coefficient = 15,500 M^{-1} cm^{-1}). (5)

References

1. Hayman, S. & Patterson, E.K. (1971) JBC, 246, 660.
2. Campbell, B.J., Lin, Y-C., Davis, R.V. & Ballew, E. (1966) BBA, 118, 371.
3. René, A.M. & Campbell, B.J. (1969) JBC, 244, 1445.
4. Plancot, M-T. & Han, K-K. (1972) EJB, 28, 327.
5. Campbell, B.J., Lin, Y-C. & Bird, M.E. (1963) JBC, 238, 3632.
6. Harper, C., René, A. & Campbell, B.J. (1971) BBA, 242, 446.

METHIONYL DIPEPTIDASE

(L-Methionyl-amino-acid hydrolase)

L-Methionyl-amino acid + H_2O = L-methionine + amino acid

Ref.

Molecular weight

source	value	conditions	Ref.
Escherichia coli B	94,000 [2]	Sephadex G 150, pH 7.4; gel electrophoresis (SDS)	(1)

Specificity and kinetic constants

substrate[(a)]	V(specific activity)	K_m (mM)	Ref.
Met-Ser	560	0.60	(1)
Met-Ala	450	0.31	(1)
Met-Thr	300	0.53	(1)
Met-Leu	270	0.45	(1)
Met-Ile	60	0.42	(1)
Met-Val	40	0.13	(1)

(a) with the enzyme from E. coli B (purified 152 x; conditions: pH 7.4, Tris, 37°). Met-Ala (1.00) could also be replaced by Met-Gly (0.59); Met-Glu (0.10); Ala-Ala (0.40); Ala-Met (0.38); Ala-Leu (0.24); Ser-Ala (0.26); Thr-Ala (0.18); Leu-Ala (0.14) and Gly-Ala (0.09) but not by Gly-Gly; formyl Met-Ala; Met (sulphoxide)-Ala; Met (sulphone)-Ala or Met-Ala-Met.

Dipeptidase M may be responsible for the removal of the NH_2-terminal methionyl residues from newly synthesised proteins in E. coli. The enzyme does not hydrolyse 4-toluenesulphonyl-L-arginine methyl ester; benzoyl-L-tyrosine ethyl ester; hippuryl-L-phenylalanine; L-leucinamide or ribonuclease. It has no esterase activity. (1)

The dipeptidase requires Mn^{2+} specifically: no other divalent ion tested was active. (1)

References

1. Brown, J.L. (1973) JBC, 248, 409.

ANGIOTENSIN-CONVERTING ENZYME

Angiotensin I + H_2O = angiotensin II + His-Leu

Ref.

Molecular properties

source	value	conditions	Ref.
Guinea-pig and hog lung	150,000	Sephadex G 150, pH 7.4; sucrose density gradient	(1)
Human lung	480,000	Sephadex G 200, pH 6.9	(2)
Hog plasma	150,000	gel filtration; sucrose gradient	(1,3)
Calf lung	300,000	Sephadex G 200, pH 8.0	(4)

Specificity and catalytic properties

Angiotensin converting enzyme catalyzes the conversion of angiotensin I to the vasoactive peptide, angiotensin II. Angiotensin I isolated from horse plasma is a decapeptide: Asp-Arg-Val-Tyr-Ile-His-Pro-Phe-His-Leu and the angiotensin converting enzyme, which is highly specific, cleaves the Phe-His peptide bond in this peptide. Angiotensin I is produced by the action of renin (EC 3.4.99.19, formerly EC 3.4.4.15) on angiotensinogen. The conformation of angiotensin II has been investigated. (1,3,4,5)

Angiotensin converting enzyme requires a divalent metal ion and NaCl for activity (e.g. Zn^{2+}, Co^{2+} or Mn^{2+}). The natural metal activator has not been identified. The enzyme is inhibited by EDTA. (3,4)

The lung (but not plasma) contains an enzyme (molecular weight = 80,000 daltons) which hydrolyses His-Leu (Km = 0.2 mM). This enzyme does not cleave the His-Leu bond in angiotensin I. It was completely inhibited by EDTA. (1)

source	substrate	ko(sec^{-1})	K_m(mM)	conditions	Ref.
Guinea-pig and hog lung (purified 200 x)	angiotensin I	-	0.02	pH 7.4, Pi, 37°	(1)
Calf lung (purified 700 x)	CbzPhe(NO_2)-His-Leu[a]	13.9	0.23	pH 8.0, Tris, 25°	(4)
	CbzPhe(NO_2)-Gly-Gly	56.7	1.5		
	Hip(NO_2)-Gly-Gly	99.7	4.8		

(a) in the presence of 5% methanol. Angiotensin I was converted to angiotensin II at 10% the rate of CbzPhe (NO_2)-His-Leu cleavage.

The enzyme from calf lung was inhibited by Pyr-Lys-Trp-Ala-Pro (Ki = 0.5 µM); Gly-Gly (Ki = 1.5 mM) and Phe-Arg (Ki = 0.81 mM). His-Leu (Ki = 0.88 mM) was also inhibitory; the hog plasma enzyme, however, was not inhibited by this compound. (4,3)

Ref.

Specificity and catalytic properties continued

The angiotensin converting enzyme from rabbit lung was inhibited by bradykinin and related compounds. The venoms of several species of snake contain converting enzyme inhibitors. (6)

Renin (EC 3.4.99.19) has been highly purified from a number of sources. The molecular weight of hog renin is ∿45,000 (gel filtration). At pH 7.0, the Km with hog angiotensinogen is 0.71 ng ml^{-1}(∿0.5 μM) and with rat angiotensinogen, 2.5 ng ml^{-1}(∿2 μM). Renin is not inactivated by EDTA or diisopropylphosphofluoridate. (7)

Streptomyces sp. contains a powerful inhibitor of renin, pepstatin A (isovaleryl-L-Val-L-Val-4-amino-3-hydroxy-6-methylheptanoyl-L-Ala-4-amino-3-hydroxy-6-methylheptanoate). (8)

Light absorption data

The light absorption properties of synthetic substrates of the angiotensin converting enzyme are known. (4)

Abbreviations

Cbz benzyloxycarbonyl group.

References

Handbook of Experimental Pharmacology (1973) Vol 37: Angiotensin (ed. Page, I.H. & Bumpus, F.M.) Springer-Verlag: Heidelberg.

1. Lee, H-J., Larue, J.N. & Wilson, I.B. (1971) BBA, 250, 549.
2. Fitz, A. & Overturf, M. (1972) JBC, 247, 581.
3. Angus, C.W., Lee, H-J. & Wilson, I.B. (1972) BBA, 276, 228.
4. Stevens, R.L., Micalizzi, E.R., Fessler, D.C. & Pals, D.T. (1972) B, 11, 2999.
5. Fermandjian, S., Fromageot, P., Tistchenko, A-M., Leicknam, J-P. & Lutz, M. (1972) EJB, 28, 174.
6. Sander, G.E., West, D.W. & Huggins, C.G. (1972) BBA, 289, 392.
7. Smeby, R.R. & Bumpus, F.M. (1970) Methods in Enzymology, 19, 699.
8. Corvol, P., Devaux, C. & Menard, J. (1973) FEBS lett., 34, 189.

Angus, C.W., Lee, H.J. & Wilson, I.B. (1973) BBA, 309, 169.
Poulsen, K., Burton, J. & Haber, E. (1973) B, 12, 3877.

KALLIKREIN

The trivial name kallikrein designates a number of enzymes from different sources and of different specificities with the common property of liberating vasoactive peptides (kinins) from their inactive kininogen precursors.

Ref.

Molecular properties

Kallikrein is present in mammalian blood plasma in an inactive form (prekallikrein). Human plasma has at least three protein prekallikrein activators. Bovine plasma has also a complex activation system. The activation is accompanied by the specific cleavage of the prekallikrein molecule, probably at a single arginyl residue. Prekallikrein is also activated by trypsin, casein, acetone and glass powder. (2,3)

Kallikrein is a serine protease. Thus, the enzyme from pig pancreas is inactivated by the active site titrants 4-nitrophenyl-4'-guanidinobenzoate, cinnamoyl imidazole and indoleacryloyl imidazole. It is also inactivated by diisopropylfluorophosphate but not by EDTA. (5,6)

source	value	conditions	
Bovine plasma prekallikrein	90,000 [1]	pH 8.4; sucrose density gradient; Sephadex G 100, pH 8.0; gel electrophoresis (SDS).	(3)
Pig pancreas [(a)]	26,200	ultracentrifugation.	(4)
Pig pancreas [(a)]	25,700	active site titration.	(5,8)

(a) pig pancreas contains two prekallikreins each giving rise to a kallikrein (kallikreins A and B). The two kallikreins have identical molecular weights but they differ in carbohydrate content: kallikrein contains about 7% carbohydrate. The amino acid and carbohydrate compositions of the enzyme have been reported.

Specific activity

Bovine plasma (1300 x)	52	N-α-tosyl-L-arginine methyl ester (pH 8.5, Tris, 37°)	(3)
Porcine pancrease (kallikrein A or B)	300	N-α-benzoyl-L-arginine ethyl ester (pH 8.0, 25°)	(4,5)

Specificity and catalytic properties

Kallikrein is highly specific for kininogen which could not be replaced by several other proteins or peptides tested. Kallikrein also hydrolyzes (at decreasing rates) esters of basic; aromatic and neutral aliphatic amino-acids. It was most active with L-arginine esters. The kinetic constants obtained with several esters have been reported and the specificity properties of kallikrein (EC 3.4.21.8; formerly EC 3.4.4.21); trypsin (EC 3.4.21.4; formerly EC 3.4.4.4) and clostripain (EC 3.4.22.8; formerly EC 3.4.4.20) have been compared. (4,8)

Ref.

Kininogen

Human and bovine kinin (kallidin II) has the structure Lys-Arg-Pro-Pro-Gly-Phe Ser Pro Phe Arg and its precursor, kininogen, is identical to α-globulin of serum. Bradykinin (kallidin I), which is identical to kinin except that the amino terminal Lys residue is missing, is formed by the action of amino peptidase, trypsin or snake venom (but not kallikrein) on kininogen. (7)

References

1. Colman, R.W., Girey, G.J.D., Zacest, R. & Talamo, R.C. (1971) in Progress in Haematology (eds. Brown, E.B. & Moore, C.V.) Vol 7, Grune and Stratton, New York. p 255.
2. Bagdasarian, A., Talamo, R.C. & Colman, R.W. (1973) JBC, 248, 3456.
3. Takahashi, H., Nagasawa, S. & Suzuki, T. (1972) J. Biochem. (Tokyo), 71, 471.
4. Kutzbach, C. & Schmidt-Kastner, G. (1972) Hoppe-Seylers' Z. Physiol. Chem., 353, 1099.
5. Fiedler, F., Müller, B. & Werle, E. (1972) FEBS lett., 24, 41.
6. Fiedler, F., Müller, B. & Werle, E. (1972) FEBS lett., 22, 1.
7. Dayhoff, M.O. (1972) Atlas of Protein Sequence and Structure, National Biomedical Research Foundation: Washington. p D-178.
8. Fiedler, F., Leysath, G. & Werle, E. (1973) EJB, 36, 152.

ENTEROPEPTIDASE

Selective cleavage of Lys^{6}-Ile^{7} bond in trypsinogen

Ref.

Molecular properties

source	value	conditions	Ref.
Porcine duodenal mucosa	195,000 [2]	pH 6.0; gel electrophoresis (SDS); Sephadex G 200, pH 6.0; amino acid composition	(1)

Enteropeptidase is composed of two polypeptide chains of molecular weights 134,000 and 62,000. Both chains are glycoproteins and the dimer contains 20% neutral sugar, 15% amino sugar and 2% sialic acid. The light chain has a serine residue which reacts irreversibly with diisopropylfluorophosphate. The enzyme also reacts irreversibly with N-α-tosyl-L-lysine chloromethylketone. (1)

Specific activity

			Ref.
Porcine duodenal mucosa (1000 x)	1.3	bovine trypsinogen (pH 5.6, succinate + Ca^{2+}, 25°)	(1)
	8	BAEE (pH 7.9, 25°)	(1)

Specificity and catalytic properties

source	substrate	V relative	K_m (mM)	Ref.
Porcine duodenal mucosa (purified 650 x)	bovine trypsinogen(a)	1.0	0.07	(2)
	BAEE	5.9	-	(2)
	TAME	1.7	-	(2)

(a) Ca^{2+} had little effect on the activation. Bovine chymotrypsinogens A and B were not activated and 4-nitrophenylacetate was not hydrolyzed. Enteropeptidase also cleaved the Lys^{6}-Ile^{7} bond in S-carboxymethylated trypsinogen but no bond was cleaved in S-carboxymethylated chymotrypsinogen:- There are 3 Lys-Ile bonds in chymotrypsinogen A.

The catalytic properties of trypsin and enteropeptidase have been compared. Enteropeptidase was not inhibited by the trypsin protein inhibitors. (2)

Light absorption data

See trypsin (EC 3.4.21.4; formerly EC 3.4.4.4)

Abbreviations

BAEE N-benzoyl-L-arginine ethyl ester
TAME 4-toluenesulphonyl-L-argine methyl ester

References

1. Baratti, J., et al (1973) BBA, 315, 147.
2. Maroux, S., Baratti, J. & Desnuelle, P. (1971) JBC, 246, 5031.

PROINSULINASE

Proinsulin + $3H_2O$ = insulin + connecting peptide + arginine

Ref.

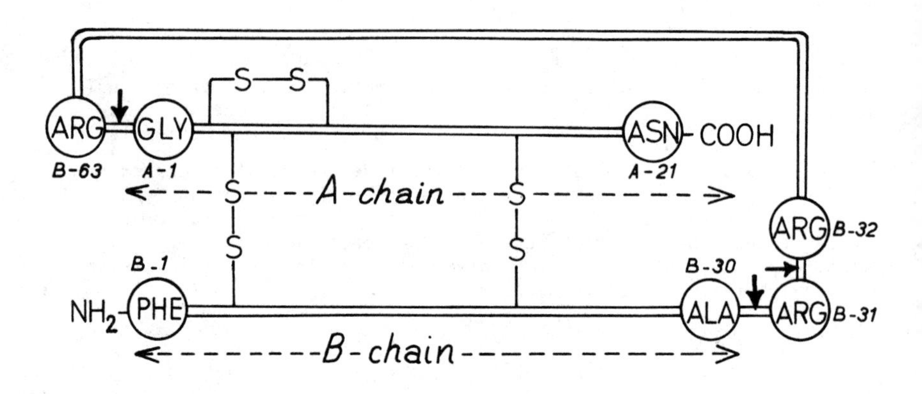

Fig. 1. The structure of porcine proinsulin

Proinsulinase cleaves 3 peptide bonds in proinsulin; these are indicated by arrows in Fig. 1. In this way 3 fragments are released: insulin (A- and B-chains), the connecting peptide (ARG B-32 - ARG B-63) and ARG B-31. (1,2,3)

Proinsulinase has been highly purified from bovine pancreas. The enzyme is a single polypeptide chain of molecular weight 70,000 daltons (gel electrophoresis, SDS). It does not have carboxypeptidase B activity (EC 3.4.12.3; formerly EC 3.4.2.2). (2)

The conversion of proinsulin to insulin is a 2-step reaction: a fast hydrolysis to an intermediate which is then slowly cleaved to yield insulin. The slow step, but not the fast step, is inhibited by Zn^{2+}. The connecting peptide is released intact together with ARG B-31. Neither ALA B-30 nor any other amino acid is released. Proinsulinase has many enzymic properties in common with trypsin (EC 3.4.21.4; formerly EC 3.4.4.4); thus it hydrolyses α-N-benzoyl-DL-arginine-4-nitroanilide (this reaction is not affected by Zn^{2+}) and it is inhibited by the trypsin protein inhibitors from porcine pancreas, lima bean or soybean. Proinsulinase is also inhibited by diisopropylfluorophosphate but not by L-1-tosylamido-2-phenylethyl chloromethyl ketone. (2)

The sequences of proinsulins and insulins isolated from several sources are known. (4)

Ref.

References

1. Behrens, O.K. & Grinnan, E.L. (1969) Ann. Rev. Biochem, 38, 96.
2. Yip, C.C. (1971) PNAS, 68, 1312.
3. Chance, R.E., Ellis, R.M. & Bromer, W.W. (1968) Science, 161, 165.
4. Dayhoff, M.O. (1972) Atlas of Protein Sequence and Structure, National Biomedical Research Foundation: Washington, p D-208.

Tager, H.S., Emdin, S.O., Clark, J.L. & Steiner, D.F. (1973) JBC, 248, 3476.

Yu, S.S. & Kitabchi, A.E. (1973) JBC, 248, 3753.

STREPTOMYCES ALKALINE PROTEINASE

Ref.

Molecular weight

source	value	conditions	
Streptomyces rectus var. proteolyticus	21,500	pH 5.6; Sephadex G 100, pH 5.6; amino acid composition	(1)

Specificity and catalytic constants

source	substrate	ko (sec^{-1})	K_m (mM)	
S. rectus[a]	acetyl-L-phenylalanine methyl ester	310	8	(2)
	N-benzoyl-L-arginine ethyl ester	5.7	16	(2)

(a) purified 3 x. The assay conditions were: pH 8.0, 0.1 M NaCl and 5 mM borate, 25°. Casein was also hydrolyzed.

Streptomyces alkaline proteinase is a serine proteinase. It also has an exposed sulphydryl group which is involved in maintaining the organization of the active site. (2)

Light absorption data

See trypsin (EC 3.4.21.4; formerly EC 3.4.4.4).

References

1. Mizusawa, K. & Yoshida, F. (1972) JBC, 247, 6978.
2. Mizusawa, K. & Yoshida, F. (1973) JBC, 248, 4417.

PREMONOPHENOL MONOOXYGENASE ACTIVATING ENZYME

Converts premonophenol monooxygenase to monophenol monooxygenase (EC 1.14.18.1; formerly EC 1.10.3.1 and EC 1.10.3.2)

Ref.

Molecular properties

source	value	conditions	Ref.
Bombyx mori (silkworm; cuticle)	34,000 [1]	gel filtration; gel electrophoresis (SDS)	(1)

The activating enzyme is a serine proteinase and it was inactivated by diisopropyl fluorophosphate; phenylmethanesulphonyl fluoride and 4-nitrophenyl-4'-guanidinobenzoate but not by the chloromethylketone derivatives of phenylalanine and lysine. It was not inactivated by thiol group reagents or by EDTA. (1)

Specificity and catalytic properties

In haemolymph and other tissues of insects monophenol monooxygenase occurs as an inactive preenzyme which is activated by the activating enzyme or, less effectively, by α-chymotrypsin and amino peptidase but not by trypsin. The activation process is accompanied by the release of a peptide. (1)

source	substrate	k_o (sec^{-1})	K_m (μM)	conditions	Ref.
B. mori (purified 4800 x)	premonophenol monooxygenase	-	0.22	pH 7.5, Pi	(1)
	α-N-benzoyl-L-arginine ethyl ester[(a)]	107	145	pH 7.8 Tris + 0.5 M KCl 25°	(1)
	α-N-tosyl-L-arginine methyl ester	5.7	709	pH 7.8, Tris, + 0.5 M KCl 25°	(1)

(a) α-N-benzoyl-L-tyrosine ethyl ester; α-N-tosyl-L-lysine methyl ester and α-N-benzoyl-DL-arginine 4-nitroanilide were not hydrolyzed.

References

1. Dohke, K. (1973) ABB, 157, 203; 210.

CATHEPSIN B

Hydrolyses proteins with a specificity resembling that of papain (EC 3.4.22.2, formerly EC 3.4.4.10)

Ref.

Molecular weight

source	value	conditions	
Calf liver	about 25,000	Sephadex G 100, 1% NaCl	(1)
Rat liver	33,000	Sephadex G 100, pH 6.8	(2)
Beef spleen acetone powder	40,000 [2]	Biogel-P60, pH 4.5	(3)

Specificity and catalytic properties

source	substrate	ko(min^{-1})	K_m(mM)	conditions	
Calf liver	BAEE[(a)]	98	17.5	pH 6, 40°	(1)
(purified	BANA	143	20.0	pH 6, 40°	(1)
200 x)	ANA	143	20.0	pH 7, 40°	(1)
	LNA	143	28.0	pH 7, 40°	(1)
	TAME	42	26.3	pH 6, 40°	(1)
Rat liver	BANA	40	1.4	pH 7.0,Pi,37°	(2)
(purified 1500 x)	benzoylarginine amide	10	47	pH 7.0,Pi,37°	(2)

(a) Gly-Phe was not hydrolyzed

Cathepsin B is a sulphydryl enzyme: it is activated by sulphydryl-containing compounds and EDTA. It is not inhibited by diisopropyl-fluorophosphate or soybean trypsin inhibitor. It activates trypsinogen and catalyzes transamidation reactions. The specificity of cathepsin B resembles that of papain. The main sites of cleavage in the B chain of insulin (see EC 3.4.21.aa) are at the carboxyl ends of glycine, alanine, leucine and glutamate. (3)

The enzyme from rat liver is also a sulphydryl enzyme. It has different specificity properties from cathepsin B and of several compounds tested, only hydrolyzed BANA and benzoylarginine amide: benzoylarginine-4-nitro-anilide; BAEE; carbobenzoxyleucyl-β-naphthylamide; benzoylphenyl-alanyl -β-naphthylamide; ANA and LNA were not hydrolyzed. It did not activate trypsinogen and its activity on haemoglobin was low. (2)

Light absorption data

See trypsin (EC 3.4.21.4, formerly EC 3.4.4.4).

Ref.

Abbreviations

BAEE	α-N-benzoyl-L-arginine ethyl ester
BANA	α-N-benzoyl-L-arginine naphthylamide
LNA	L-leucine β-naphthylamide
ANA	L-arginine β-napthylamide
TAME	toluene-4-sulphonyl-L-arginine methyl ester

References

1. Snellman, O. (1969) BJ, 114, 673.
2. de Lumen, B.O. & Tappel, A.L. (1972) JBC, 247, 3552.
3. Keilová, H. & Keil, B. (1969) FEBS lett., 4, 295.

INSULINASE

Hydrolyses insulin. Formerly EC 3.4.99.10

Ref.

Molecular weight

source	value	conditions	
Rat liver	80,000	Sephadex G 200, pH 7.6	(1)

Insulinase is a thiol proteinase and it is inactivated by 4-hydroxymercuribenzoate and *N*-ethylmaleimide. (1)

Specificity and catalytic properties

It is thought that the major route for insulin catabolism is *via* insulinase. Insulin is also degraded, at lower rates, by trypsin, chymotrypsin or papain. (1)

Insulinase from rat liver (purified 97 x) attacked insulin (1.00), globin (0.41) and casein (0.11) but not serum albumin or human growth hormone. The enzyme required mercaptoethanol for full activity but there was no requirement for metal ions. Insulinase was not inhibited by EDTA. (1)

source	substrate[a]	K_m (nM)
Rat skeletal muscle (highly purified)	insulin	22.2
	desalanine	15.8
	monoarginine	24.4
	diarginine	24.4
	proinsulin	857.2
	cleaved	234.2
	desdipeptide	176.0
	desnonapeptide	55.0
	destridecapeptide	44.0

(a) porcine; all the substrates had the same V. The conditions were: pH 7.5, borate, 37°. (The data are from Ref. 2).

For the structures of insulin and proinsulin, see EC 3.4.21.aa.

References

1. Burghen, G.A., Kitabchi, A.E. & Brush, J.S. (1972) Endocrinology, 91,633.
2. Baskin, F.K. & Kitabchi, A.E. (1973) EJB, 37, 489.

PEPSIN C

More restricted specificity than pepsin A

(EC 3.4.23.1, formerly EC 3.4.4.1)

Ref.

Molecular properties

source	value	conditions	
Human (gastric juice)	31,500	pH 5.0; amino acid composition	(1,2)
Pig (gastric juice)	36,000	pH 6.9; amino acid composition	(3)

The amino-acid compositions of human pepsin C (gastricsin) and human pepsin A differ considerably. However, when the sequence of 19 residues at the carboxyl terminal end of pepsin C was compared with a similar sequence in pepsin A, an extensive homology was found. (1,2)

Pepsin C (pig) is derived from pepsinogen C (molecular weight = 41,400 daltons). Pepsin C contains no Pi. (3)

Specificity and catalytic properties

source	substrate	V µmoles $hr^{-1}mg^{-1}$	K_m (mM)	
Human gastric juice	Cbz-Tyr-Ala[a]	0.40	0.74	(4)
	Cbz-Tyr-Ser	0.19	4.8	(4)
	Cbz-Tyr-Thr	0.13	7.9	(4)
	Cbz-Trp-Ala	0.39	0.92	(4)

(a) the conditions were: pH 2.0, citrate, 37°. Cbz-tyr-ala (1.00) could also be replaced by Cbz-tyr-leu (0.30); Cbz-tyr-val (0.09); Cbz-tyr-tyr (0.08) and Cbz-tyr-phe (0.02) but not Cbz-tyr-gly; Cbz-ser-tyr; Cbz-ala-tyr; Cbz-phe-ser, Cbz-trp-ser or Cbz-trp-leu. The sites of cleavage by pepsin C of oxidized ribonuclease A and glucagon have been investigated.

The substrate specificities of human pepsins A and C have been compared. (4)

Abbreviations

Cbz benzyloxycarbonyl group.

References

1. Mills, J.N. &Tang, J. (1967) JBC, 242, 3093.
2. Huang, W-Y. & Tang, J. (1970) JBC, 245, 2189.
3. Ryle, A.P. & Hamilton, M.P. (1966) BJ, 101, 176.
4. Huang, W-Y. & Tang, J. (1969) JBC, 244, 1085.

APO-PYRIDOXALENZYME HYDROLASE

Ref.

Molecular weight

source	value	conditions	Ref.
Intestine of vitamin B_6-deficient rats	31,000	Sephadex G 100, pH 7.5	(1)

Specificity and catalytic properties

source	substrate	K_m (μM)	conditions	Ref.
Rat intestine (purified 500 x)	apoornithine transaminase	15.2	pH 8.6, Tris, 37°	(1)

The hydrolase splits the apo-protein of ornithine transaminase into 2 products: an homogeneous protein and an oligopeptide. The products, either separately or together, were enzymically inactive. The hydrolase inactivated all apopyridoxal enzymes tested: ornithine transaminase (1.00); homoserine dehydratase (0.84), aspartate transaminase (0.82); serine dehydratase (0.62) and tyrosine transaminase (0.08). Pyridoxal phosphate and pyridoxal (but not pyridoxamine phosphate or pyridoxol phosphate) protected the apoenzymes against the protease. The concentrations of pyridoxal phosphate and pyridoxal required for 50% protection were 6 μM and 160 μM, respectively. Non-pyridoxal enzymes (glutamate dehydrogenase, lactate dehydrogenase, urease and glutaminase), bovine serum albumin and several synthetic substrate tested (e.g. N-benzoyl-L-arginine ethyl ester; 4-toluene sulphonyl-L-arginine methyl ester and N-acetyl-L-tyrosine ethyl ester) were not cleaved by the enzyme. (1)

References

1. Kominami, E., Kobayashi, K., Kominami, S. & Katunuma, N. (1972) JBC, 247, 6848.

ω-AMIDASE

(ω-Amidodicarboxylate amidohydrolase)

An ω-amido-dicarboxylic acid + H_2O = a dicarboxylate + NH_3

Ref.

Molecular weight

source	value	conditions	Ref.
Rat liver	58,000 [2]	pH 8.0 with or without 7 M guanidine HCl; amino acid composition; peptide mapping	(1)
Bacillus subtilis	120,000	Sephadex G 200, pH 7.2	(2)
Thermus aquaticus	36,000	Sephadex G 200, pH 7.2	(2)

Specific activity

Rat liver (150 x)	11.7	2-oxoglutaramate (pH 8.5, Tris, 30°)	(1)

Specificity and Michaelis constants

In addition to its hydrolytic activity on amides and esters, ω-amidase catalyzes hydroxaminolysis and transamination reactions with esters and amides. For instance, the enzyme catalyzed the incorporation of methylamine into the amide and the monomethyl and the monoethyl esters of 2-oxoglutarate, succinate and glutarate. The enzyme has a broad specificity (see below) but it does not hydrolyze glutamine, asparagine, N-methyl-2-oxoglutaramate, dimethyl glutarate, the anilides of succinate or glutarate, N-methylsuccinamate or N-ethylsuccinamate. (1)

ω-Amidase from B. subtilis (purified 470 x) and T. aquaticus (purified 265 x) have similar catalytic properties to the enzyme from rat liver. The enzyme from B. subtilis, however, hydrolyses asparagine whereas the other enzymes do not. (2)

substrate[(a)]	V relative	K_m (μM)
5-methyl 2-oxoglutarate	1.00	30
5-ethyl 2-oxoglutarate	0.91	60
2-oxoglutaramate	0.70	3300[(b)]
glutaramate	0.19	3100
methyl glutarate	0.16	1800
succinamate	0.07	200
ethyl glutarate	0.06	9900
methyl succinate	0.03	700
ethyl succinate	0.01	8300
propyl glutarate	<0.01	6800
succinyl hydroxamate	<0.01	800

Ref.

Specificity and Michaelis constants continued

(a) in the hydrolysis reaction and with the enzyme from rat liver (pH 8.5, Tris, 30°, Ref. 1)

(b) in solution, 2-oxoglutaramate exists in equilibrium with 5-hydroxypyroglutamate. The true Km for 2-oxoglutaramate is 10 μM.

References

1. Hersh, L.B. (1971) B, 10, 2884.
2. Fernald, N.J. & Ramaley, R.F. (1972) ABB, 153, 95.

PENICILLIN AMIDASE

(Penicillin amidohydrolase)

Penicillin + H_2O = a carboxylic acid anion + penicin

Ref.

Molecular weight

source	value	conditions	
Bacillus megaterium	120,000	pH 7.0	(1)

Specific activity

B. megaterium (96 x)	32	benzylpenicillin (pH 8.7, borate, 37°)	(1)

Specificity and catalytic properties

source	substrate	K_m (mM)	conditions	
Escherichia coli (68 x)	benzylpenicillin[a]	0.8	pH 8.5, Pi, 37°	(2)
B. megaterium	benzylpenicillin[b]	4.5	pH 8.5, borate, 37°	(1)

(a) the products of the reaction were inhibitory: 6-aminopenicillanate (NC; Ki = 5.3 mM) and phenylacetate (C; Ki = 5.1 mM).

(b) the products of the reaction were inhibitory: 6-aminopenicillanate (NC; Ki = 26 mM) and phenylacetate (C; Ki = 0.45 M). Benzylpenicillin (1.00) could be replaced by phenylacetamide (0.23); Δ^2-pentenylpenicillin (0.16); DL-N-phenylacetylalanine (0.15); 3-(α-toluylamido)-propionate (0.14); allylmercaptomethylpenicillin (0.10); ethylthiomethylpenicillin (0.10); N-methylphenylacetamide (0.10); phenoxyethylpenicillin (0.04) and phenoxymethylpenicillin (0.03). The following were inactive: n-heptylpenicillin; 2(phenylthio)ethylpenicillin; dimethoxyphenylpenicillin and 5-methyl-3-phenyl-4-isoxazolyl penicillin.

Penicillin amidase (B. megaterium) does not require divalent metal ions for activity: Co^{2+} and Mn^{2+} were inhibitory. (1)

Abbreviations

penicin	6-aminopenicillanic acid; 6-amino-3,3-dimethyl-7-oxo-4-thia-1-azabicyclo[3.2.0]-heptane-2-carboxylic acid

References

1. Chiang, C. & Bennett, R.E. (1967) J. Bacteriol, 93, 302.
2. Balasingham, K., Warburton, D., Dunnill, P. & Lilly, M.D. (1972) BBA, 276, 250.

N-ACETYL-β-ALANINE DEACETYLASE

(N-Acetyl-β-alanine amidohydrolase)

N-Acetyl-β-alanine + H_2O = acetate + β-alanine

Ref.

Specificity and Michaelis constants

source	substrate	relative velocity	K_m (mM)	conditions	Ref.
Hog kidney (purified 100 x)	N-acetyl-β-alanine[a]	1.0	2.5	pH 7.6, borate-Pi, 37°	(1)
	N-acetyl-taurine	0.4	-		

(a) the N-acetyl derivatives of the following compounds were inactive:- DL-alanine; DL-aspartate; 4-aminobutyrate; DL-3-amino-n-butyrate; DL-3-amino-iso-butyrate; glycyl-glycine and glycyl-β-alanine. D-Pantothate and N-formyl-β-alanine were not hydrolyzed.

The enzyme does not require any metal ions for activity. (1)

References

1. Fujimoto, D., Koyama, T. & Tamiya, N. (1968) BBA, 167, 407.

ACYLSPHINGOSINE DEACYLASE

(N-Acylsphingosine amidohydrolase)

N-Acylsphingosine + H_2O = sphingosine + a fatty acid anion

Ref.

Equilibrium constant

The reaction is reversible. (1)

Specificity and catalytic properties

source	substrate	reaction	K_m (mM)	conditions	
Rat brain (purified 210 x)	N-palmitoyl-sphingosine	F	0.3	pH 4.8, Tris-acetate	(1)
	sphingosine	R	0.2	pH 5, Tris-acetate	(1)
	dihydrosphingosine	R	3.0	pH 5, Tris-acetate	(1)
	palmitate	R	0.04	pH 8, Tris-acetate	(1)

The enzyme hydrolyzes N-acylsphingosines or N-acyldihydrosphingosines in which the fatty acid moiety contains 16 or 18 carbon atoms. It does not catalyze the hydrolysis of N-acetylsphingosine; N-lignocerylsphingosine; cerebrosides; sphingomyelin; sphingosyl galactoside or sphingosylphosphorylcholine. In the reverse reaction, either sphingosine or dihydrosphingosine and fatty acids of 8-24 carbon atoms can be used (the highest rate was obtained with laurate). (1,2)

The enzyme does not require any added cofactor (ATP, coenzyme A or Mg^{2+}) in the forward (pH optimum = 4.8) or reverse reaction (pH optimum = 4.8); it does, however, require cholate. (1,2)

In the forward reaction, sphingosine and fatty acids were inhibitory. (1)

References

1. Gatt, S. & Yavin, E. (1969) Methods in Enzymology, 14, 139.
2. Gatt, S. (1966) JBC, 241, 3724.

CHOLOYLGLYCINE HYDROLASE

(3α,7α,12α-Trihydroxy-5β-cholan-24-oylglycine amidohydrolase)

3α,7α,12α-Trihydroxy-5βcholan-24-oylglycine + H_2O =
3α,7α,12α-trihydroxy-5β-cholanate + glycine

Ref.

Specificity and catalytic properties

source	substrate[(a)]	K_m (mM)	conditions	
Clostridium perfringens (purified 15 x)	glycocholate	3.6	pH 5.8, Pi, 37°	(1)
	glycodeoxycholate	1.2	pH 5.8, Pi, 37°	(1)
	taurochenodeoxycholate	3.0	pH 5.8, Pi, 37°	(1)
	taurodeoxycholate	3.5	pH 5.8, Pi, 37°	(1)
	glycochenodeoxycholate	14.0	pH 5.8, Pi, 37°	(1)
	taurocholate	37.0	pH 5.8, Pi, 37°	(1)

(a) glycodehydrocholate was an inhibitor but not a substrate. Cholate, a product of the hydrolysis of glycocholate, was a competitive inhibitor (Ki = 90 μM).

The enzyme does not have a metal ion requirement. (1)

Abbreviations

choloylglycine or glycocholate	3α,7α,12α-trihydroxy-5β-cholan--24-oylglycine
cholate	3α,7α,12α-trihydroxy-5β-cholanate

For other abbreviations see Ref. 1.

References

1. Nair, P.P., Gordon, M. & Reback, J. (1967) JBC, 242, 7.

ASPARTYLGLUCOSYLAMINASE

(1-L-β-Aspartamido-2-acetamido-1,2-dideoxy-β-D-glucose amidohydrolase)

1-L-β-Aspartamido-2-acetamido-1,2-dideoxy-β-D-glucose + H_2O =
1-amino-2-acetamido-1,2-dideoxy-β-D-glucose + L-aspartate

Ref.

Equilibrium constant

1-Amino-2-acetamido-1,2-dideoxy-β-D-glucose decomposes nonenzymically to 2-acetamido-2-deoxy-β-D-glucose and NH_3.

Molecular weight

source	value	conditions	Ref.
Hog kidney[a]	70,000 [1]	Sephadex G 150, pH 7.0; gel electrophoresis (SDS); amino acid composition.	(2)
Rat kidney	31,000	Sephadex G 200.	(3)
Hen oviduct	101,000-110,000	sucrose density gradient; Sephadex G 200, pH 7.5.	(4)

(a) multiple forms have been observed.

Specific activity

Hog kidney (20,000 x) 1.55 Asn-Glc NAc (pH 5.5, Pi-citrate, 37°) (2)

Specificity and catalytic properties

The amidase is specific for substituted amides of aspartic acid but less specific for the substituent which can vary considerably in size or type. A further requirement is that the amino- and carboxyl-groups of the aspartate moiety are non-substituted. (1,4)

source	substrate	K_m (mM)	conditions	Ref.
Hog kidney	Asn-GlcNAc[a]	0.77	pH 5.5, Pi-citrate, 37°	(2)
	ovalbumin glycopeptide[a]	0.77	pH 5.5, Pi-citrate, 37°	(2)
Hen oviduct (purified 1500 x)	Asn-GlcNAc	0.3	pH 7.2, Pi, 30°	(4)

(a) the Km value obtained depended on the enzyme concentration; the value given was obtained at 5 μg ml^{-1} enzyme.

Ref.

Specificity and catalytic properties continued

The amidase (hog kidney) was inhibited by EDTA, by Cu^{2+}, Ni^{2+}, Zn^{2+} and Mn^{2+} and by aspartate but not by N-acetylglucosamine or NH_3. The mechanism of the reaction has been investigated. (2)

The enzyme (rat kidney, purified 430 x) was neither activated nor inhibited by Ca^{2+}, Mg^{2+}, Co^{2+} or Zn^{2+}. (3)

The enzyme (hen oviduct) was inhibited irreversibly by 5-diazo-4-oxo-L-norleucine (Ki = 9.5 μM). (4)

Aspartylglucosylaminase was not inhibited by N-acetylglucosaminono-(1,5)-lactone. (5)

Abbreviations

Asn-Glc NAc	1-L-β-Aspartamido-2-acetamido-1,2-dideoxy-β-D-glucose

References

1. Yamashina, I. (1972) in Glycoproteins volume 5B (ed. Gottschalk, A.) Elsevier Publishing Co; Amsterdam, p 1187.
2. Kohno, M. & Yamashina, I. (1972) BBA, 258, 600.
3. Mahadevan, S. & Tappel, A.L. (1967) JBC, 242, 4568.
4. Tarentino, A.L. & Maley, F. (1969) ABB, 130, 295.
5. Conchie, J. & Strachan, I. (1969) BJ, 115, 709.

N-ACETYLMURAMOYL-L-ALANINE AMIDASE

(Mucopeptide amidohydrolase)

Hydrolyses the link between N-acetylmuramoyl residues and L-amino acid residues in certain cell wall glycopeptides

Ref.

Molecular weight

source	value	conditions	
Bacillus megaterium	20,000 [1]	sucrose density gradient; gel electrophoresis (SDS)	(1)
Staphylococcus aureus	800,000	sucrose density gradient; Sepharose 4B, pH 6.8	(2)

Specificity and kinetic properties

Amidase from B. megaterium (purified 540 x) removes L-alanine from B. subtilis cell wall from which teichoic acid had been removed: no other amino acid was released. The enzyme does not hydrolyze the products of lysozyme digestion of alkali treated B. subtilis. It requires a certain ion strength and maximum activity was obtained at pH 6.8 with 20 mM KCl or 4 mM $MgCl_2$. It was not inhibited by N-acetylmuramate, DL-Ala-DL-Ala or a combination of these. (1)

References

1. Chan, L. & Glaser, L. (1972) JBC, 247, 5391.
2. Singer, H.J., Wise, E.M. & Park, J.T. (1972) J. Bact, 112, 932.

FORMYLMETHIONINE DEFORMYLASE

(*N*-Formyl-L-methionine amidohydrolase)

N-Formyl-L-methionine + H_2O = formate + L-methionine

Ref.

Specificity and Michaelis constants

source	substrate	K_m (mM)	conditions	
Euglena gracilis (purified)	*N*-formylmethionine[a]	3.8	pH 7.2, Pi, 37°	(1)

(a) the following compounds were inactive: *N*-acetyl-L-methionine; *N*-acetyl-L-alanine, α-*N*-acetyl-L-ornithine or *N*-formyl-L-methionyl-L-alanine. The reaction mixture contained 0.4 mM Co^{2+}.

References

1. Aronson, J.N. & Lugay, J.C. (1969) BBRC, 34, 311.

ACETYLHISTIDINE DEACETYLASE

(*N*-Acetyl-L-histidine amidohydrolase)

N-Acetyl-L-histidine + H_2O = L-histidine + acetate

Ref.

Molecular weight

source	value	conditions	Ref.
Katsuwonus pelamis, brain (skipjack tuna)	83,000	Sephadex G 100, pH 7.4	(1)

Specificity and catalytic properties

The deacetylase has been purified (14 x) from *K. pelamis*. The enzyme is highly specific for *N*-acetylhistidine (1.00) which could be replaced by *N*-acetyl-DL-methionine (0.09) but not by the *N*-acetyl derivatives of the following: histamine; L-tryptophan; L-tyrosine; L-alanine, L-valine; L-glutamate; L-aspartate; L-cysteine; L-ornithine or glycine (pH 7.2, Pi 30°). (1)

References

1. Baslow, M.H. & Lenney, J.F. (1967) Can. J. Biochem. 45, 337.

N-METHYL-2-OXOGLUTARAMATE HYDROLASE

(N-Methyl-2-oxoglutaramate methylamidohydrolase)

N-Methyl-2-oxoglutaramate + H_2O = 2-oxoglutarate + methylamine

Ref.

Equilibrium constant

N-Methyl-2-oxoglutaramate cyclizes non-enzymically to HMPG (Keq = 1 x 10^5 at pH 8 and 30°). In the overall reaction, the synthesis of HMPG is strongly favoured:-

$$\frac{[\text{2-oxoglutarate}]\ [\text{methylamine}]}{[\text{HMPG}]} = 2\ \text{mM (pH 8.0, tricine, 30°)}$$ (1)

Molecular weight

source	value	conditions	
Pseudomonas MA	90,000	Bio-Gel P-150, pH 8. The buffer contained mercaptoethanol, EDTA and 2-oxoglutarate.	(1)

Specificity and catalytic properties

N-Methyl-2-oxoglutaramate hydrolase catalyzes the synthesis and hydrolysis of 5-amides of 2-oxoglutarate. The mechanism of the reaction has been investigated. The enzyme does not require any added cofactor for activity. (1,2,3)

substrate[a]	reaction	V[b]	K_m (mM)	conditions	
2-oxoglutarate[c]	synthesis	25	20	pH 8, tricine, 30°	(1)
methylamine[d]	synthesis	25	59	pH 8, tricine, 30°	(1)
HMPG	hydrolysis	12	50	pH 8, tricine, 30°	(1)
2-oxoglutaramate	hydrolysis[e]	22	3	pH 8, tricine, 30°	(1)

(a) with the enzyme from Pseudomonas MA (purified 710 x).
(b) µmoles min^{-1} mg^{-1} enzyme protein.
(c) the following are neither substrates nor inhibitors: glutarate; glutamate; laevulinate; 2-hydroxyglutarate; 2-oxoleucine or 2-oxonorvaline. Oxalate is an inhibitor but not a substrate.
(d) methylamine could be replaced by n-butylamine; isobutylamine; aniline; propylamine; ethylamine; sec-butylamine or isopropylamine but not by tert-butylamine; dimethylamine; diethylamine or trimethylamine.
(e) to 2-oxoglutarate and NH_3.

Abbreviations

HMPG 5-hydroxy-N-methylpyroglutamate

References

1. Hersh, L.B. (1970) JBC, 245, 3526.
2. Hersh, L.B. (1971) JBC, 246, 6803.
3. Hersh, L.B., Tsai, L. & Stadtman, E.R. (1969) JBC, 244, 4677.

5-AMINOVALERAMIDASE

(5-Aminovaleramide amidohydrolase)

5-Aminovaleramide + H_2O = 5-aminovalerate + NH_3

Ref.

Molecular weight

source	value	conditions	Ref.
Pseudomonas putida	67,000	Sephadex G 100, pH 7.0; sucrose density gradient	(1)

Specificity and catalytic properties

substrate[a]	V relative	K_m (mM)	conditions	Ref.
5-aminovaleramide	1.00	2.0	pH 7.6, Tris, 37°	(1)
4-aminobutyramide	0.81	4.1	pH 7.6, Tris, 37°	(1)
6-aminocaproamide	0.01	12.0	pH 7.6, Tris, 37°	(1)

(a) with the enzyme from *P. putida* (purified 400 x).

The following compounds were inactive as substrates or inhibitors:- acetamide; acetamide + propylamine; *n*-butyramide; *n*-valeramide; glycinamide; *N*-methylamide of 5-aminovalerate; 5-aminovalerate hydrazide; 5-benzyloxycarbonyl-5-aminovalerate, asparagine and glutamine. The enzyme required EDTA and dithioerythritol for activity; all divalent metal ions tested (including Mg^{2+}) were inhibitory. It was not inhibited by sulphydryl alkylating reagents or by diisopropylfluorophosphate. (1)

References

1. Reitz, M.S. & Rodwell, V.W. (1970) JBC, 245, 3091.

ALKYLAMIDASE

(N-Methyl caproamide amidohydrolase)

N-Methyl caproamide + H_2O = caproate + methylamine

Ref.

Molecular weight

source	value	conditions	Ref.
Sheep liver	240,000	Sephadex G 200, pH 7.4	(1)

Specificity and catalytic properties

substrate[a]	$V_{relative}$	K_m (mM)
N-methyl caproamide	1.00	0.66
N-methyl valeramide	2.16	5.0
N-methyl heptylamide	1.65	1.25
N-methyl caprylamide	0.28	3.7
N-ethyl caproamide	1.31	0.83
N-propyl caproamide	2.63	1.8
N-butyl caproamide	0.07	2.2
N-phenyl caproamide	101.00	2.5
N,N-dimethyl caproamide	0.44	0.19

(a) with the enzyme from sheep liver (purified 59 x). The conditions were: pH 7.8, 37°. (Ref. 1).

The enzyme (sheep liver) hydrolyzed N-monosubstituted and N,N-disubstituted amides and there was some activity towards primary amides such as caproamide. It had little or no activity towards short chain substrates: it was inactive towards N-methylformamide; N-methyl acetamide; N-methyl butyramide and also N,N-diethyl caproamide. The enzyme did not require added divalent metal ions for activity. It was inhibited by paraoxon (O,O-diethyl-4-nitrophenyl phosphate) and also by high substrate concentrations. (1)

References

1. Chen, P.R.S. & Dauterman, W.C. (1971) BBA, 250, 216.

CEPHALOSPORINASE

(Cephalosporin amido-β-lactamhydrolase)

Hydrolyses the CO.NH bond in the lactam ring of cephalosporin

Molecular properties

source	value	conditions	Ref.
Enterobacter cloacea	14,000	Sephadex G 50; amino acid composition	(1)
Bacillus cereus	35,000	ultracentrifugation	(2,3)

Specificity and catalytic properties

source	substrate	V(a)	K_m(μM)	Ref.
E. cloacea (highly purified)	cephalosporin C	365	-	(1)
	7-(phenylacetamido) cephalosporinate	93	-	(1)
	cephaloridine	414	-	(1)
	benzylpenicillin	5	53	(1)
	ampicillin	0.6	-	(1)
	phenoxymethyl penicillin	4.5	-	(1)
	cephalothin	59	-	(1)
B. cereus (purified 332 x)	cephalosporin C	∿2200	-	(2)

(a) as specific activities. With the enzyme from E. cloacea the conditions were pH 5.9, 30° and with that from B. cereus, pH 7.0, 30°.

The enzyme (B. cereus) requires Zn^{2+} for activity. (2)

Abbreviations

cephalosporin C — 7-(D-5-amino-5-carboxyvaleramido)-3-(hydroxymethyl)-8-oxo-5-thia-1-azabicyclo[4.2.0]oct-2-ene-2-carboxylic acid acetate

References

1. Hennessey, T.D. & Richmond, M.H. (1968) BJ, 109, 469.
2. Kuwabara, S. (1970) BJ, 118, 457.
3. Lloyd, P.H. & Peacocke, A.R. (1970) BJ, 118, 467.

ALLANTOICASE

(Allantoate amidinohydrolase)

Allantoate + H_2O = (-)-ureidoglycollate + urea

Ref.

Equilibrium constant

Allantoicase catalyzes the reversible conversions of allantoate (diureidoacetate) to (-)-ureidoglycollate and urea (reaction 1) and of (+)-ureidoglycollate to glyoxylate and urea (reaction 2). (1,2,3)

Molecular properties

Purified allantoicase (*Pseudomonas aeruginosa*) consists of 75% protein of molecular weight 154,000 daltons, 15% of 11,000 daltons and 10% of smaller particles and larger aggregates. The forms of molecular weights 154,000 and 11,000 are readily interconvertible and they have very similar specific activities towards allantoate. The amino acid composition of the enzyme, which is a glycoprotein, has been determined. (1)

Allantoicase contains tightly bound Mn^{2+} (1 gm atom per 11,400 daltons) which is essential for enzymic activity. The Mn^{2+} could be replaced by 13 other divalent metal ions tested but not all of the metallo-enzymes obtained were as active as the Mn^{2+}-enzyme. The dissociation constant of the Mn^{2+} -enzyme complex is 30μM. (1,2)

Specific activity

				Ref.
P. aeruginosa	(70 x)	590	allantoate (pH 7.9, triethanolamine, 30°)	(1)

Specificity and catalytic properties

substrate or inhibitor	reaction[(a)]	K_m (mM)	K_i (mM)	Ref.
allantoate	1	25	-	(3)
allantoate	2	-	22	(3)
(±)-ureidoglycollate	2	26	-	(3)
glyoxylate	2	32	-	(3)
glyoxylate	1	-	30	(3)

(a) see Equilibrium constant section. The enzyme used was that from *P. aeruginosa* and the assay conditions were: pH 7.9, ethanolamine, 30°.

The enzyme (*P. aeruginosa*) is highly specific for its substrates. Thus, neither urea nor NH_3 was released from the *N*-carbamoyl derivatives of glycine; L- or D-alanine; L-aspartate; L- or D-asparagine or D-glutamine or from hydroxyurea, oxonate, hydroxonate or oxamate.

Ref.

Specificity and catalytic properties continued

In the synthetic reaction, no disappearance of urea or NH_3 occurred in the presence of glycollate; oxalate; pyruvate; L-lactate; tartrate; malate or DL-threonine. Allantoicase did not catalyze a reaction between glyoxylate and guanidine; formylurea; pyruvate; 2-oxoglutarate; succinate; malate or tartrate. A number (150) of substances have been tested as inhibitors of allantoicase. These included carboxylic, hydroxy, D-amino and N-carbamoyl-D-amino acids of which several were inhibitory. The mechanism of the reaction has been investigated. (2,3)

Allantoicase has also been purified from *Candida utilis* (purified 97 x) and its catalytic properties have been studied. (4)

References

1. 'S-Gravenmade, E.J., van der Drift, C. & Vogels, G.D. (1971) BBA, 251, 393.
2. van der Drift, C. & Vogels, G.D. (1970) BBA, 198, 339.
3. Vogels, G.D. (1969) BBA, 185, 186.
4. Choi, K.S., Lee, K.W., Hico, S.C.Y. & Roush, A.H. (1968) ABB, 126, 261.

GUANIDINOBUTYRASE

(4-Guanidinobutyrate amidinohydrolase)

4-Guanidinobutyrate + H_2O = 4-aminobutyrate + urea

Ref.

Molecular weight

source	value	conditions	
Pseudomonas putida	184,000	sucrose density gradient; Sephadex G 200, pH 8.0	(1)
Lizard liver	252,000	sucrose density gradient	(2)

Specificity and catalytic properties

source	substrate	V[a]	K_m (mM)	
P. putida (purified 68 x)	4-guanidinobutyrate[b]	606	32	(1)
	5-guanidinovalerate	27	206	(1)
	6-guanidinocaproate	11	163	(1)

(a) specific activity (pH 10, glycine, 30°)

(b) 4-guanidinobutyrate could not be replaced by guanidinoacetate; 3-guanidinopropionate; L-2-amino-4-guanidinobutyrate; L- or D-arginine; L-homoarginine or agmatine. Of these compounds only agmatine was inhibitory.

Guanidinobutyrase requires Mn^{2+} for activity. (1,2)

Guanidinobutyrase from Streptomyces griseus could not utilize any of the following compounds: L- or D-arginine; 4-guanidinobutyramide; guanidinoacetate; 3-guanidinopropionate; 5-guanidinovalerate; agmatine; guanidinobutane; streptomycin or streptidine. (3)

References

1. Chou, C-S., & Rodwell, V.W. (1972) JBC, 247, 4486.
2. Mora, J., Tarrab, R., Martuscelli, J. & Sóberon, G. (1965) BJ, 96, 588.
3. van Thoai, N., Thome - Beau, F. & Olomucki, A. (1966) BBA, 115, 73.

FORMIMINOGLUTAMASE

(N-Formimino-L-glutamate formiminohydrolase)

N-Formimino-L-glutamate + H_2O = L-glutamate + formamide

Ref.

Molecular weight

source	value	conditions	Ref.
Bacillus subtilis	220,000	pH 7.4. The buffer contained Mn^{2+}	(1)

Specificity and catalytic properties

source	substrate	V(a)	K_m(M)	conditions	Ref.
B. subtilis (purified 140 x)	N-formiminoglutamate	4400	0.53	pH 8.7, Tris, 37°	(1)
	N-formiminoglutamate	460	0.039	pH 7.4, Tris, 37°	(1)
Aerobacter aerogenes (purified 40 x)	N-formiminoglutamate	-	0.40	pH 9.0, Tris, 37°	(2)

(a) specific activity

The enzyme (B. subtilis) has a requirement for Mn^{2+} (1.00) which could be replaced by Co^{2+} (0.20); Mg^{2+} (0.18); Zn^{2+} (0.15); Cd^{2+} (0.08) or Ni^{2+} (0.07). Ba^{2+}, Ca^{2+}, Cu^{2+} and Fe^{2+} exhibited less than 5% the activity of Mn^{2+}. The enzyme from A. aerogenes also required Mn^{2+} which could not be replaced by any other metal ion tested. (1,2)

The following compounds were inhibitory: high concentrations of N-formiminoglutamate, L-glutamate and barbiturate and borate buffers. Histidine, tetrahydrofolate and formamide did not inhibit. (1)

References

1. Kaminskas, E., Kimhi, Y. & Magasanik, B. (1970) JBC, 245, 3536.
2. Lund, P. & Magasanik, B. (1965) JBC, 240, 4316.

AMIDINO-ASPARTATE HYDROLASE

(N-Amidino-L-aspartate amidinohydrolase)

N-Amidino-L-aspartate + H_2O = L-aspartate + urea

Ref.

Molecular weight

source	value	conditions	
Pseudomonas chlororaphis	300,000	Sephadex G 200	(1)

Specificity and catalytic properties

source	substrate	relative velocity	K_m (mM)	conditions	
P. chlororaphis (purified 21 x)	N-amidino-aspartate[(a)]	1.00	8.3	pH 8.0, HEPES	(1)
	N-amidino-glutamate	0.04	-	pH 8.0, HEPES	(1)

(a) N-amidino-aspartate (L-guanidinosuccinate) could not be replaced by D-guanidinosuccinate; 3-guanidinopropionate; 4-guanidinobutyrate; 2-amino-4-guanidinopropionate; agmatine; arginine; N-acetylarginine; creatinine or guanidinoacetate.

The enzyme (P. chlororaphis) requires Co^{2+} for activity. The following were ineffective as activators: Zn^{2+}; Cu^{2+} (both were strong inhibitors); Mn^{2+}; Fe^{2+} (both were less effective inhibitors) and Ca^{2+} (ineffective as inhibitor). (1)

The enzyme was inhibited by urea (C; Ki = 0.25 mM) and aspartate (C; Ki = 0.25 mM). (1)

References

1. Milstien, S. & Goldman, P. (1972) JBC, 247, 6280.

FORMIMINOGLUTAMATE DEIMINASE

(N-Formimino-L-glutamate iminohydrolase)

N-Formimino-L-glutamate + H_2O = N-formyl-L-glutamate + NH_3

Ref.

Molecular weight

source	value	conditions	Ref.
Pseudomonas sp.	100,000 [2]	ultracentrifugation; gel electrophoresis (SDS)	(1)

Specific activity

			Ref.
Pseudomonas sp. (750 x)	12.8	N-formiminoglutamate (pH 7.2, Pi, 37°)	(1)

References

1. Wickner, R.B. & Tabor, H. (1971) Methods in Enzymology, 17B, 80.

dCTP DEAMINASE

(dCTP aminohydrolase)

$$dCTP + H_2O = dUTP + NH_3$$

Ref.

dCTP deaminase has been purified (10 x) from *Bacillus subtilis* infected with phage PBS 1. The enzyme is specific for dCTP which could not be replaced by dCDP, dCMP, deoxycytidine or CTP. It requires Mn^{2+} (1.0) for full activity; Mg^{2+} (0.9) and Ca^{2+} (0.7) were also effective as activators. The deaminase was inhibited by dTTP (NC(dCTP)) and to a less extent by dTDP and dTMP but the following did not inhibit: deoxythymidine; dUDP; dUMP and deoxyuridine. (1)

References

1. Tomita, F. & Takahashi, I. (1969) BBA, 179, 18.

GUANOSINE DEAMINASE

(Guanosine aminohydrolase)

Guanosine + H_2O = xanthosine + NH_3

Ref.

Molecular weight

source	value	conditions	Ref.
Pseudomonas convexa	~150,000	Sephadex G 100, pH 7.4	(1)

Specificity and catalytic properties

source	substrate	K_m (μM)	conditions	Ref.
P. convexa (purified 180 x)	guanosine[a]	36	pH 7.0, Pi, 37°	(1)
	deoxyguanosine	62	pH 7.0, Pi, 37°	(1)
	8-azaguanosine	122	pH 7.0, Pi, 37°	(1)

(a) the following were neither substrates nor inhibitors: cytidine; adenosine; guanine; 5'- and 3'- guanylate; 5'-deoxyguanylate and a number of other analogous compounds. 4-Amino-5-imidazole-carboxamide ribonucleoside was an inhibitor.

The enzyme did not require any metal ions for activity. It was not inhibited by EDTA. (1)

Light absorption data

The transformation of guanosine into xanthosine is accompanied by a decrease in absorption at 260 nm (molar extinction coefficient = 3,900 M^{-1} cm^{-1}). (1)

References

1. Ishida, Y., Shirafuji, H., Kida, M. & Yoneda, M. (1969) Agr. Biol. Chem. (Tokyo), 33, 384.

GTP CYCLOHYDROLASE

(GTP 7,8-8,9-dihydrolase)

GTP + H_2O = formate + 2-amino-4-hydroxy-6-(erythro-1',2',3'-trihydroxypropyl)-dihydropteridine triphosphate

Ref.

Molecular weight

source	value	conditions	
Escherichia coli	>300,000	Sephadex G 200	(1)

Specificity and catalytic properties

The GTP cyclohydrolase reaction involves 4 steps all of which may be catalyzed by one protein molecule. The mechanism of the reaction has been investigated. (1,2)

source	substrate	K_m (μM)	conditions	
E. coli (purified 700 x)	GTP(a)	22	pH 8.5, Tris, 42°	(1,2)

(a) GTP (1.00) could be replaced by 7-methyl-GTP (0.08) but not by GDP; GMP; guanosine; guanine or any other purine nucleotide.

The cyclohydrolase does not require any added cofactor (tetrahydrofolate, NAD, NADP, CoA) or metal activator (Mg^{2+}, Mn^{2+}, Fe^{2+}, Fe^{3+} or Cu^{2+}) for activity. All of the metal ions tested (except for Mg^{2+}) and ascorbate were inhibitory. (1)

References

1. Burg, A.W. & Brown, G.M. (1968) JBC, 243, 2349.
2. Wolf, W.A. & Brown, G.M. (1969) BBA, 192, 468.

ADENINE NUCLEOTIDE DEAMINASE

(Adenosine (phosphate) aminohydrolase)

$$5'\text{-AMP} + H_2O = 5'\text{-IMP} + NH_3$$

Ref.

Equilibrium constant

The reaction was irreversible [F] with the enzyme from *Desulfovibrio desulfuricans* or *Microsporum audouini* as catalyst. (2,3)

Molecular weight

source	value	conditions	Ref.
D. desulfuricans	∿45,000	Bio-Rad chromatography	(2)

Specificity and catalytic properties

source	substrate	V relative	K_m(µM)	conditions	Ref.
D. desulfuricans (purified 800 x)	ATP[a]	1.00	285	pH 5.8, Pi	(2)
	ADP	0.89	300	pH 5.8, Pi	(2)
	AMP	0.97	250	pH 5.8, Pi	(2)
	NAD	2.11	59,000	pH 5.8, Pi	(2)
M. audouini (purified 4000 x)	ATP[b]	1.00	33	pH 5.0, acetate, 25°	(1,3)
	ADP	∿0.8	47	pH 5.0, acetate, 25°	(1,3)
	AMP	0.84	83	pH 5.0, acetate, 25°	(1,3)
Porphyra crispata (red alga; purified 1240 x)	ATP[c]	1.00	66	pH 6.0, Pi, 22°	(4)
	ADP	1.04	47	pH 6.0, Pi, 22°	(4)
	AMP	1.45	47	pH 6.8, Pi, 22°	(4)
	NAD	0.71	72	pH 5.6, Pi, 22°	(4)

(a) ATP (V relative = 1.00) could also be replaced by dATP (0.09); dADP (0.09); dAMP (0.08); adenosine (0.07); deoxyadenosine (0.05); ADP-ribose (2.75) and by adenosine 5'-phosphosulphate, CoA and 3',5'-cyclic AMP. The following were inactive: adenine; 2'-AMP; 3'-AMP; 2',3'-cyclic AMP and NADP. The enzyme was not activated by metal ions and EDTA did not inhibit.

(b) ATP (1.00) could also be replaced by dATP (0.30); dADP (0.20); dAMP (0.08); adenosine (0.07); deoxyadenosine (0.01); ATPP (0.91); ADP-ribose (0.50); ADP-glucose (0.36); 3'-AMP (0.06); 3',5'-cyclic AMP (0.10); NAD (0.10); FAD (0.11) and CoA (0.12) but not by 2'-AMP, adenine, NADP, GTP or CMP. There was no evidence for a metal ion requirement.

(c) ATP (V relative = 1.00) could also be replaced by adenosine (0.41) but not by 2'-AMP, 3'-AMP, NADP or adenine. This enzyme required Ca^{2+} (Mg^{2+} and Ba^{2+} were less effective) for the hydrolysis of ATP, ADP, AMP and NAD but not of adenosine.

Ref.

Specificity and catalytic properties continued

The deaminase (D. desulfuricans) was inhibited by a number of substances including urea, sulphydryl reagents, citrate, maleate, succinate, Pi, PPi and several metal ions. The enzyme from M. audouini was also inhibited by metal ions and sulphydryl reagents. (2,3)

Light absorption data

The deamination of AMP, ADP or ATP is accompanied by a decrease in absorption at 265 nm (molar extinction coefficient = 7250 $M^{-1}cm^{-1}$). (5)

References

1. Zielke, C.L. & Suelter, C.H. (1971) The Enzymes, 4, 75.
2. Yates, M.G. (1969) BBA, 171, 299.
3. Chung, S-T. & Aida, K. (1967) J. Biochem. (Tokyo), 61, 1.
4. Su, J-C., Li, C-C. & Ting, C.C. (1966) B, 5, 536.
5. Kalckar, H.M. (1947) JBC, 167, 461.

PHOSPHORIBOSYL-AMP CYCLOHYDROLASE

(1-N-(5'-Phospho-D-ribosyl)-AMP 1,6-hydrolase)

1-N-(5'-phospho-D-ribosyl)-AMP + H_2O =

5-(5'-phospho-D-ribosyl-aminoformimino)-1-(5''-phosphoribosyl)-imidazole-4-carboxamide

Ref.

Molecular and catalytic properties

Phosphoribosyl-AMP cyclohydrolase from *Neurospora crassa* (purified ∿1000 x) is a trifunctional enzyme which catalyzes three steps (reactions 2,3 and 10) of the histidine pathway:-

[2] 1-N-(5'-phosphoribosyl)-ATP + H_2O =
1-N-(5'-phosphoribosyl)-AMP + PPi

[3] EC 3.5.4.19 (see above)

[10] EC 1.1.1.23 (histidinol dehydrogenase)

With 1-N-(5'-phosphoribosyl)-ATP as the substrate a specific activity of 8.6 was obtained (pH 8.6, Tris, 37°) and with L-histidinol, 12.4 (pH 9.8, glycine, 37°). All the three activities reside on one enzyme protein of molecular weight 126,000. The amino acid composition of the protein has been determined. (1)

References

1. Minson, A.C. & Creaser, E.H. (1969) BJ, 114, 49.

Δ'-PYRROLINE 4-HYDROXY 2-CARBOXYLATE DEAMINASE

(Δ'-Pyrroline-4-hydroxy-2-carboxylate aminohydrolase (decyclizing))

Δ'-Pyrroline 4-hydroxy 2-carboxylate + H_2O = 2,5-dioxovalerate + NH_3

Ref.

Equilibrium constant

The reaction is irreversible [F] with the enzyme from *Pseudomonas striata* as catalyst. (1)

Molecular weight

source	value	conditions	
P. striata	62,000	ultracentrifugation	(1)

Specific activity

P. striata (60 x) 4.8 HPC (pH 7.0, Pi, 25°) (1)

Specificity and catalytic properties

source	substrate	K_m (mM)	conditions	
P. striata	HPC [a]	1.4	pH 7.0, Pi, 25°	(1)

(a) HPC (1.00) could be replaced by the D-isomer (0.02) but not by Δ'-pyrroline-3-hydroxy-2-carboxylate; Δ'-pyrroline-3-hydroxy-5-carboxylate; Δ'-pyrroline-2-carboxylate; 2-oxo-4,5-dihydroxyvalerate or DL-3-hydroxy-4-aminobutyrate.

The deaminase does not require any added cofactor for activity. It was not inhibited by EDTA or by pyridoxal enzyme or cobamide enzyme inhibitors. (1)

Abbreviations

HPC	Δ'-pyrroline 4-hydroxy 2-carboxylate (L-isomer).
2,5 dioxo-valerate	2-oxoglutarate semialdehyde

References

1. Singh, R.M.M. & Adams, E. (1965) JBC, 240, 4344, 4352.

THIAMINASE

(Thiamine hydrolase)

Thiamine + H_2O = 2-methyl-4-amino-5-hydroxymethyl-pyrimidine + 4-methyl-5-(2'-hydroxyethyl)-thiazole

Molecular weight

source	value	Ref.
Bacillus aneurinolyticus	100,000	(1)

Specific activity

B. aneurinolyticus (25 x)	4[a]	thiamine (pH 8.6, 37°)	(1)

(a) this is a maximum value.

Specificity and Michaelis constants

source	substrate	K_m (μM)	conditions	
B. aneurinolyticus	thiamine	3.0	pH 8.6, 37°	(1)

Thiaminase hydrolyzes thiamine and thiamine derivatives with the side chain intact at the 5- position of the thiazole moiety. It will hydrolyze thiothiamine but not thiamine pyrophosphate. The enzyme requires cysteine and EDTA for full activity. (1)

References

1. Wittliff, J.L. & Airth, R.L. (1970) Methods in Enzymology, 18A, 234.

ADENOSINETETRAPHOSPHATASE

(Adenosinetetraphosphate phosphohydrolase)

Adenosine 5'-tetraphosphate + H_2O = ATP + Pi

Ref.

Specificity and catalytic properties

source	substrate	V relative	K_m (μM)
Rabbit muscle (purified 1000 x)	adenosine tetraphosphate[a]	1.0	27
	inosine tetraphosphate	0.7	34

(a) adenosine tetraphosphate (1.00) could also be replaced by PPPi (0.39) but not by PPi or by the common nucleotides.

The enzyme exhibits an absolute requirement for a divalent metal ion. With adenosine tetraphosphate, Co^{2+} was the most effective activator and with PPPi, Ni^{2+}. (1)

References

1. Small, G.D. & Cooper, C. (1966) B, 5, 14.

NUCLEOSIDETRIPHOSPHATE PYROPHOSPHATASE

(Nucleosidetriphosphate pyrophosphohydrolase)

A nucleoside triphosphate + H_2O = a nucleotide + PPi

Ref.

Molecular properties

source	value	conditions	
Rabbit reticulocytes	37,000	sucrose density gradient	(1)

Specific activity

Rabbit reticulocytes (2300 x) 26.5 ITP (pH 9.5, β-alanine, 37°) (1)

Specificity and kinetic properties

The enzyme hydrolyzed ITP (1.00); dITP (1.03); XTP (0.71); dUTP (0.13); UTP (0.12); GTP (0.10); dGTP (0.06); dTTP (0.03); CTP (0.005); dCTP (0.005) and dATP (0.002). ATP, IDP, IMP, glucose 6-phosphate; ribose 5-phosphate; glycerol 3-phosphate; 2,3-diphosphoglycerate and 4-nitrophenol phosphate were resistant to hydrolysis. (1)

The enzyme requires Mg^{2+} for activity. Mn^{2+} was less effective as activator and Ca^{2+}, Fe^{2+}, Zn^{2+} and Cu^{2+} were inactive in the place of Mg^{2+}. All monovalent cations (e.g. Na^+, K^+, NH_4^+) tested were inhibitory. (1)

Several nucleotide derivatives were inhibitory and of these IDP was the most potent. High substrate concentrations were also inhibitory. (1)

References

1. Chern, C.J., MacDonald, A.B. & Morris, A.J. (1969) JBC, 244, 5489.

NUCLEOSIDE PHOSPHOACYLHYDROLASE

(Nucleoside-5'-phosphoacylate acylhydrolase)

Hydrolyses mixed phospho-anhydride bonds

Ref.

Molecular weight

source	value	conditions	
Escherichia coli	23,000	ultracentrifugation; Sephadex G 50, pH 7.5; amino acid composition	(1)

Specificity and catalytic properties

source	substrate(a)	V(b)	K_m (mM)
E. coli (purified 1700 x)	adenosine 5'-NP	750	0.056
	guanosine 5'-NP	980	0.11
	uridine 5'-NP	760	0.65
	cytidine 5'-NP	1235	4.0
	thymidine 5'-NP	32	770

(a) NP = 4-nitrophenyl phosphate. The conditions were pH 7.5, Tris, 30°.

(b) specific activity.

Nucleoside phosphoacylhydrolase cleaves the mixed phosphoanhydride type of bond and it is without activity on phosphomono esters or diesters, on pyrophosphates or on mixed anhydrides not involving a phosphoryl residue as one of the partners. In addition to being a mixed phosphoanhydride, an active substrate must have its phosphoryl group linked to a nucleoside: acetylphosphate and di-4-nitrophenyl phosphate were inactive. The nucleosides and nucleotides were therefore inactive as substrates but several were competitive inhibitors of uridine 5'-NP utilization, for example:- alaninyl adenylate (Ki = 0.22 mM); lysinyl adenylate (Ki = 0.34 mM); AMP (Ki = 0.60 mM); GMP (Ki = 1.3 mM); UMP (Ki = 8.5 mM) and CMP (Ki = 150 mM). (1)

The enzyme does not require the addition of a cofactor or metal ion. It was not inhibited by EDTA. (1)

Light absorption data

4-Nitrophenol (formed by the action of the hydrolase on uridine-5'-4-nitrophenol phosphate) has an absorption band at 400 nm (molar extinction coefficient = 18,200 $M^{-1}cm^{-1}$ in strong alkali). (1)

References

1. Spahr, P.F. & Gesteland, R.F. (1970) EJB, 12, 270.

ADENYLYLSULPHATASE

(Adenylylsulphate sulphohydrolase)

Adenylylsulphate + H_2O = AMP + sulphate

Ref.

Molecular weight

source	value	conditions	Ref.
Bovine liver	68,500 [1]	Sephadex G 100, pH 7.2; gel electrophoresis (SDS); amino acid composition.	(1)

Specific activity

Bovine liver (1200 x) 6 APS (pH 5.2, acetate, 37°) (1)

Specificity and catalytic properties

source	substrate	K_m (mM)	conditions	Ref.
Rat liver (purified 137 x)	APS[(a)]	1	pH 5.2, acetate, 37°	(2,3)

(a) ATP was inactive

The enzyme (bovine liver) did not hydrolyze ATP, ADP or 3'-phosphoadenosine 5'-phosphosulphate. (1)

The enzyme (rat liver) was inhibited by Co^{2+}, Pi, PPi, ADP, ATP, GTP and ITP. AMP and sulphate were poor inhibitors. (2,3)

Abbreviations

APS adenylylsulphate; adenosine 5'-phosphosulphate

References

1. Stokes, A.M., Denner, W.H.B., Rose, F.A. & Dodgson, K.S. (1973) BBA, 302, 64.
2. Bailey-Wood, R., Dodgson, K.S. & Rose, F.A. (1970) BBA, 220, 284.
3. Bailey-Wood, R., Dodgson, K.S. & Rose, F.A. (1969) BJ, 112, 257.

PHOSPHONATASE

(2-Phosphonoacetaldehyde phosphohydrolase)

2-Phosphonoacetaldehyde + H_2O = acetaldehyde + Pi

Ref.

Molecular weight

source	value	conditions	Ref.
Bacillus cereus	75,000 [2]	Sephadex G 150; gel electrophoresis; amino acid composition; peptide mapping	(1)

Mg^{2+}, which is required for activity, maintains the enzyme in its dimer form. Mg^{2+} could be replaced to a certain extent by Mn^{2+} but Zn^{2+} and Ca^{2+} were antagonistic. The enzyme had no metal content. An orthophosphite-enzyme complex has been prepared but orthophosphite does not bind covalently to the phosphonatase. (1)

Specific activity

B. cereus (200 x) 13.9 2-phosphonoacetaldehyde (pH 8.5, ammediol, 25°) (1)

Specificity and catalytic properties

source	substrate	K_m (μM)	conditions	Ref.
B. cereus	2-phosphonoacetaldehyde(a)	40	pH 8.5, ammediol, 25°	(1)

(a) 2-phosphonoacetaldehyde (1.00) could be replaced by 4-nitrophenylphosphate (0.03) but not by α- or β-glycerophosphate; ethanolamine phosphate; 2-aminoethylphosphonate; 2-aminomethylphosphonate; 2-hydroxy 5-nitrobenzylphosphonate or phosphonomycin.

The phosphonatase was inhibited by orthophosphite in the presence of 2-phosphonoacetaldehyde or acetaldehyde. (1)

References

1. La Nauze, J.M., Rosenberg, H. & Shaw, D.C. (1970) BBA, 212, 332.

ACETOLACTATE DECARBOXYLASE

(2-Hydroxy-2-methyl-3-oxobutyrate carboxy-lyase)

(+)-2-Hydroxy-2-methyl-3-oxobutyrate = (-)-2-acetoin + CO_2

Ref.

Specific activity

Aerobacter aerogenes (145 x)	72	acetolactate (pH 6.2, Pi, 37°)	(1)

Specificity and catalytic properties

source	substrate	V relative	K_m (mM)	conditions	
A. aerogenes	D-2-acetolactate[a]	1.0	3.4	pH 6.5, Pi, 37°	(1,2)
	D-2-acetohydroxybutyrate	1.0	10	pH 6.5, Pi, 37°	(1,2)

(a) acetoacetate and oxaloacetate were inactive.

The enzyme does not require divalent metal ions for activity. There was no inhibition by EDTA. (1)

Abbreviations

acetolactate (+)-2-hydroxy-2-methyl-3-oxobutyrate
acetoin acetylmethylcarbinol

References

1. Løken, J.P. & Størmer, F.C. (1970) EJB, 14, 133.
2. Juni, E. (1952) JBC, 195, 715.

OXALYL-CoA DECARBOXYLASE

(Oxalyl-CoA carboxy-lyase)

Oxalyl-CoA = formyl-CoA + CO_2

Ref.

Equilibrium constant

The reaction is irreversible [F] with the enzyme from *Pseudomonas oxalaticus* as the catalyst. (1,2)

Specificity and Michaelis constants

source	substrate	K_m (mM)	conditions	
P. oxalaticus (purified 10 x)	oxalyl-CoA[(a)]	1	pH 6.5, Pi, 30°	(1,2)

(a) oxalate, malonate, succinate, malonyl-CoA and succinyl-CoA were inactive.

The enzyme (*P. oxalaticus*) requires thiamine pyrophosphate and possibly Mg^{2+} or Mn^{2+} for activity. (1,2)

Light absorption data

Formyl-CoA has an absorption band at 235 nm (molar extinction coefficient = 4100 $M^{-1}cm^{-1}$). (2)

References

1. Quayle, J.R. (1969) *Methods in Enzymology*, 13, 369.
2. Quayle, J.R. (1963) *BJ*, 89, 492.

ORNITHINE DECARBOXYLASE

(L-Ornithine carboxy-lyase)

L-Ornithine = putrescine + CO_2

Ref.

Molecular properties

source	value	conditions	Ref.
Rat prostate	65,000-85,000[a]	sucrose density gradient; Bio-Gel P-200, pH 7.5	(1)
Rat liver	100,000	Sephadex G 150, pH 7.0. The buffer contained pyridoxal phosphate, dithiothreitol and EDTA.	(2)

(a) in the presence of dithiothreitol; in its absence the enzyme dimerized and became inactive.

Specific activity

Rat liver (5400 x) 0.2 L-ornithine (pH 7.0, Pi, 37°) (2)

Specificity and catalytic properties

source	substrate	K_m (mM)	conditions	Ref.
Rat prostate (purified 290 x)	L-ornithine[a]	0.1	pH 7.2, Gly-Gly, 37°	(1)
Rat liver	L-ornithine[b]	0.2	pH 7.0, Pi, 37°	(2)
Escherichia coli (purified 30 x)	L-ornithine	3.9	pH 7.6, HEPES, 37°	(4)

(a) D-ornithine was neither a substrate nor an inhibitor (Ref. 3).
(b) D-ornithine was a weak inhibitor but not a substrate.

Ornithine decarboxylase has an absolute requirement for pyridoxal phosphate and a thiol compound was required for maximum activity. The enzyme did not require divalent metal ions; thus that from rat prostate was not inhibited by EDTA. (1,2,3,4)

Ornithine decarboxylases isolated from different sources differ in their response to polyamines. The enzyme from *E. coli* was subject to strong feed-back inhibition (of a complex type) by putrescine. Spermine and spermidine were also potent inhibitors of this enzyme. The rat prostate enzyme was inhibited by putrescine (C(L-ornithine); Ki = 1 mM) but spermidine and spermine were poor inhibitors. With the rat liver enzyme, putrescine was a poor inhibitor and spermidine, spermine, cadaverine and histamine were only slightly inhibitory. (4,1,3,2)

References

1. Jänne, J. & Williams-Ashman, H.G. (1971) JBC, 246, 1725.
2. Ono, M., Inoue, H., Suzuki, F. & Takeda, Y. (1972) BBA, 284, 285.
3. Pegg, A.E. & Williams-Ashman, H.G. (1968) BJ, 108, 533.
4. Morris, D.R., Wu, W.H., Applebaum, D. & Koffron, K.L. (1970) Ann. New York Acad. Sci., 171, 968.

PHOSPHOPANTOTHENOYL-CYSTEINE DECARBOXYLASE

(4'-Phospho-N-(L-pantothenoyl)-L-cysteine carboxy-lyase)

4'-Phospho-N-(L-pantothenoyl)-L-cysteine =

pantotheine 4'-phosphate + CO_2

Ref.

Specificity and catalytic properties

source	substrate	K_m (mM)	conditions	Ref.
Rat liver (purified 112 x)	4'-phospho-N-(L-pantothenoyl)-L-cysteine[a]	0.14	pH 8.0, Tris, 37°	(1,2)

(a) dephospho-α-carboxy-CoA and α-carboxy-CoA were decarboxylated at low rates but pantothenoyl-L-cysteine was neither a substrate nor an inhibitor.

The decarboxylase required cysteine or reduced glutathione for full activity but it did not require the addition of any cofactor or divalent metal ion (e.g. Mg^{2+}). (1,2)

The reaction product, pantotheine 4'-phosphate, was an inhibitor [C(4'-phospho-N-(L-pantothenoyl)-L-cysteine) Ki = 0.43 mM]. The following compounds were also inhibitory: ATP, UTP, pyridoxal phosphate and pyridoxal. ADP and AMP did not inhibit. (1,2)

References

1. Abiko, Y. (1970) Methods in Enzymology, 18A, 354.
2. Abiko, Y. (1967) J. Biochem (Tokyo), 61, 300.

METHYLMALONYL-CoA DECARBOXYLASE

(S-Methylmalonyl-CoA carboxy-lyase)

S-Methylmalonyl-CoA = propionyl-CoA + CO_2

Ref.

Equilibrium constant

The reaction is essentially irreversible [F] with the enzyme from Micrococcus lactilyticus as catalyst. (1)

Molecular properties

source	value	conditions	
M. lactilyticus	275,000-300,000	sucrose density gradient	(1)

The preparation obtained contained 0.1 mole biotin per 300,000 daltons. The enzyme was inhibited by avidin. (1)

Specificity and catalytic properties

source	substrate	K_m (μM)	conditions	
M. lactilyticus (purified 5 x)	S-methylmalonyl-CoA[a]	1	pH 7.0, Pi, 23°	(1)

(a) S-methylmalonyl-CoA could not be replaced by R-methylmalonyl-CoA, succinyl-CoA or malonyl-CoA.

The decarboxylase did not require any cofactor or metal ion for activity. It was not inhibited by EDTA. (1)

source	inhibitor[a]	K_i (μM)	
M. lactilyticus	methylmalonyl-CoA	80	(1)
	malonyl-CoA	100	(1)
	succinyl-CoA	300	(1)
	butyryl-CoA	500	(1)
	glutaryl-CoA	3000	(1)
	reduced CoA	4000	(1)

(a) NC(methylmalonyl-CoA). The conditions were: pH 7.0, Pi, 23°. Succinate and malonate did not inhibit.

References

1. Galivan, J.H. & Allen, S.H.G. (1968) ABB, 126, 838.

CARNITINE DECARBOXYLASE

(Carnitine carboxy-lyase)

Carnitine = 2-methylcholine + CO_2

Ref.

Equilibrium constant

The reaction is essentially irreversible [F] with the enzyme from rat heart as catalyst. (1)

Specificity and catalytic properties

source	substrate	K_m (mM)	conditions	
Rat heart mitochondria (purified 30 x)	DL-carnitine[(a)]	0.24	pH 7.4, Pi, 37°	(1)

(a) the enzyme appears to be specific for the l isomer of carnitine. Palmitoylcarnitine and acetylcarnitine had about 30% the activity of DL-carnitine.

The enzyme requires ATP and Mg^{2+} for activity and concomitant with the decarboxylation of carnitine, ADP and Pi are formed. NAD, biotin and pyridoxal phosphate are not required by this enzyme. (1)

References

1. Khairallah, E.A. & Wolf, G. (1967) JBC, 242, 32.

PHENYLPYRUVATE DECARBOXYLASE

(Phenylpyruvate carboxy-lyase)

Phenylpyruvate = phenylacetaldehyde + CO_2

Ref.

Specificity and catalytic properties

source	substrate	K_m (μM)	conditions
Achromobacter eurydice (purified 26 x)	phenylpyruvate[a]	51	pH 7.3, Pi, 30° (1,2)
	thiamine pyrophosphate	1.5	pH 7.3, Pi, 30° (1,2)

(a) phenylpyruvate (1.00) could be replaced by indolepyruvate (0.48); 2-oxo-4-ethio-butyrate (0.30); 2-oxocaproate (0.31); 2-oxo-4-methio-butyrate (0.23) and 2-oxo-isocaproate (0.05) but not by the following: 4-hydroxyphenylpyruvate; 2-oxo-3-methylvalerate; 2-oxo-isovalerate; 2-oxoglutarate; hydroxypyruvate or pyruvate.

The enzyme requires thiamine pyrophosphate and a divalent metal ion (Mg^{2+} (1.00); Mn^{2+} (0.55); Co^{2+} (0.45) or Ni^{2+} (0.45). Zn^{2+} or Cu^{2+} are not activators, instead they are potent inhibitors). It was inhibited by EDTA. 4-Hydroxyphenylpyruvate was a competitive inhibitor of phenylpyruvate decarboxylation (Ki = 20 μM). (1)

References

1. Asakawa, T., Wada, H. & Yamano, T. (1968) BBA, 170, 375.
2. Fujioka, M., Morino, Y. & Wada, H. (1970) Methods in Enzymology, 17A, 589.

4-CARBOXYMUCONOLACTONE DECARBOXYLASE

(4-Carboxymuconolactone carboxy-lyase)

4-Carboxymuconolactone = 3-oxoadipate enol-lactone + CO_2

Ref.

Molecular weight

source	value	conditions	
Pseudomonas putida	93,000 [~8]	Sephadex G 200, pH 7.1; gel electrophoresis (SDS); amino acid composition.	(1,2,3)

Specific activity

P. putida (668 x) 2139 4-carboxymuconolactone (pH 8.0, Tris, 25°) (3)

Catalytic properties

source	substrate	K_m (mM)	conditions	
P. putida (purified 437 x)	4-carboxymucono-lactone[a]	0.82	pH 8.0, Tris, 25°	(1,2)

(a) 4-carboxymuconolactone decarboxylates spontaneously (at 30° and in neutral solutions the half life is 30 min) to 3-oxoadipate enol-lactone. It was generated from 3-carboxy-cis,cis-muconate with 3-carboxy-cis,cis muconate-lactonizing enzyme as the catalyst.

The product of 4-carboxymuconolactone decarboxylation tautomerizes. (1,2)

Light absorption data

4-Carboxymuconolactone has an absorption band at 230 nm and the reaction can be followed at this wavelength. (1,2)

Abbreviations

4-carboxymuconolactone	4-carboxymethyl-4-carboxy-Δ^2-butenolide
3-oxoadipate enol-lactone	4-carboxymethyl-Δ^3-butenolide

References

1. Ornston, L.N. (1966) JBC, 241, 3787.
2. Ornston, L.N. (1970) Methods in Enzymology, 17A, 543.
3. Parke, D., Meagher, R.B. & Ornston, L.N. (1973) B, 12, 3537.

AMINOCARBOXYMUCONATE-SEMIALDEHYDE DECARBOXYLASE

((3'-Oxo-prop-2'-enyl)-2-amino-but-2-ene-dioate carboxy-lyase)

2-Amino-3-carboxymuconate semialdehyde =

2-aminomuconate semialdehyde + CO_2

Ref.

Equilibrium constant

The reaction is irreversible [F]. 2-Aminomuconate semialdehyde rearranges non-enzymically to picolinate. (1)

Specificity and catalytic properties

source	substrate	K_m (µM)	conditions	
Cat liver (purified 200 x)	2-amino-3-carboxy-muconate semialdehyde	1	pH 8.0, Tris, 24°	(1)

The enzyme did not require metal ions for activity and it was not inhibited by EDTA. 2-Hydroxymuconate semialdehyde was a powerful inhibitor but the following compounds did not inhibit: picolinate; quinolinate; niacin; nicotinamide; NAD; 2-oxoadipate; glutarate or 2-hydroxymuconate. (1)

Light absorption data

2-Amino-3-carboxymuconate semialdehyde has an absorption band at 360 nm (molar extinction coefficient = 45,000 $M^{-1}cm^{-1}$). (1)

Abbreviations

2-Amino-3-carboxymuconate semialdehyde — (3'-oxo-prop-2'-enyl)-2-amino-but-2-ene-dioate

References

1. Ichiyama, A., Nakamura, S., Kawai, H., Honjo, T., Nishizuka, Y., Hayaishi, O. & Senoh, S. (1965) JBC, 240, 740.

o-PYROCATECHUATE DECARBOXYLASE

(2,3-Dihydroxybenzoate carboxy-lyase)

2,3-Dihydroxybenzoate = catechol + CO_2

Ref.

Specificity and catalytic properties

source	substrate	K_m(mM)	conditions	Ref.
Aspergillus niger (purified 815 x)	2,3-dihydroxybenzoate[a]	0.3	pH 5.2, citrate-Pi, 30°	(1,2)

(a) the following were inactive: 3-hydroxybenzoate; salicylate; 4-hydroxybenzoate; 2,4-dihydroxybenzoate; 2,5-dihydroxybenzoate; 2,6-dihydroxybenzoate; 3,4-dihydroxybenzoate; 3,6-dihydroxybenzoate; 2,3,4-trihydroxybenzoate; anthranilate; 3-hydroxyanthranilate and 4-aminobenzoate. The enzyme did not decarboxylate amino- or 2-oxo-acids.

The decarboxylase had no cofactor (e.g. thiamine pyrophosphate or pyridoxal phosphate) or metal ion requirement. Various structural analogues of 2,3-dihydroxybenzoate were competitive inhibitors (e.g. salicylate; 2,4-dihydroxybenzoate and 2,6-dihydroxybenzoate). (1,2)

Light absorption data

Catechol has an absorption band at 276 nm and the reaction can be followed at this wavelength. (1,2)

Abbreviations

o-pyrocatechuate	2,3-dihydroxybenzoate
salicylate	2-hydroxybenzoate

References

1. Subba Rao, P.V., Moore, K. & Towers, G.H.N. (1970) Methods in Enzymology, 17A, 514.
2. Subba Rao, P.V., Moore, K. & Towers, G.H.N. (1967) ABB, 122, 466.

INDOLE-3-GLYCEROL-PHOSPHATE SYNTHASE

(1-(2'-Carboxyphenylamino)-1-deoxyribulose-5-phosphate carboxy-lyase (cyclizing))

1-(2'-Carboxyphenylamino)-1-deoxyribulose 5-phosphate =
1-C-(3'-indolyl)-glycerol 3-phosphate + CO_2 + H_2O

Ref.

Molecular properties

source	value	conditions	Ref.
Escherichia coli	45,000 [1]	ultracentrifugation at pH 7.0 or in guanidine-HCl; amino acid composition; peptide mapping.	(1)

Indole-3-glycerol-phosphate synthase is an enzyme of the tryptophan pathway (see EC 4.1.3.27) and in many organisms it is associated with anthranilate synthase (EC 4.1.3.27). Unless otherwise stated the properties given here are for the unaggregated enzyme.

The In GP synthase and PRA isomerase reactions of E. coli are catalyzed by a single polypeptide chain. (1,2)

Specific activity

E. coli (37.5 x) 7 CDRP (pH 7.8, Tris, 37°; In GP synthesis) (1)

Specificity and catalytic properties

source	substrate	K_m (μM)	conditions	Ref.
E. coli (purified 37.5 x)	CDRP (a)	5	pH 7.8, Tris, 37°	(1)
E. coli (purified 17 x)	CDRP (b)	17.5	pH 8.8, Tris, 37°	(3)

(a) In In GP synthesis

(b) In In GP synthesis. The following compounds are good inhibitors: anthranilate; 3-, 4-, and 5-methyl anthranilate. Other derivatives of anthranilate were less inhibitory.

The enzyme requires dithiothreitol for activity. (1,2)

The In GP synthase of Neurospora crassa has also been studied; this enzyme does not require the addition of any cofactor for activity and it was not inhibited by EDTA. (4)

Ref.

Light absorption data

The transformation of CDRP to In GP is accompanied by an increase in absorption at 280 nm (molar extinction coefficient = 448 $M^{-1}cm^{-1}$). (1)

Abbreviations

CDRP 1-(2'-carboxyphenylamino)-1-deoxyribulose 5-phosphate
In GP 1-C-(3'-indolyl)-glycerol 3-phosphate
PRA N-(5'-phosphoribosyl) anthranilate

References

1. Creighton, T.E. & Yanofsky, C. (1966) JBC, 241, 4616.
2. Creighton, T.E. & Yanofsky, C. (1970) Methods in Enzymology, 17A, 372.
3. Gibson, F. & Yanofsky, C. (1960) BBA, 43, 489.
4. Wegman, J. & de Moss, J.A. (1965) JBC, 240, 3781.

S-ADENOSYLMETHIONINE DECARBOXYLASE

(S-Adenosyl-L-methionine carboxy-lyase)

S-Adenosyl-L-methionine =
5'-deoxyadenosyl-(5'), 3-aminopropyl-(1), methylsulphonium salt + CO_2

Molecular properties

source	value	conditions	Ref.
Escherichia coli W	113,000 [~8][a]	pH 7.35; gel electrophoresis (SDS)	(1)
Rat liver	52,000	Sephadex G 100, pH 7.2; sucrose density gradient	(2)

(a) the enzyme contains 1-2 moles pyruvate (but no pyridoxal phosphate) per 113,000 daltons.

Specific activity

			Ref.
E. coli (800 x)	1.4[a]	adenosylmethionine (pH 7.4, triethanolamine 37°)	(1)

(a) this is a maximum value.

Specificity and catalytic properties

source	substrate	K_m (μM)	conditions	Ref.
E. coli	adenosylmethionine	90	pH 7.4, triethanolamine, 37°	(1)
Rat liver (350 x)	adenosylmethionine	36	pH 7.2, Pi; at saturating putrescine	(2)
	putrescine	330	pH 7.2, Pi	(2)
Human prostate (cytosol)	adenosylmethionine[a]	40	pH 7.5, Pi, 37°	(3)
	putrescine	131	pH 7.5, Pi, 37°	(3)

(a) the following compounds were inactive: S-pentosyl-L-methionine; S-methyl-L-methionine; L-methionine; S-inosyl-L-methionine and S-adenosyl-L-homocysteine.

The decarboxylase from rat liver requires pyridoxal phosphate but that from E. coli does not: this enzyme has covalently linked pyruvate which may function as a prosthetic group. The enzyme from human prostate may have either pyridoxal phosphate or pyruvate as a prosthetic group; this enzyme was activated by putrescine (other polyamines such as spermidine or spermine did not activate). The enzyme from rat liver also requires putrescine (or spermidine) for full activity. The enzyme from E. coli requires Mg^{2+} for activity but those from rat liver and human prostate do not. (1,2,3)

Adenosylmethionine decarboxylase from rat prostate (purified 500 x) had similar properties to the human prostate enzyme. The rat enzyme was activated by putrescine or spermidine. (4)

Ref.

References

1. Wickner, R.B., Tabor, C.W. & Tabor, H. (1970) JBC, 245, 2132.
2. Feldman, M.J., Levy, C.C. & Russell, D.H. (1972) B, 11, 671.
3. Zappia, V., Carteni-Farina, M. & della Pietra, G. (1972) BJ, 129, 703.
4. Jänne, J. & Williams-Ashman, H.G. (1971) BBRC, 42, 222.

6-METHYLSALICYLATE DECARBOXYLASE

(6-Methylsalicylate carboxy-lyase)

6-Methylsalicylate = m-cresol + CO_2

Ref.

Molecular weight

source	value	conditions	Ref.
Penicillium patulum	∿60,000	Bio-Gel A 5 m, pH 7.2. The buffer contained EDTA and dithiothreitol. Ultracentrifugation.	(1,2)

Specificity and catalytic properties

source	substrate	K_m (μM)	conditions	Ref.
P. patulum (extract)	6-methylsalicylate[a]	7	pH 7.2, Pi	(1)

(a) with a highly purified enzyme preparation (see Ref. 2), 6-methylsalicylate could not be replaced by 3- or 4-methylsalicylate.

References

1. Light, R.J. (1969) BBA, 191, 430.
2. Vogel, G. & Lynen, F. (1970) Naturwissenschaften, 57, 664.

ORSELLINATE DECARBOXYLASE

(Orsellinate carboxy-lyase)

Orsellinate = orcinol + CO_2

Ref.

Equilibrium constant

The reaction is essentially irreversible [F] with the enzyme from Gliocladium roseum (a mould) as catalyst. (1)

Molecular weight

source	value	conditions	Ref.
Lasallia pustulata (a lichen)	72,000 [1]	gel electrophoresis (SDS)	(2)

Specificity and catalytic properties

source	substrate	K_m (mM)	conditions	Ref.
L. pustulata (purified 400 x)	orsellinate(a)	0.211	pH 6.8, Pi, 25°	(2)
G. roseum (purified 41 x)	orsellinate(b)	0.26	pH 5.2, Pi, 25°	(1)

(a) orsellinate could be replaced by 3- or 5-chloro-orsellinate; homo-orsellinate or everninate but not by isoavernniate; salicylate; 4-aminosalicylate; 6-methylsalicylate or 2,4-dihydroxybenzoate.

(b) orsellinate (1.00) could be replaced by 2,4-dihydroxy-5,6-dimethylbenzoate (0.25) but not by the following: benzoate; 2-, 3-, and 4-hydroxybenzoate; 2- and 4-hydroxy-6-methylbenzoate; 2,3-, 2,4-, 2,5-, 2,6- and 3,4-dihydroxybenzoate; 2,4-dihydroxy-3,6-dimethylbenzoate; 2,3,4- and 3,4,5-trihydroxybenzoate. A number of other aromatic substances tested were neither substrates nor inhibitors.

The enzyme (G. roseum) had no cofactor or metal ion requirement. It was not inhibited by EDTA. (1)

The decarboxylase (G. roseum) was inhibited competitively by 2-hydroxybenzoate (Ki = 4.6 mM); 2-hydroxy-6-methylbenzoate (Ki = 1 mM); 2,3-dihydroxybenzoate (Ki = 1.5 mM); 2,4-dihydroxybenzoate (Ki = 5.2 mM) and 2,5-dihydroxybenzoate (Ki = 1.1 mM). The enzyme from L. pustulata was also inhibited by 2,4-dihydroxybenzoate (C(orsellinate); Ki = 0.8 mM). (1,2)

Light absorption data

Orsellinate has an absorption band at 250 nm (molar extinction coefficient = 8,360 $M^{-1}cm^{-1}$ at pH 6.8). (2)

Abbreviations

orsellinate 2,4-dihydroxy-6-methyl benzoate

References

1. Pettersson, G. (1965) Acta. Chem. Scand, 19, 2013.
2. Mosbach, K. & Schultz, J. (1971) EJB, 22, 485.

DIALKYLAMINO-ACID DECARBOXYLASE (PYRUVATE)

(2,2-Dialkyl-L-amino-acid carboxy-lyase (aminotransferring))

2-Aminoisobutyrate + pyruvate = dimethyl ketone + CO_2 + L-alanine

Ref.

Equilibrium constant

The overall reaction is catalyzed in two steps. In the first, AIB reacts with pyridoxal phosphate - enzyme to form pyridoxamine phosphate - enzyme, CO_2 and acetone. This reaction is irreversible [F]. In the second, a reversible exchange transamination occurs by which pyruvate is converted to L-alanine and pyridoxal phosphate - enzyme is regenerated. The overall reaction is, therefore, irreversible [F]. (1,2)

Molecular properties

source	value	conditions	Ref.
Pseudomonas cepacia[(a)]	188,000 [4 or 8]	ultracentrifugation with or without urea or guanidine-HCl; amino acid analysis; peptide mapping.	(1)
Arthrobacter sp.	188,000 [8]	ultracentrifugation with or without guanidine-HCl	(1)
P. fluorescens[(a)]	150,000	pH 7.5	(2)

(a) the highly purified enzyme contained bound vitamin B_6.

Specific activity

P. cepacia (23 x)	9	AIB[(a)]	(pH 7.5, Pi, 37°)	(1)
P. fluorescens (166 x)	10	AIB[(a)]	(pH 7.5, Pi, 30°)	(2)

(a) the second substrate was pyruvate.

Specificity and catalytic properties

source	substrate	V relative	K_m (mM)	conditions	Ref.
P. cepacia	AIB	-	8.7	pH 7.5, Pi, 37°	(1)
	pyruvate	-	0.17	pH 7.5, Pi, 37°	(1)
	pyridoxal phosphate	-	0.014	pH 7.5, Pi, 37°	(1)
P. fluorescens	pyruvate[(a)]	-	2	pH 7.5, Pi, 30°	(2)
	pyridoxal phosphate	-	0.006	pH 7.5, Pi, 30°	(2)
	AIB[(b)]	1.0	8	pH 7.5, Pi, 30°	(2)
	L-alanine (transamination)	2.4	33	pH 7.5, Pi, 30°	(2)

Ref.

Specificity and catalytic properties continued

(a) pyruvate (1.00) could be replaced by 2-oxobutyrate (0.95); 2-oxovalerate (0.49); glyoxylate (0.12) or 2-oxoisocaproate (0.06) but not by 2-oxophenylpyruvate or 2-oxoglutarate. The amino acid substrate was DL-isovaline.

(b) AIB could be replaced by L-isovaline or 1-aminocyclopentane-carboxylate. The 2-oxoacid substrate was pyruvate.

The mechanism of the reaction has been investigated. Dialkylamino acid decarboxylase has been shown to catalyze, slowly, the following reaction:

D-alanine + pyruvate = CO_2 + acetaldehyde + L-alanine (3)

The enzyme (P. fluorescens) is competitively inhibited by L-cycloserine (Ki = 2.5 μM) and D-cycloserine (Ki = 8 mM). (3)

Abbreviations

AIB 2-aminoisobutyrate

References

1. Lamartiniere, C.A., Itoh, H. & Dempsey, W.B. (1971) B, 10, 4783.
2. Bailey, G.B. & Dempsey, W.B. (1967) B, 6, 1526.
3. Bailey, G.B., Chotamangsa, O. & Vuttivej, K. (1970) B, 9, 3243.

2-KETO-3-DEOXY-L-ARABONATE ALDOLASE

(2-Keto-3-deoxy-L-arabonate glycolaldehyde-lyase)

2-Keto-3-deoxy-L-arabonate = pyruvate + glycolaldehyde

Ref.

Equilibrium constant

			Ref.
[pyruvate] [glycolaldehyde] / [2-keto-3-deoxy-L-arabonate]	= 0.37 M	(pH 7.4, HEPES, 28°; in the presence of Mn^{2+})	(1)
[pyruvate] [lactaldehyde] / [2-keto-3-deoxy-D-fuconate]	= 0.12 M	(pH 7.5, HEPES, 28°; in the presence of Mn^{2+})	(2)

Specificity and catalytic properties

source	substrate	V relative	K_m (mM)	
Pseudomonad MSU-1 (purified 50 x)	2-keto-3-deoxy-L-arabonate[a]	1.00	1.8	(2)
	2-keto-3-deoxy-D-fuconate	0.47	2.9	(2)

(a) at pH 7.5, HEPES, 28°. The following were inactive: 2-keto-3-deoxy-D-galactonate and its 6-phosphoester; 2-keto-3-deoxy-D-gluconate and its 6-phosphoester; N-acetylneuraminate and 2-keto-4-hydroxy-DL-glutarate. The two D-galactonate and two gluconate derivatives did not inhibit 2-keto-3-deoxy-D-fuconate cleavage.

The enzyme requires a divalent metal ion for activity and it was inhibited by EDTA. The following were activators: Mn^{2+} (1.00); Co^{2+} (0.95); Mg^{2+} (0.54) and Ni^{2+} (0.45). (2)

References

1. Dahms, A.S. & Anderson, R.L. (1972) JBC, 247, 2238.
2. Dahms, A.S. & Anderson, R.L. (1969) BBRC, 36, 809.

DIMETHYLANILINE N-OXIDE ALDOLASE

(N,N-Dimethylaniline-N-oxide formaldehyde-lyase)

N,N-Dimethylaniline N-oxide = N-methylaniline + formaldehyde

Ref.

Specificity and catalytic properties

source	substrate	V relative	K_m (mM)	
Pig liver microsomes	N,N-dimethylnaphthylamine N-oxide[a]	1.00	7	(1)
	N,N-dimethyl-4-toluidine N-oxide	1.43	20	(1)
	N,N-dimethylaniline N-oxide	0.84	80	(1)

(a) N,N-dimethylnaphthylamine N-oxide (1.00) could be replaced by 4-chlorodimethylaniline N-oxide (0.43) or N-ethyl-N-methylaniline N-oxide (0.10) but not by N,N-dimethylbenzylamine N-oxide; N,N-dimethylcyclohexylamine N-oxide; N,N-dimethyloctylamine N-oxide or morphine N-oxide.

The enzyme does not require any added cofactor or oxygen for activity. Potent competitive inhibitors were pyridine and SKF-525 A; both had Ki = 2 μM.

Abbreviations

SKF-525A diethylaminoethanol ester of diphenylpropylacetic acid

References

1. Machinist, J.M., Orme-Johnson, W.H. & Ziegler, D.M. (1966) B, 5, 2939.

DIHYDRONEOPTERIN ALDOLASE

(2-Amino-4-hydroxy-6-(D-*erythro*-1',2',3'-trihydroxypropyl)-7,8-dihydropteridine glycolaldehyde-lyase)

2-Amino-4-hydroxy-6-(D-*erythro*-1',2',3'-trihydroxypropyl)-7,8-dihydropteridine = 2-amino-4-hydroxy-6-hydroxymethyl-7,8-dihydropteridine + glycolaldehyde

Ref.

Equilibrium constant

The reaction is essentially irreversible [F] with the enzyme from *E. coli* as catalyst. (1)

Molecular weight

source	value	conditions	
Escherichia coli B	100,000 [1]	gel electrophoresis (SDS)	(1)

Specific activity

E. coli B (1100 x) 4×10^{-3} dihydroneopterin (pH 9.6, ammonium acetate 37°) (1)

Specificity and Michaelis constants

source	substrate	K_m (μM)	conditions	
E. coli B	dihydroneopterin	9	pH 9.6, ammonium acetate, 37°	(1)

The enzyme was active with an epimer of dihydroneopterin, namely 2-amino-4-hydroxy-6-(L-*threo*-trihydroxypropyl)-7,8-dihydropteridine, but the corresponding D-*threo*- and L-*erythro* compounds were inactive. The following compounds were also inactive: neopterin; tetrahydroneopterin; dihydroneopterin monophosphate and dihydroneopterin triphosphate. (1)

The aldolase does not require divalent metal ions for activity. It was not inhibited by EDTA. (1)

One of the products of the reaction, dihydropterin-CH_2OH, is a powerful inhibitor. Other dihydropteridines, including dihydropteroate and dihydrofolate, are less potent inhibitors. (1)

Abbreviations

dihydroneopterin	2-amino-4-hydroxy-6-(D-*erythro*-1',2',3-trihydroxypropyl)-7,8-dihydropteridine.
dihydropterin-CH_2OH	2-amino-4-hydroxy-6-hydroxymethyl-7,8-dihydropteridine.

References

1. Mathis, J.B. & Brown, G.M. (1970) JBC, 245, 3015.

ACETOLACTATE SYNTHASE

(Acetolactate pyruvate-lyase (carboxylating))

2-Acetolactate + CO_2 = 2 pyruvate

Ref.

Molecular properties

source	value	conditions	
Aerobacter aerogenes	220,000 [4]	pH 6.0, in the presence of Mg^{2+}, thiamine pyrophosphate, pyruvate and ammonium sulphate; gel electrophoresis (SDS); amino acid composition; electron microscopy.	(1,2)

The enzyme contains three moles of TPP (which are required for activity) and 6 moles of Pi per 200,000 daltons. (2)

Neurospora crassa has two forms of the synthase; one has molecular weight about 115,000 but the other form is much larger. (6)

Specific activity

A. aerogenes (120 x) 530 2-acetolacetate formed (pH 5.8, acetate, 37°) (3)

Specificity and catalytic properties

source	substrate	K_m (mM)	conditions	
A. aerogenes	pyruvate	6.3	pH 5.8, acetate, 37°	(4)
	2-oxobutyrate[(a)]	5	pH 5.8, acetate, 37°	(5)

(a) 2-oxobutyrate condenses with pyruvate to form acetohydroxybutyrate. The activity is low.

Although the enzyme was active in the absence of divalent metal ions stimulation occurred in their presence. It was also activated by acetate (1.00), monochloroacetate (0.96), formate (0.70) or trichloroacetate (0.52), (no activator = 0.43). It was inhibited by several inorganic anions and of these sulphate, Pi and PPi were particularly effective. Further, the use of Pi and PPi buffers in kinetic experiments resulted in sigmoid kinetics whereas in acetate buffer Michaelis-Menten kinetics were observed. The enzyme was also inhibited by glyoxylate (C(pyruvate), Ki = 51 µM) and phenylpyruvate (C(pyruvate), Ki = 0.11 mM) and to a lesser extent by oxamate, 2-oxobutyrate (C (pyruvate), Ki = 5mM), propionate and oxalate. The following compounds did not inhibit: oxaloacetate; acetamide; acetylphosphate; phenylalanine or 2,3-butanediol. (2,4,5)

Ref.

Specificity and catalytic properties continued

The enzyme from N. crassa (purified 30-60 x) has a pH optimum of about 7.5. The enzyme in intact mitochondria requires TPP, FAD and Mn^{2+} (or Mg^{2+}) for activity and it is sensitive to end product inhibition by L-valine (NC(pyruvate); Ki = 0.51 mM). The Km for pyruvate was 3.2 mM. High concentrations of pyruvate were inhibitory. (6)

References

1. Huseby, N-E., Christensen, T.B., Olsen, B.R. & Størmer, F.C. (1971) EJB, 20, 209.
2. Størmer, F.C., Solberg, Y. & Hovig, T. (1969) EJB, 10, 251.
3. Størmer, F.C. (1967) JBC, 242, 1756.
4. Størmer, F.C. (1968) JBC, 243, 3735.
5. Huseby, N-E. & Størmer, F.C. (1971) EJB, 20, 215.
6. Glatzer, L., Eakin, E. & Wagner, R.P. (1972) J. Bact, 112, 453.
   Arfin, S.M. & Koziell, D.A. (1973) BBA, 321, 348; 356.

CITRAMALATE LYASE

(L-(+)-Citramalate pyruvate-lyase)

Citramalate = acetate + pyruvate

Ref.

Equilibrium constant

$$\frac{[\text{acetate}]\ [\text{pyruvate}]}{[\text{citramalate}]} = 8.3\ \text{M} \quad (\text{pH } 7.4,\ 25°)$$ (1)

Specificity and Michaelis constants

source	substrate	K_m (mM)	conditions	Ref.
Clostridium tetanomorphum (extract)	L-(+)-citramalate[a]	0.6	pH 7.4, Pi, 24°	(1)
	Mg^{2+}[b]	0.1	pH 7.4, Pi, 24°	(1)

(a) D-(-)-citramalate, DL-isocitrate, DL-malate, citrate and mesaconate were inactive. A higher Km was obtained at high substrate concentrations. Very high substrate concentrations were inhibitory.

(b) Mg^{2+} was required for activity. Mn^{2+} or Co^{2+} were also active but no activation took place with Ca^{2}, Fe^{2+} or monovalent metal ions. The enzyme was inhibited by EDTA and PPi.

References

1. Barker, H.A. (1969) Methods in Enzymology, 13, 344.

DECYLCITRATE SYNTHASE

((-)-Decylcitrate oxaloacetate-lyase (CoA-acylating))

(-)-Decylcitrate + CoA = lauroyl-CoA + H_2O + oxaloacetate

Ref.

Molecular properties

source	value	conditions	
Penicillium spiculisporum	90,000 [2]	pH 7.8	(1,2)

A stable enzyme-lauroyl-CoA complex has been prepared. (1)

Specificity and catalytic properties

source	substrate	K_m (μM)	conditions	
P. spiculisporum (purified 20-40 x)	oxaloacetate[a]	35	pH 8.1, Pi, 25°	(1,2)
	lauroyl-CoA[b]	<0.1	pH 8.1, Pi, 25°	(1,2)
	11-formamido-undecanoyl-CoA	30	pH 8.1, Pi, 25°	(1)

(a) 2-oxoglutarate and pyruvate had less than 10% the activity of oxaloacetate.

(b) the enzyme was subject to potent inhibition by this substrate. Lauroyl-CoA could be replaced by several acyl-CoA derivatives of chain lengths less than C-12. Thus, CoA derivatives of the following acids were active: decanoic; octanoic; butyric; acetic and phenylacetic. Myristoyl-CoA and palmitoyl-CoA were poor substrates. With acetyl-CoA as substrate, citrate was formed.

The enzyme does not have any metal ion requirement. It was inhibited by palmitoyl-CoA, acetyl-CoA and CoASSCoA (C(lauroyl-CoA)). (1,2)

Light absorption data

The cleavage of the thioester bond of lauroyl-CoA can be followed at 232 nm (for acetyl-CoA the molar extinction coefficient = 4,500 $M^{-1}cm^{-1}$). (1,3)

References

1. Måhlén, A. (1971) EJB, 22, 104.
2. Måhlén, A. & Gatenbeck, S. (1968) Acta Chem. Scand., 22, 2617.
3. Dixon, G.H. & Kornberg, H.L. (1959) BJ, 72, 3P.

ANTHRANILATE SYNTHASE

(Chorismate pyruvate-lyase (amino-accepting))

Chorismate + L-glutamine = anthranilate + pyruvate + L-glutamate

Ref.

Molecular properties

Anthranilate synthase catalyzes the first specific reaction of the tryptophan pathway:

CHORISMATE $\xrightarrow{\text{anthranilate synthase}}$ ANTHRANILATE $\xrightarrow{\text{PR transferase}}$ *N*-(5'-PHOSPHO-D-RIBOSYL)-ANTHRANILATE $\xrightarrow{\text{PRA isomerase}}$ 1-(2'-CARBOXY-PHENYLAMINO)-1-DEOXYRIBULOSE 5-PHOSPHATE $\xrightarrow{\text{In GP synthase}}$ 1-*C*-(3'-INDOLYL)-GLYCEROL 3-PHOSPHATE $\xrightarrow{\text{tryptophan synthase}}$ TRYPTOPHAN

Chorismate is also a precursor of tyrosine and phenylalanine (see EC 5.4.99.5)

Except for tryptophan synthase (EC 4.2.1.20) the enzymes of the pathway are, depending on the organism, associated to varying degrees. Thus, there are three types of anthranilate synthase:

Type 1, unassociated (found in several bacteria); type 2, associated with PR transferase (found in enteric bacteria) and type 3, associated with PRA isomerase - In GP synthase or In GP synthase (found in certain fungi and plants).

All anthranilate synthases so far studied are oligomeric proteins containing nonidentical subunits: components I and II and in certain cases, component III. On its own, component I catalyzes the following reaction:

$$\text{chorismate} + NH_3 = \text{anthranilate} + \text{pyruvate} + H_2O$$

In type 1 anthranilate synthase, component II provides the glutamine binding site. In type 2, component II provides both the glutamine binding site and the PR transferase activity and in type 3 an additional subunit, component III, provides PRA isomerase - In GP synthase or In GP synthase activities. (1)

Anthranilate synthases have been highly purified from several sources and the molecular properties of some of these are given below. The reader is referred to Ref. 1 for the properties of anthranilate synthases from other sources and to Ref. 9 for the properties of the enzyme from *Acinetobacter calcoaceticus*.

Molecular properties continued — Ref.

source	type	subunit composition	subunit molecular weight (x 10^{-3}) I	II	III	Ref.
Pseudomonas putida	1	I_1II_1	64	18	-	(5)
Serratia marcescens	1	I_2II_2	60	21	-	(7)
Escherichia coli	2	I_2II_2	60	60	-	(1,8)
Salmonella typhimurium	2	I_2II_2	64	63	-	(2)
Neurospora crassa	3	$I_2II_2III_4$	40	30	40	(6)

Catalytic properties

The catalytic properties of the enzyme from S. typhimurium have been investigated in detail.

Each component I has one site for chorismate (K_D = 3.6 µM) and probably a separate site for the end product inhibitor, tryptophan (K_D = 40 µM). The NH_3 site is also on component I. Each component II has one site for glutamine to which the glutamine analogue DON binds irreversibly. On its own component I has normal Michaelis-Menten kinetics with either chorismate or NH_3 as the variable substrate and in the presence or absence of tryptophan. The complex I_2II_2 has normal kinetics with either chorismate or glutamine as the variable substrate in the absence of tryptophan and with glutamine in the presence of tryptophan. In the presence of tryptophan, there was positive cooperativity for chorismate or Mg^{2+} binding and negative cooperativity for NH_3 binding. Anthranilate synthase is subject to inhibition by the products anthranilate, pyruvate and glutamate. (1,2)

Abbreviations

PR transferase	anthranilate phosphoribosyltransferase	EC 2.4.2.18
PRA isomerase	N-(5'-phosphoribosyl) anthranilate isomerase	see EC 4.1.1.48
In GP synthase	indole 3-glycerol -phosphate synthase	EC 4.1.1.48
DON	6-diazo-5-oxonorleucine	

References

1. Zalkin, H. (1973) Advances in Enzymology, 38, 1.
2. Henderson, E.J. & Zalkin, H. (1971) JBC, 246, 6891.
3. Nagano, H. & Zalkin, H. (1970) JBC, 245, 3097.
4. Henderson, E.J., Nagano, H., Zalkin, H. & Hwang, L.H. (1970) JBC, 245, 1416.
5. Queener, S.W., Queener, S.F., Meeks, J.R. & Gunsalus, I.C. (1973) JBC, 248, 151.
6. Arroyo-Begovich, A. & de Moss, J.A. (1973) JBC, 248, 1262.
7. Zalkin, H. & Hwang, L.H. (1971) JBC, 246, 6899.
8. Ito, J., Cox, E.C. & Yanofsky, C. (1969) J. Bacteriol, 97, 725.
9. Sawula, R.V. & Crawford, I.P. (1973) JBC, 248, 3573.

DECYLHOMOCITRATE SYNTHASE

(Decylhomocitrate 2-oxoglutarate-lyase (CoA-acylating))

Decylhomocitrate + CoA = lauroyl-CoA + H_2O + 2-oxoglutarate

Ref.

Molecular properties

source	value	conditions	
Penicillium spiculisporum	43,000	gel electrophoresis (SDS)	(1)

The enzyme contains no bound lipid. It has a strong tendency to aggregate. (1)

Specificity and catalytic properties

source	substrate	K_m (μM)	conditions	
P. spiculisporum (purified 50 x)	lauroyl-CoA[(a)]	~1	pH 8.1, Pi, 22°	(1)
	2-oxoglutarate[(b)]	750	pH 8.1, Pi, 22°	(1)

(a) lauroyl-CoA (1.0) could be replaced by decanoyl-CoA (0.2) but with the following low or no activities were obtained: myristoyl-CoA; palmitoyl-CoA; 11-formamidoundecanoyl-CoA and 11-aminoundecanoyl-CoA.

(b) oxaloacetate and pyruvate were inactive as substrates.

The enzyme does not require metal ions. (1)

The synthase was inhibited by oxaloacetate (C(2-oxoglutarate); Ki = 0.6 mM) and by palmitoyl-CoA. Pyruvate, citrate and ATP were not inhibitory. (1)

Light absorption data

See EC 4.1.3.23.

References

1. Måhlén, A. (1973) EJB, 38, 32.

TYROSINE PHENOL-LYASE

(L-Tyrosine phenol-lyase (deaminating))

L-Tyrosine + H_2O = phenol + pyruvate + NH_3

Ref.

Equilibrium constant

The reaction is reversible. (1)

Molecular properties

source	value	conditions	
Escherichia intermedia	170,000	pH 6.0	(2)

The enzyme contains two moles of pyridoxal phosphate per 170,000 daltons (K_D = 1.3 μM). (2)

Specific activity

E. intermedia	7.2[a]	pyruvate (pH 9.0, NH_4Cl, 30°; in reaction (a))	(1)
	1.94	L-tyrosine (pH 8.0, Pi, 30°; in reaction (a))	(2)

(a) this is a maximum value.

Specificity and catalytic properties

Tyrosine phenol-lyase catalyzes a series of α,β-elimination, β-replacement and racemization reactions.

(a) α,β-elimination

L-tyrosine + H_2O = phenol + pyruvate + NH_3

(b) β-replacement

L-tyrosine + pyrocatechol = phenol + 3,4-dihydroxyphenyl-L-alanine

(c) racemization

L-alanine = D-alanine

source	substrate	reaction	V relative	K_m (mM)	conditions	
E. intermedia	L-tyrosine[a]	(a)	0.28	0.23	pH 8.0, NH_4Cl, 30°	(2)
	phenol[b]	(a)	1	4.4	pH 9.0, NH_4Cl, 30°	(1)
	pyruvate	(a)	1	5.1	pH 9.0, NH_4Cl, 30°	(1)
	L-alanine	(c)	0.007	26.0	pH 7.3, Pi, 30°	(3)

(a) L-tyrosine (1.00) could be replaced by S-methyl-L-cysteine (0.31); D-tyrosine (0.26); L-serine (0.16); D-serine (0.15); 3,4-dihydroxyphenyl-L-alanine (0.15) or L-cysteine. With each substrate pyruvate was formed. The following were inactive: L-phenylalanine; L-histidine; L-tryptophan; L-aspartate; L-valine; L-methionine; L-alanine and D-alanine.

Ref.

Specificity and catalytic properties continued

(b) phenol (1.00) could be replaced by pyrocatechol (0.63) or resorcinol (0.32) but not by H_2S or CH_3SH.

The lyase requires added pyridoxal phosphate for activity. There was also a requirement for K^+ or NH_4^+ for full activity but Na^+ was not an activator. (2)

L-Alanine inhibited pyruvate formation from L-tyrosine, S-methyl-L-cysteine or L-serine. The inhibition was competitive (K_i = 6.5 mM). Phenol and pyrocatechol also inhibited the forward reaction (both NC, Ki = 35.6 μM and 0.46 mM, respectively). (2)

Tyrosine phenol-lyase has also been purified (85 x) from *Clostridium tetanomorphum*. This enzyme also requires pyridoxal phosphate for activity and, in addition, Mg^{2+}. (4)

References

1. Yamada, H., Kumagai, H., Kashima, N. & Torii, H. (1972) BBRC, 46, 370.
2. Kumagai, H., Yamada, H., Matsui, H., Ohkishi, H. & Ogata, K. (1970) JBC, 245, 1767; 1773.
3. Kumagai, H., Kashima, N. & Yamada, H. (1970) BBRC, 39, 796.
4. Brot, N. & Weissbach, H. (1970) Methods in Enzymology, 17A, 642.

ARABONATE DEHYDRATASE

(D-Arabonate hydro-lyase)

D-Arabonate = 2-keto-3-deoxy-D-arabonate + H_2O

Ref.

Equilibrium constant

The reaction is essentially irreversible [F] with the enzyme from pork liver as the catalyst. (1)

Specificity and catalytic properties

source	substrate	relative velocity	K_m(mM)	conditions	
Pork liver	L-fuconate[(a)]	1.0	1.0	pH 7.0, TAES, 37°	(1)
(purified 13 x)	D-arabonate	0.6	0.6	pH 7.0, TAES, 37°	(1)

(a) L-fuconate could not be replaced by D-fuconate; L-arabonate; D-gluconate; D-galactonate; D-gluconate; L-mannonate or D-glucuronate.

The enzyme requires Mg^{2+} for activity. Mn^{2+} or Zn^{2+} could not replace Mg^{2+} and EDTA was inhibitory. (1)

References

1. Yuen, R. & Schachter, H. (1972) Can. J. Biochem, 50, 798.

INDOLEACETALDOXIME DEHYDRATASE

(3-Indoleacetaldoxime hydro-lyase)

3-Indoleacetaldoxime = 3-indoleacetonitrile + H_2O

Ref.

Specificity and catalytic properties

source	substrate	K_m (mM)	conditions	Ref.
Gibberella fujikuroi (a fungus; purified 14 x)	3-indoleacetaldoxime[a]	0.17	pH 7.0, Pi, 30°	(1)

(a) the following compounds were inactive: *cis*-benzaldoxime; phenylacetaldoxime; phenylpropionaldoxime; mandelaldoxime; isonitrosoacetophenone; salicylaldoxime or pyridine-3'-aldoxime.

The enzyme requires pyridoxal phosphate, L-dehydroascorbate and Fe^{3+} for full activity. It was not inhibited by EDTA. (1,2)

The dehydratase was inhibited by a number of substrate analogues. The inhibition was competitive and with phenylacetaldoxime Ki = 2.2×10^{-8}M. High substrate concentrations were also inhibitory. (1)

References

1. Shukla, P.S. & Mahadevan, S. (1968) ABB, 125, 873.
2. Kumar, S.A. & Mahadevan, S. (1963) ABB, 103, 516.

GLYCEROL DEHYDRATASE

(Glycerol hydro-lyase)

Glycerol = 3-hydroxypropionaldehyde + H_2O

Ref.

Molecular properties

source	value	conditions	
Aerobacter aerogenes	188,000 [2]	sucrose density gradient; amino acid composition	(2)

The molecular weight given above is that of an apoenzyme-hydroxycobalamin complex. The physiological cofactor of the enzyme is cobalamin (vitamin B_{12}). The apoenzyme - cobalamin complex is very unstable and difficult to isolate but when cobalamin is replaced by hydroxycobalamin a stable but enzymically inactive complex results. This complex is composed of 2 different types of protein subunit (one has a molecular weight of about 22,000) and 1 mole of hydroxycobalamin. The active dehydratase is regenerated by incubating the inactive apoenzyme - hydroxycobalamin complex with cobalamin. This exchange process requires sulphite and Mg^{2+}. (1,2)

A molecular weight of 240,000 for glycerol dehydratase (*A. aerogenes*) has been reported. (1)

Specific activity

A. aerogenes (110 x) 3 glycerol (pH 8.6, Pi, 30°) (2)

The apoenzyme - cobalamin complex is unstable in the presence or absence of the substrate, glycerol and kinetic constants have not been obtained. Mg^{2+} is required for apoenzyme - cobalamine complex formation but not in the dehydratase reaction. (1,2)

Abbreviations

cobalamin (vitamin B_{12})	α-(5,6-benzimidazolyl)-5'-deoxyadenosyl cobamide.

References

1. Abeles, R.H. (1971) The Enzymes, 5, 481.
2. Schneider, Z., Larsen, E.G., Jacobson, G., Johnson, B.C. & Pawelkiewicz, J. (1970) JBC, 245, 3388.

CITRACONATE HYDRATASE

((-)-Citramalate hydro-lyase)

(-)-Citramalate = citraconate + H_2O

Ref.

Equilibrium constant

The reaction is essentially irreversible [R] with the enzyme from Pseudomonad sp. as catalyst. (1)

Specificity and catalytic properties

source	substrate	K_m (mM)	conditions	
Pseudomonad sp. (purified 25 x)	citraconate[(a)]	1.1	pH 7.5, Tris, 30°	(1)
	Fe^{2+}[(b)]	0.12	pH 7.5, Tris, 30°	(1)

(a) citraconate could not be replaced by mesaconate, maleate, fumarate, *cis*-aconitate, or *trans*-aconitate.

(b) Fe^{2+} was an activator in the presence of glutathione. The enzyme was inhibited by EDTA or PPi. No other metal ion tested could replace Fe^{2+}.

References

1. Subramanian, S.S. & Raghavendra Rao, M.R. (1968) JBC, 243, 2367.

trans-EPOXYSUCCINATE HYDRATASE

(meso-Tartrate hydro-lyase)

meso-Tartrate = trans-2,3-epoxysuccinate + H_2O

Ref.

Molecular weight

source	value	conditions	Ref.
Pseudomonas putida	66,000 [1]	ultracentrifugation in pH 7.0 buffer and in 6 M-guanidine + 0.1 M mercaptoethanol; amino acid composition.	(1)

Specific activity

			Ref.
P. putida (290 x)	229	l-trans-2,3-epoxysuccinate (pH 8.6, glycylglycine).	(1)

Specificity and Michaelis constants

source	substrate[a]	V relative	K_m (mM)	Ref.
P. putida	l-trans-2,3-epoxysuccinate	1.00	2.15	(1)
	d-trans-2,3-epoxysuccinate	0.26	0.53	(1)

(a) the conditions were: pH 8.8, glycylglycine. cis-2,3-Epoxysuccinate was inactive as a substrate or inhibitor.

The enzyme was not inhibited by EDTA. (1)

References

1. Allen, R.H. & Jakoby, W.B. (1969) JBC, 244, 2078.

GLUCONATE DEHYDRATASE

(D-Gluconate hydro-lyase)

D-Gluconate = 2-keto-3-deoxy-D-gluconate + H_2O

Ref.

Molecular weight

source	value	conditions	
Achromobacter-Alcaligenes group	270,000	Sephadex G 200 (thin layer)	(1,2)

Specific activity

Achromobacter-Alcaligenes group (100 x)	10.2	D-gluconate (pH 8.5, Tris, 30°)	(2)

Specificity and Michaelis constants

Gluconate dehydratase could utilize D-gluconate (1.00; Km = 21 mM); D-xylonate (0.88); D-fuconate (0.13); D-galactonate (0.13) and L-arabonate (0.02) but not 6-phospho-D-gluconate; 6-phospho-D-galactonate; D-glucose; D-glucitol; 5-keto-D-gluconate; D-glucarate; D- or L-gulonate; L-galactonate; D-mannonate; D-mannarate; D-galactarate; D-talonate; α-D-glucoheptonate; α-D-galactoheptonate; D-arabonate; D-ribonate; D-xylarate; D-, L- or *meso*-tartrate; D-glycerate; L-rhamnitol or L-fucitol. The enzyme requires Mg^{2+} or Mn^{2+} for optimal activity. (1)

References

1. Wood, W.A. (1971) *The Enzymes*, *5*, 578.
2. Kersters, K., Khan-Matsubara, J., Nelen, L. & DeLey, J. (1971) *Antonie van Leeuwenhoek* (*J. Microbiol. Serol.*), *37*, 233.

2-KETO-3-DEOXY-L-ARABONATE DEHYDRATASE

(2-Keto-3-deoxy-L-arabonate hydro-lyase)

2-Keto-3-deoxy-L-arabonate = 2-oxoglutarate semialdehyde + H_2O

Ref.

Molecular properties

source	value	conditions	Ref.
Pseudomonas saccharophilia	85,000	Sephadex G 200, pH 7.4	(1)

The enzyme does not contain any B_{12} derivative. (1)

Specific activity

P. saccharophilia (415 x) 54 2-keto-3-deoxy-L-arabonate (pH 7.2, Pi, 30°) (1)

Specificity and catalytic properties

source	substrate	K_m (µM)	conditions	Ref.
P. saccharophilia	2-keto-3-deoxy-L-arabonate[a]	70	pH 7.2, Pi, 30°	(1)

(a) the following compounds were inactive: glycerol; 1,2-propanediol and ethylene glycol.

The dehydratase does not require any added cofactor (e.g. B_{12} derivatives) or metal ions for activity. Thus, it was not inhibited by hydroxycobalamin or by EDTA. (1)

The mechanism of the reaction has been investigated. (1)

References

1. Stoolmiller, A.C. & Abeles, R.H. (1966) JBC, 241, 5764.

CDP GLUCOSE 4,6-DEHYDRATASE

(CDPglucose 4,6-hydro-lyase)

CDPglucose = CDP-4-keto-6-deoxy-D-glucose + H_2O

Ref.

Equilibrium constant

The reaction is essentially irreversible [F] with the enzyme from *Pasteurella pseudotuberculosis* as the catalyst. (1)

Specificity and catalytic properties

source	substrate	K_m (μM)	conditions	
P. pseudotuberculosis (purified 217 x)	CDPglucose[a]	17	pH 7.5, Tris, 38°	(1)
Salmonella typhi (purified 50 x)	CDPglucose	400	pH 7.0, Pi, 37°	(2)

(a) CDPglucose (1.0) could be replaced by dCDP glucose (1.0) but not by CDPmannose; CDPgalactose; GDPglucose; GDPmannose; dTDPglucose; dTDPgalactose; dTDPmannose; UDPglucose or ADPglucose.

The dehydratase requires NAD for activity. With the enzyme from *P. pseudotuberculosis*, NAD (1.0) could be replaced by reduced NAD (0.7) but NADP or reduced NADP were much less effective cofactors. (1,2)

The mechanism of the reaction has been investigated. (3)

References

1. Matsuhashi, S., Matsuhashi, M., Brown, J.G. & Strominger, J.L. (1966) JBC, 241, 4283.
2. Hey, A.E. & Elbein, A.D. (1966) JBC, 241, 5473.
3. Melo, A., Elliott, W.H. & Glaser, L. (1968) JBC, 243, 1467.

dTDP GLUCOSE 4,6-DEHYDRATASE

(dTDPglucose 4,6-hydro-lyase)

dTDPglucose = dTDP-4-keto-6-deoxy-D-glucose + H_2O

Ref.

Molecular properties

source	value	conditions	
Escherichia coli	80,000 [2]	pH 8.0; gel filtration.	(1)

The dehydratase contains one mole of NAD per 80,000 daltons. When the NAD is removed, the enzyme dissociates into subunits of 40,000. The process is reversible. Several NAD analogues bind to the apoenzyme but the resulting holoenzymes are catalytically inactive. The affinity for NAD of the apoenzyme is dependent on temperature: at 37°, the K_D = 0.5 mM and at 16°, 5 µM. (1,4,5)

Specific activity

E. coli B (1260 x)	13 dTDPglucose (pH 8.0, Tris, 37°; the reaction mixture contained EDTA)	(2)
E. coli	5.5 dTDPglucose (pH 8.0, Tris, 37°)	(5)

Specificity and catalytic properties

The removal of H_2O from dTDPglucose takes place in the following way:

dTDPglucose + [enzyme - NAD] =
dTDP-4-ketoglucose + [enzyme - reduced NAD]

dTDP-4-ketoglucose =
dTDP-4-keto-5,6-glucoseen + H_2O

dTDP-4-keto-5,6-glucoseen + [enzyme - reduced NAD] =
dTDP-4-keto-6-deoxyglucose + [enzyme-NAD]

The reactions mediated by enzyme-bound NAD have B stereospecificity with respect to the nicotinamide moiety. The enzyme - NAD complex does not require added NAD or any other cofactors or any metal ions for activity. EDTA did not inhibit. (1,2,3)

Ref.

Specificity and catalytic properties continued

substrate or inhibitor (a)	V relative	K_m (mM)	K_i (mM)	conditions	
dTDPglucose[b]	1.00	70	-	pH 8.0, Tris, 37°	(2,5)
dUDPglucose	1.00	2000	-	pH 8.0, Tris, 37°	(4,5)
dTDP-6-deoxyglucose[c]	-	-	0.22	pH 8.0, Tris, 37°	(1,2)
dTDP-6-deoxygalactose[d]	-	-	0.70	-	(1)
dTDP or dTTP[e]	-	-	0.20	pH 8.0, Tris, 37°	(1,4)

(a) C(dTDPglucose)

(b) UDPglucose and other ribonucleotide sugars; dTMP; dTTP-L-rhamnose and dTDP-4-keto-6-deoxyglucose were neither substrates nor inhibitors.

(c) dTDP-6-deoxyglucose reacts with enzyme-NAD to produce dTDP-4-keto-6-deoxy-glucose and enzyme-reduced NAD.

(d) dTDP-6-deoxygalactose has been reported not to inhibit in Ref. 2.

(e) dTMP, UDP, UDPglucose, GDP, ADP and CDP did not inhibit.

The mechanism of the reaction has been investigated. (1)

References

1. Glaser, L. & Zarkowsky, H. (1971) The Enzymes, 5, 467.
2. Wang, S.F. & Gabriel, O. (1969) JBC, 244, 3430.
3. Wang, S.F. & Gabriel, O. (1970) JBC, 245, 8.
4. Zarkowsky, H., Lipkin, E. & Glaser, L. (1970) JBC, 245, 6599.
5. Zarkowsky, H. & Glaser, L. (1969) JBC, 244, 4750.

GDP MANNOSE 4,6-DEHYDRATASE

(GDPmannose 4,6-hydro-lyase)

GDPmannose = GDP-4-keto-6-deoxy-D-mannose + H_2O

Ref.

Molecular properties

source	value	conditions	
Phaseolus vulgaris	120,000	pH 7.0	(1)

The enzyme from *Escherichia coli* had tightly bound NAD the presence of which was essential for the dehydratase activity. There was no evidence that the enzyme from *P. vulgaris* contained bound NAD. (2,1)

Specificity and Michaelis constants

source	substrate	K_m(µM)	conditions	
P. vulgaris (purified 47 x)	GDPmannose	28	pH 7.0, Pi, 37°	(1)
E. coli (purified 84 x)	GDPmannose[a]	550	pH 7.0, Pi, 37°	(2)

(a) GDPglucose was inactive.

The enzyme has no cofactor or metal ion requirement. The enzyme from *E. coli* was not inhibited by EDTA. (1,2)

Inhibitors

source	inhibitor	type	K_i(µM)	conditions	
P. vulgaris	GDPglucose	C(GDPmannose)	84	pH 7.0, Pi, 37°	(1)
	GTP[a]	C(GDPmannose)	19	pH 7.0, Pi, 37°	(1)

(a) GMP and ITP were also inhibitory. ATP, UTP and CTP were poor inhibitors.

References

1. Liao, T-H. & Barber, G.A. (1972) BBA, 276, 85.
2. Elbein, A.D. & Heath, E.C. (1965) JBC, 240, 1926.

D-GLUTAMATE CYCLASE

(D-Glutamate hydro-lyase (cyclizing))

D-Glutamate = D-2-pyrrolidone 5-carboxylate + H_2O

Ref.

Equilibrium constant

[D-pyrrolidone carboxylate] / [D-glutamate]	=		Ref.
	16	(pH 8.3, 37°)	(1)
	26.8	(pH 7.9, 30°)	(2)

Specificity and catalytic properties

The cyclization of D-glutamate to D-pyrrolidone carboxylate (also called D-pyroglutamate) involves the formation of a *cis*-peptide bond. The reaction, which occurs nonenzymically at an appreciable rate, is catalyzed by D-glutamate cyclase (EC 4.2.1.48) and also by glutamine synthetase (EC 6.3.1.2). The cyclase requires Mg^{2+} (or Mn^{2+}) for activity and EDTA was inhibitory. (1,2)

source	substrate	V relative	K_m (mM)	
Mouse liver (purified 180 x)[(a)]	D-glutamate	1.00	200	(2)
	D-2-pyrrolidone 5-carboxylate	0.017	90	(2)
	cis-D-3-fluoro 2-pyrrolidone 5-carboxylate	0.95	200	(2)
	trans-D-3-fluoro 2-pyrrolidone 5-carboxylate	0.95	1500	(2)
Mouse kidney (purified 60 x)[(b)]	D-glutamate	-	15	(1)

(a) conditions: pH 7.9, 2-methylimidazole, 30°.
(b) conditions: pH 8.3, ammediol, 37°. D-Glutamate (1.00) could be replaced by DL-γ-hydroxyglutamate, isomer A (0.69); DL-γ-hydroxyglutamate, isomer B (0.21); DL-β-hydroxyglutamate, isomer B (0.47); DL-β-methylglutamate (0.84); D-α-methylglutamate (0.21); D-γ-methylglutamate (0.78); DL-allo-α-aminotricarbalylate (0.95) or by DL-α-aminotricarbalylate (0.47) but not by L-glutamate; D- or L-glutamine; D-aspartate; D-α-aminoadipate or D-homoglutamine.

References

1. Meister, A., Bukenberger, M.W. & Strassburger, M. (1963) BZ, 338, 217.
2. Unkeless, J.C. & Goldman, P. (1971) JBC, 246, 2354.

UROCANATE HYDRATASE

(3-(4'-Hydroxyimidazole-5'-yl)-propionate hydro-lyase)

3-(4'-Hydroxyimidazole-5'-yl)-propionate = trans-urocanate + H_2O

Ref.

Equilibrium constant

The reaction is irreversible [R]. The product of the reaction, 3-(4'-hydroxyimidazole-5'-yl)-propionate, spontaneously tautomerizes to racemic 3-(imidazole-4'-one-5'-yl)-propionate. (1)

Molecular properties

source	value	conditions	Ref.
Pseudomonas putida[a]	110,000 [2]	pH 7.5; gel electrophoresis (SDS); amino acid composition; peptide mapping.	(2,3)
Bacillus subtilis[b]	120,000 [at least 2]	sucrose density gradient; gel electrophoresis (SDS).	(4)
Cat liver[c]	127,000	Sephadex G 200.	(6)

(a) the enzyme contains two moles of 2-oxobutyrate per 110,000 daltons. It contains no pyridoxal phosphate, biotin, FAD or FMN.

(b) pyridoxal phosphate was required to stabilize the enzyme but it was not required in the reaction.

(c) the enzyme had a light-absorption maximum at 279 nm but there was no maximum above this wavelength.

Specific activity

P. putida (65 x)	5.1[a]	urocanate (pH 7.5, Pi, 25°)	(2)
Beef liver (300 x)	0.13	urocanate (pH 7.2, Pi, 25°)	(5)
Cat liver (300 x)	0.2	urocanate (pH 7.4, Pi, 30°)	(6)

(a) this is a maximum value.

Ref.

Specificity and catalytic properties

source	substrate	K_m (μM)	conditions	
P. putida	urocanate	240	pH 7.5, Pi, 25°	(2)
B. subtilis (purified 38 x)	urocanate	77	pH 7.5, Pi, 37°	(4)
Beef liver	urocanate	5	pH 7.2, Pi, 25°	(5)
Cat liver	urocanate[a]	7.1	pH 7.4, Pi, 30°	(6)

(a) no other substance tested could replace trans-urocanate e.g. cis-urocanate; vinyl imidazole or sorbic acid.

Urocanase does not require added cofactors (e.g. pyridoxal phosphate) or metal ions for activity. It was not inhibited by EDTA. (4,5,6)

Urocanase was inhibited by 4-imidazolone 5-propionate (C(urocanate); Ki = 0.03 - 0.8 mM depending on the source of the enzyme). The enzyme from P. putida was also inhibited by succinate and fumarate. (5,6,7)

Light absorption data

Urocanate has an absorption band at 277 nm (molar extinction coefficient = 18,800 $M^{-1}cm^{-1}$). (2)

References

1. Kaeppeli, F. & Retey, J. (1971) EJB, 23, 198.
2. George, D.J. & Phillips, A.T. (1970) JBC, 245, 528.
3. Lynch, M.C. & Phillips, A.T. (1972) JBC, 247, 7799.
4. Kaminskas, E., Kimhi, Y. & Magasanik, B. (1970) JBC, 245, 3536.
5. Hassall, H. & Greenberg, D.M. (1971) Methods in Enzymology, 17B, 84.
6. Swaine, D. (1969) BBA, 178, 609.
7. Hug, D.H., Roth, D. & Hunter, J. (1968) J. Bact., 96, 396.

DIHYDRODIPICOLINATE SYNTHASE

(L-Aspartate-β-semialdehyde hydro-lyase (adding pyruvate and cyclizing)

L-Aspartate β-semialdehyde + pyruvate = dihydrodipicolinate + $2H_2O$

Ref.

Equilibrium constant

The reaction is probably irreversible [F]. 2,5-Dihydrodipicolinate is in equilibrium with 2,3-dihydrodipicolinate and 4-hydroxy-2,3,4,5-tetrahydrodipicolinate and it is not clear which of these is the immediate product of the reaction. (1)

Molecular properties

source	value	conditions	Ref.
Escherichia coli W[(a)]	134,000 [4]	Sephadex G 200, pH 7.5; sucrose density gradient; amino acid composition.	(1)
E. coli K_{12}	112,000	Sephadex G 200, pH 8.1. The buffer contained Mg^{2+}, EDTA and L-lysine.	(2)

(a) when the enzyme is reduced with sodium borohydride in the presence of pyruvate, a covalent enzyme-substrate complex is formed.

Specific activity

Escherichia coli W	(5000 x)	100	aspartate semialdehyde (pH 7.4, imidazole, 25°)	(1)

Specificity and catalytic properties

source	substrate	K_m (mM)	conditions	Ref.
E. coli W	pyruvate[(a)]	0.25	pH 7.4, Tris, 25°	(3,4)
	aspartate semialdehyde[(b)]	0.13	pH 7.4, Tris, 25°	(3,4)

(a) phospho*enol*pyruvate and oxaloacetate were inactive.

(b) *N*-acetyl aspartate semialdehyde and succinate semialdehyde were inactive.

The enzyme (E. coli W) was subject to feedback inhibition by L-lysine (C(aspartate semialdehyde); Ki = 0.21 mM). (3,4)

The enzyme from E. coli K_{12} was also inhibited by L-lysine and in this case the inhibition was cooperative: in the presence of L-lysine, sigmoid substrate dependence curves were obtained whereas in its absence Michaelis Menten kinetics were observed. (2)

Ref.

Light absorption data

The condensation reaction can be followed at 270 nm when it is carried out in imidazole (but not Pi or Tris) buffers. (1,3)

References

1. Shedlarski, J.G. & Gilvarg, C. (1970) JBC, 245, 1362.
2. Truffa-Bachi, P., Patte, J-C & Cohen, G.N. (1967) Comp. Rend. Ser. D, 265, 928.
3. Yugari, Y. & Gilvarg, C. (1965) JBC, 240, 4710.
4. Shedlarski, J.G. (1971) Methods in Enzymology, 17B, 129.

CROTONOYL ACYL CARRIER PROTEIN HYDRATASE

(D-3-Hydroxybutyryl-acyl carrier protein hydro-lyase)

D-3-Hydroxybutyryl-ACP = crotonoyl-ACP + H_2O

Ref.

Equilibrium constant

$$\frac{[\text{crotonoyl-ACP}]\ [H_2O]}{[\text{D-3-hydroxybutyryl-ACP}]} = 19\ M\ (\text{pH } 8.5, \text{Tris}, 25^o) \quad (1)$$

Molecular properties

source	value	conditions	Ref.
Escherichia coli	26,000	Sephadex G 100, pH 7.0. The buffer contained mercaptoethanol	(1)

Crotonoyl-ACP hydratase is a component of the fatty acid synthetase complex (see EC 2.3.1.38).

Specific activity

			Ref.
E. coli	(525 x)	5 crotonoyl-ACP (pH 8.5, Tris, 25°)	(1)

Specificity and Michaelis constants

source	substrate	V relative	K_m (μM)	conditions	Ref.
E. coli	crotonoyl-ACP[a]	1.00	10	pH 8.5, Tris, 25°	(1)
	2-hexenoyl-ACP	0.56	10	pH 8.5, Tris, 25°	(1)
	2-octenoyl-ACP	0.05	10	pH 8.5, Tris, 25°	(1)

(a) 2-decenoyl-ACP and crotonoyl-CoA were inactive.

There are conflicting reports as to whether E. coli has a single 3-hydroxyacyl-ACP dehydratase which acts on C_4 to C_{16} substrates or whether there are 4 enzymes, namely, a short chain dehydratase (EC 4.2.1.58), a medium chain dehydratase (EC 4.2.1.59) and 2 long chain dehydratases (EC 4.2.1.60 and 61). The properties of a single, non specific dehydratase from E. coli (purified 2900 x) has been described; this enzyme has a molecular weight of 170,000 daltons. It catalyzes the hydration of cis-5-trans-2-dodecadienoyl-ACP (an intermediate in the synthesis of unsaturated fatty acids) in addition to its catalytic role in the fatty acid synthetase complex. (2,3)

Light absorption data

2,3-Dehydroacyl thioesters (e.g. crotonoyl-ACP) have an absorption band at 263 nm (molar extinction coefficient = 6700 $M^{-1}cm^{-1}$). (2)

Abbreviations

ACP acyl carrier protein (see EC 2.3.1.39)

References

1. Mizugaki, M., Weeks, G., Toomey, R.E. & Wakil, S.J. (1968) JBC, 243, 3661.
2. Birge, C.H. & Vagelos, P.R. (1972) JBC, 247, 4930.
3. Prescott, D.J. & Vagelos, P.R. (1972) Advances in Enzymology, 36, 299.

3-HYDROXYDECANOYL ACYL CARRIER PROTEIN DEHYDRATASE

(D-3-Hydroxydecanoyl acyl carrier protein hydro-lyase)

D-3-Hydroxydecanoyl-ACP = 2,3-*trans*- or 3,4-*cis*-decenoyl-ACP + H_2O

Ref.

Equilibrium constant

At equilibrium (pH 7.0, Pi, 30°): (1,2)

D-3-hydroxydecanoyl-NAC = 71.8%
2,3-*trans*-decenoyl-NAC = 25.8%
3,4-*cis*-decenoyl-NAC = 2.4%

Molecular properties

source	value	conditions	Ref.
Escherichia coli	36,000 [2]	pH 7.0; Sephadex G 150, pH 8.0; gel electrophoresis (SDS); amino acid composition.	(3)
	28,000	sucrose density gradient; Sephadex G 100, pH 7.0	(2,4)

The dehydratase is a component of the fatty acid synthetase complex (see EC 2.3.1.38).

Specific activity

E. coli	(1200 x)	6	3,4-*cis*-decenoyl-NAC isomerized to 2,3-*trans*-decenoyl-NAC (pH 8.0, Tris, 25°)	(3,4)

Specificity and catalytic properties

The dehydratase catalyzes the reversible interconversionsof 3 substrates (the numbers in brackets are relative velocities):

D-3-hydroxydecanoyl-NAC
(1) ↓↑ (36) (10) ↑↓ (5)
3,4-*cis*-decenoyl-NAC ⇌ (140) / (11) 2,3-*trans*-decenoyl-NAC

The dehydratase was inactive with 3,4-*trans*-decenoyl-NAC. It could utilize a number of thioesters of D-3-hydroxydecanoate: -ACP (1.00); -NAC (0.17); -pantetheine (0.10) and -CoA (0.10). It was most active with 3-hydroxydecanoyl-NAC (1.00) which could be replaced by 3-hydroxyoctanoyl-NAC (0.07) or 3-hydroxydodecanoyl-NAC (0.05). The 3-hydroxydecanoyl-thioester substrates were, however, the only substrates to

Specificity and catalytic properties continued

produce the corresponding 3,4-unsaturated products: the C_8 and C_{12} 3-hydroxyacyl thioesters yielded only the corresponding 2,3-unsaturated products. (1,2,4)

The mechanism of action of the dehydratase has been investigated. The enzyme does not require any added cofactors or metal ions for activity. (5)

There are conflicting reports as to whether *E. coli* has a single 3-hydroxyacyl ACP dehydratase which acts on C_4 to C_{16} substrates or whether there are 4, more specific, enzymes. (see EC 4.2.1.58).

Inhibitors

3-Decynoyl-NAC is an active-site directed inhibitor (all the reactions are inhibited) which forms a covalent bond with an active-site histidyl residue. 2 Moles of inhibitor are bound per 36,000 daltons. The following are inactive as inhibitors: 3-decynoate and its methylester; 2-decynoyl-NAC and 3-octynoyl-NAC. (2,3)

Light absorption data

See EC 4.2.1.58.

Abbreviations

ACP acyl carrier protein (see EC 2.3.1.39)
NAC *N*-acetylcysteamine

References

1. Bloch, K. (1969) Accounts Chem. Res., 2, 193.
2. Kass, L.R. (1969) Methods in Enzymology, 14, 73.
3. Helmkamp, G.M. & Bloch, K. (1969) JBC, 244, 6014.
4. Kass, L.R., Brock, D.J.H. & Bloch, K. (1967) JBC, 242, 4418, 4432.
5. Rando, R.R. & Bloch, K. (1968) JBC, 243, 5627.

ARENE-OXIDE HYDRATASE

(Dihydrodiol hydro-lyase (arene-oxide-forming))

Dihydrodiol = arene oxide + H_2O

Ref.

Specificity and catalytic properties

source	substrate	K_m (mM)	conditions	
Guinea-pig liver microsomes (purified 41 x)	styrene oxide[a]	0.53	pH 9.0, Tris, 37°	(1,2,3)

(a) styrene oxide (1.00) could be replaced by naphthalene-1,2-oxide (4.15); indene-1,2-oxide (3.07); octene-1,2-oxide (2.79); phenanthrene-9,10-oxide (2.51); 4-chlorophenyl-2,3-epoxypropyl ether (0.95); benzene oxide (0.26); cyclohexene oxide (0.08); bisnorsqualene-1,2-oxide (0.03) or squalene-2,3-oxide (very low activity). The guinea-pig liver may have several arene-oxide hydratases with overlapping specificity properties since crude homogenates had different relative activities.

Arene-oxide hydratase does not have any cofactor or metal ion requirement. Thus, the guinea-pig liver enzyme was not inhibited by EDTA. (1)

References

1. Oesch, F. & Daly, J.W. (1971) BBA, 227, 692.
2. Oesch, F., Jerina, D.M. & Daly, J.W. (1971) ABB, 144, 253.
3. Oesch, F., Kaubisch, N., Jerina, D.M. & Daly, J.W. (1971) B, 10, 4858.

3-CYANOALANINE HYDRATASE

(L-Asparagine hydro-lyase)

L-Asparagine = 3-cyanoalanine + H_2O

Ref.

Equilibrium constant

The reaction is irreversible [R] with the enzyme from *Lupinus angustifolius* as catalyst. (1)

Molecular weight

source	value	conditions	
L. angustifolius seed (Blue lupine)	400,000-500,000	Sepharose 6B, pH 8	(1)

Specificity and catalytic properties

source	substrate	K_m (mM)	conditions	
L. angustifolius (purified 300 x)	3-cyanoalanine[a]	2	pH 8.0, Tris, 37°	(1)

(a) the following compounds were neither substrates nor inhibitors: asparagine; glutamine; 5-diazo-4-oxo-L-norvaline; urea; 3-alanylamide; citrulline; arginine; nicotinamide; 3-aminopropionitrile; 2-aminopropionitrile; 2-amino-4-cyanobutyrate; ricinine; toyocamycin; indole-3-acetamide and indole-3-acetonitrile.

References

1. Castric, P.A., Farnden, K.J.F. & Conn, E.E. (1972) ABB, 152, 62.

CHONDROITIN ABC LYASE

(Chondroitin ABC lyase)

Elimination of Δ-4,5-D-glucuronate residues from polysaccharides containing 1,4-β-hexosaminyl and 1,3-β-D-glucuronosyl or 1,3-α-L-iduronosyl linkages, thus bringing about depolymerization

Ref.

Molecular weight

source	value	conditions	
Proteus vulgaris	~150,000	sucrose density gradient	(1)

Specific activity

P. vulgaris (134 x)	130 μmoles of ΔDi-6S released from chondroitin sulphate C(pH 8.0, Tris, 37°)	(1)

Specificity and catalytic properties

Chondroitin ABC lyase catalyzes the following reactions:

chondroitin sulphate A = n Δ Di-4S
chondroitin sulphate B = n Δ Di-4S
chondroitin sulphate C = n Δ Di-6S

It also catalyzes, at low rates, the degradation of chondroitin and hyaluronate but not of keratosulphate, heparin or heparin sulphate. (1)

Light absorption data and abbreviations

compound	abbreviation	molar extinction coefficient at 232 nm	
2-acetamido-2-deoxy-3-O-(β-D-gluco-4-enepyranosyluronate)-D-galactose	Δ Di-0S	5700 $M^{-1}cm^{-1}$	(1)
2-acetamido-2-deoxy-3-O-(β-D-gluco-4-enepyranosyluronate)-4-O-sulpho-D-galactose	Δ Di-4S	5100 $M^{-1}cm^{-1}$	(1)
2-acetamido-2-deoxy-3-O-(β-D-gluco-4-enepyranosyluronate)-6-O-sulpho-D-galactose	Δ Di-6S	5500 $M^{-1}cm^{-1}$	(1)

References

1. Yamagata, T., Saito, H., Habuchi, O. & Suzuki, S. (1968) JBC, 243, 1523; 1536.

ETHANOLAMINE PHOSPHATE PHOSPHO-LYASE

(Ethanolamine phosphate phospho-lyase (deaminating))

Ethanolamine phosphate + H_2O = acetaldehyde + NH_3 + Pi

Ref.

Molecular weight

source	value	conditions	
Rabbit liver	168,000	Sephadex G 200, pH 7.8. In the presence or absence of pyridoxal phosphate	(1)

Specificity and catalytic properties

source	substrate	K_m (μM)	conditions	
Rabbit liver (purified 107 x)	ethanolamine phosphate[(a)]	610	pH 7.8, Tris, 37°	(1,3)
	pyridoxal phosphate[(b)]	0.27	pH 7.8, Tris, 37°	(1,3)

(a) the following were inactive: 2-aminoethylphosphonate; 2-aminoethanethiol; 2-aminoethanol; 2-aminoethanol-O-sulphate; ethylamine; ethylenimine; β-glycerophosphate; glycollate; glyoxal; glyoxalate; O-phosphocholine; O-phospho-DL-serine; O-phospho-L-threonine; taurine or ethanolamine.

(b) the enzyme contains tightly bound pyridoxal phosphate (essential for activity) which was removed during the purification procedure. The pyridoxal phosphate could not be replaced by vitamin B_{12}.

The phospho-lyase was inhibited by ethylphosphate (Ki = 29 μM) and Pi (C(ethanolamine phosphate); Ki = 1.3 mM) and also by cyanide. (1)

The mechanism of the reaction has been investigated. (1,2)

References

1. Fleshood, H.L. & Pitot, H.C. (1970) JBC, 245, 4414.
2. Sprinson, D.B. & Weliky, I. (1969) BBRC, 36, 866.
3. Fleshood, H.L. & Pitot, H.C. (1969) BBRC, 36, 110.

CYSTEINE SYNTHASE

(O-Acetyl-L-serine acetate-lyase (adding hydrogen-sulphide))

O-Acetyl-L-serine + H_2S = L-cysteine + acetate

Ref.

Molecular properties

source	value	conditions	Ref.
Salmonella typhimurium[a]	68,000 [2]	ultracentrifugation with or without guanidine-HCl; amino acid analysis; peptide mapping	(1)

(a) the enzyme contains two moles of pyridoxal phosphate (which are required for activity) per 68,000 daltons.

Cysteine synthase is a component of the cysteine synthetase complex of S. typhimurium. This complex (molecular weight = 309,000 daltons) has been purified and is composed of two molecules of cysteine synthase and one molecule of serine acetyltransferase (EC 2.3.1.30). The complex dissociates in the presence of O-acetyl-L-serine. (2,3)

Specific activity

S. typhimurium (55 x) 1100 O-Acetyl-L-serine (pH 7.4, Tris, 25°) (1)

Specificity and catalytic properties

source	substrate	K_m (mM)	conditions	Ref.
S. typhimurium	O-acetyl-L-serine[a]	5	pH 7.4, Tris, 25°	(1,3)
	H_2S[b]	<0.1	pH 7.4, Tris, 25°	(1,3)

(a) the Ks for O-acetyl-L-serine = 0.6 μM. The following compounds were inactive: L-serine; N-acetyl-L-serine; O-acetyl-L-threonine; O-succinyl-DL-serine; O-acetyl-DL-homoserine and O-succinyl-L-homoserine.

(b) the following were inactive: β-mercaptopyruvate; L-cysteine and 2-mercaptoethanol. L-Cysteine was non-inhibitory. With methyl mercaptan, S-methylcysteine was produced.

The enzyme was not inhibited by EDTA. (1,3)

The cysteine synthase reaction is also catalyzed by β-cyanoalanine synthase (EC 4.4.1.9).

References

1. Becker, M.A., Kredich, N.M. & Tomkins, G.M. (1969) JBC, 244, 2418.
2. Kredich, N.M., Becker, M.A. & Tomkins, G.M. (1969) JBC, 244, 2428.
3. Kredich, N.M. & Becker, M.A. (1971) Methods in Enzymology, 17B, 459.

CYSTATHIONINE γ-SYNTHASE

(O-Succinyl-L-homoserine succinate-lyase (adding L-cysteine))

O-Succinyl-L-homoserine + L-cysteine = L-cystathionine + succinate

Ref.

Molecular properties

source	value	conditions	
Salmonella typhimurium	160,000 [4]	ultracentrifugation with or without guanidine-HCl; amino acid analysis.	(1,2)

The enzyme contains 4 moles of tightly bound pyridoxal phosphate per 160,000 daltons. The Km for pyridoxal phosphate is 4×10^{-8}M. (1,2,3)

Specificity and catalytic properties

Cystathionine γ-synthase catalyzes the following types of reaction:

Reaction 1 A γ-replacement reaction

O-Succinyl-L-homoserine + L-cysteine =
L-cystathionine + succinate

Reaction 2 A γ-elimination reaction

O-Succinyl-L-homoserine + H_2O =
2-oxobutyrate + succinate + NH_3

Reaction 3 A β-replacement reaction

O-Succinyl-L-serine + L-homocysteine =
succinate + L-cystathionine

Reaction 4 A β-elimination reaction

O-Succinyl-L-serine + H_2O =
pyruvate + succinate + NH_3

Reaction 3 proceeds at about 0.2% of the rate obtained with reaction 1. (1,3,4)

The enzyme also catalyzes rapid exchanges of the α- and β-hydrogen atoms in several amino acids. (3)

Ref.

Specificity and catalytic properties continued

substrate	reaction	V(a)	K_m	conditions	Ref.
L-cysteine	1 (b)	94	7×10^{-5}	pH 7.8, PPi, 37°	(1,2)
H_2S	1 (b)	69	3×10^{-3}	pH 7.8, PPi, 37°	(1,2)
CH_3SH	1 (b)	13	1×10^{-1}	pH 7.8, PPi, 37°	(1,2)
O-succinyl-L-homoserine	1 (c)	94	4×10^{-3}	pH 7.8, PPi, 37°	(1,2)
O-succinyl-L-homoserine (d)	2	19	3×10^{-4}	pH 8.2, PPi, 37°	(1,2)
cystathionine	2	0.25	4×10^{-3}	pH 8.2, PPi, 37°	(1,2)
O-acetyl-L-homoserine	2	3	3×10^{-2}	pH 8.2, PPi, 37°	(1,2)

(a) specific activity

(b) second substrate = O-succinyl-L-homoserine. With H_2S, homocysteine was the product and with CH_3SH, methionine.

(c) second substrate = L-cysteine.

(d) the following were inactive as substrates or inhibitors: O-acetyl-L-serine; DL-homoserine; L-serine; O-phosphoryl-L-homoserine (slightly inhibitory) DL-threonine and L-methionine. β-Mercaptopropionate was NC(Ki = 1 mM) and cystathionine C(Ki = 5 mM); D-cysteine and L-homocysteine also inhibited the γ-elimination reaction.

Abbreviations

cystathionine — 2-amino-4-[(2-amino-2-carboxyethyl)thio]-butyric acid

References

1. Kaplan, M.M. & Guggenheim, S. (1971) Methods in Enzymology, 17B, 425.
2. Kaplan, M.M. & Flavin, M. (1966) JBC, 241, 5781; 4463.
3. Guggenheim, S. & Flavin, M. (1969) JBC, 244, 6217; 3722.
4. Guggenheim, S. & Flavin, M. (1971) JBC, 246, 3562.

METHIONINE SYNTHASE

(O-Acetyl-L-homoserine acetate-lyase (adding methanethiol))

O-Acetyl-L-homoserine + methanethiol = L-methionine + acetate

Ref.

Specificity and catalytic properties

Methionine synthase (reaction 1) also catalyzes the following reaction (reaction 2):-

O-Acetyl-L-homoserine + H_2S = L-homocysteine + acetate

The purified enzyme (Neurospora crassa) requires pyridoxal phosphate for activity. (1)

source	substrate	reaction	V relative	K_m (mM)	conditions	
N. crassa (purified 500 x)	O-acetylhomo-serine[a]	2	-	7	pH 8.0, Tris, 30°	(1)
	methanethiol	1	1.0	0.8		
	H_2S	2	0.5	0.7		

(a) O-acetylhomoserine could not be replaced by L-homoserine; O-succinyl-DL-homoserine; O-phosphoryl-L-homoserine; L-serine; O-acetyl-L-serine or O-phosphoryl-L-serine. The enzyme does not form cystathionine from cysteine nor does it have γ-elimination activity with O-acetylhomoserine or cystathionine (see EC 4.2.99.9 for these activities).

The enzyme (N. crassa) was not inhibited by methionine or S-adenosylmethionine. (1,2)

References

1. Kerr, D. (1971) Methods in Enzymology, 17B, 446.
2. Kerr, D. & Flavin, M. (1969) BBA, 177, 177.

METHYLGLYOXAL SYNTHASE

(Dihydroxyacetone-phosphate phospho-lyase)

Dihydroxyacetone phosphate = methylglyoxal + Pi

Ref.

Equilibrium constant

The reaction is essentially irreversible [F] with the enzyme from *Escherichia coli* K_{12} as the catalyst. (1)

Molecular weight

source	value	conditions	Ref.
E. coli	67,000	Sephadex G 100 (1mM KH_2PO_4)	(2)

Specific activity

E. coli (1650 x) 530 dihydroxyacetone phosphate (pH 7.0, imidazole, 30°) (2)

Specificity and catalytic properties

source	substrate	K_m (mM)	conditions	Ref.
E. coli	dihydroxyacetone phosphate[(a)]	0.47	pH 7.0, imidazole, 30°	(2)

(a) dihydroxyacetone; DL-glyceraldehyde 3-phosphate and fructose 1,6-diphosphate were inactive. The Km given was obtained in the absence of the inhibitor, Pi. In the presence of Pi, substrate dependence curves were sigmoidal and the results obtained suggest that there are at least three binding sites for dihydroxyacetone phosphate and two for Pi per mole of enzyme.

The enzyme had no cofactor or metal ion requirement. It was not inhibited by EDTA. (1,2)

The synthase was inhibited by Pi (see above); PPi (C(dihydroxyacetone phosphate); Ki = 95 µM), phospho*enol*pyruvate, 3-phosphoglycerate and arsenate. (2)

References

1. Hopper, D.J. & Cooper, R.A. (1971) FEBS lett. 13, 213.
2. Hopper, D.J. & Cooper, R.A. (1972) BJ, 128, 321.

ETHANOLAMINE AMMONIA-LYASE

(Ethanolamine ammonia-lyase)

Ethanolamine = acetaldehyde + NH_3

Ref.

Equilibrium constant

The reaction is essentially irreversible [F] with the clostridial enzyme as catalyst. (1,4)

Molecular properties

source	value	conditions	Ref.
Clostridium sp.	520,000 [8-10]	ultracentrifugation in pH 7.4 + ethanolamine + Pi or guanidine-HCl; amino acid composition.	(4)

When the enzyme was isolated under mild conditions it contained 1.35 to 3.1 moles of α-(adenylyl) cobamide per 520,000 daltons. The apo-enzyme has two active cobamide binding sites per 520,000 daltons. Incubation of the deaminase with cyanocobalamin, hydroxycobalamin or methylcobalamin resulted in a concentration and time dependent inhibition phenomenon which was not reversed by α-(adenylyl) cobamide. (4,6)

Specific activity

Clostridium sp. (45 x) 15.9 ethanolamine (pH 7.4, Pi) (4)

Specificity and catalytic properties

source	substrate	K_m (μM)	conditions	Ref.
Clostridium sp.	ethanolamine[(a)]	22	pH 7.4, Pi	(4)
	α-(adenylyl)cobamide	7.7	pH 7.4, Pi	(4)
	α-(dimethylbenzimidazolyl)cobamide	1.5	pH 7.4, Pi	(4)
	α-(benzimidazolyl)cobamide	0.2	pH 7.4, Pi	(4)
	potassium ion[(b)]	470	pH 7.4, Tris	(4)

(a) the following were neither inhibitors nor substrates: glycerol; ethylene glycol; propanediol; 2-amino-2-methylpropanediol; glycine; sarcosine; DL-serine and choline. The following were inactive as substrates but they were competitive inhibitors of ethanolamine utilization: DL-2-amino 1-propanol (Ki = 7.1 μM); 1-amino 2-propanol (Ki = 20 μM) and 2-amino 1-butanol (Ki = 27 μM).

(b) the enzyme required K^+ for activity. K^+ could be replaced by $NH_4{}^+$ or Rb^+ but Na^+ and Li^+ were inhibitory.

Ref.

Specificity and catalytic properties continued

The mechanism of the reaction has been investigated. (3,5)

References

1. Barker, H.A. (1972) Ann. Rev. Biochem, 41, 66.
2. Stadtman, T.C. (1972) The Enzymes, 6, 540.
3. Abeles, R.H. (1971) The Enzymes, 5, 481.
4. Kaplan, B.H. & Stadtman, E.R. (1968) JBC, 243, 1787; 1794.
5. Carty, T.J., Babior, B.M. & Abeles, R.H. (1971) JBC, 246, 6313.
6. Babior, B.M. & Li, T.K. (1969) B, 8, 154.

UROPORPHYRINOGEN I SYNTHASE

(Porphobilinogen ammonia-lyase (polymerizing))

4 Porphobilinogen = uroporphyrinogen I + 4 NH_3

Ref.

Molecular weight

source	value	conditions	Ref.
Rhodopseudomonas spheroides	36,000 [1]	Sephadex G 100, pH 7.2; sucrose density gradient and gel electrophoresis (SDS)	(1)

Specific activity

R. spheroides (700 x) 2.1 porphobilinogen (pH 7.2, Tris, 37°) (1)

Specificity and Michaelis constants

source	substrate	K_m(M)	conditions	Ref.
R. spheroides	porphobilinogen	1.6×10^{-5}	pH 7.6, Tris, 37°	(1)
Spinach leaf (purified 67 x)	porphobilinogen[(a)]	7.2×10^{-5}	pH 8.2, Tris, 37°	(2)

(a) haemopyrroledicarboxylate; cryptopyrroledicarboxylate and carboxyporphobilinogen were inactive.

The mechanism of the reaction is discussed in Refs. 3 and 4.

Inhibitors

The enzyme (spinach leaf) was inhibited by opsopyrrole dicarboxylate and isoporphobilinogen (C(porphobilinogen)). (2,4)

References

1. Jordan, P.M. & Shemin, D. (1973) JBC, 248, 1019.
2. Bogorad, L. (1962) Methods in Enzymology, 5, 885.
3. Radmer, R. & Bogorad, L. (1972) B, 11, 904.
4. Pluscec, J. & Bogorad, L. (1970) B, 9, 4736.

SERINE SULPHATE AMMONIA-LYASE

(L-Serine-O-sulphate ammonia-lyase (pyruvate forming))

L-Serine O-sulphate + H_2O = pyruvate + NH_3 + sulphate

Ref.

Specific activity

Rat liver (520 x) 5 L-serine sulphate (pH 7.0, Tris, 37°) (1,3)

Specificity and catalytic properties

source	substrate	K_m (mM)	conditions	
Rat liver	L-serine sulphate	2.3[a]	pH 7.0, Tris, 37°	(1)

(a) a value of 23 mM is reported in Ref. 2. L-Serine sulphate (1.00) could be replaced by D-serine sulphate (0.03); β-chloro-L-alanine (1.35; the products were pyruvate, NH_3 and chloride); L-seryl-L-alanine-O-sulphate (0.29); L-serylglycine-O-sulphate (0.24); L-seryl-L-valine-O-sulphate (0.11); L-seryl-L-leucine-O-sulphate (0.05); L-threonine-O-sulphate (0.02) and L-seryl-L-phenylalanine-O-sulphate (0.01) but not the following:- DL-α-methylserine-O-sulphate; glycyl-L-serine-O-sulphate; L-hydroxyproline-O-sulphate; DL-homoserine-O-sulphate; ethanolamine O-sulphate; glycollic acid sulphate; 3-hydroxypropionic acid sulphate; L-serine-N-sulphate; O-methyl-DL-serine; O-acetyl-L-serine; O-phospho-L-serine; L-serine; L-threonine or DL-homoserine.

The lyase neither contained pyridoxal phosphate nor was it activated by this cofactor. In the presence of cupric chloride, pyridoxal phosphate catalyzes the nonenzymic breakdown of serine sulphate to pyruvate, NH_3 and sulphate. (1,3)

The cleavage of L-serine sulphate was competitively inhibited by D-serine sulphate (Ki = 18 mM) and DL-homoserine-O-sulphate (Ki = 56 mM) and noncompetitively inhibited by O-phospho-L-serine (Ki = 6 mM) and O-acetyl-L-serine (Ki = 1.7 mM). (1,2)

References

1. Thomas, J.H. & Tudball, N. (1967) BJ, 105, 467.
2. Tudball, N. & Thomas, J.H. (1971) Methods in Enzymology, 17B, 361.
3. Tudball, N., Thomas, J.H. & Fowler, J.A. (1969) BJ, 114, 299.

ORNITHINE CYCLODEAMINASE

(L-Ornithine ammonia-lyase (cyclizing))

L-Ornithine = L-proline + NH_3

Ref.

Equilibrium constant

The reaction is essentially irreversible [F] with the enzyme from *Clostridium* PA 3679 as the catalyst. (1)

Molecular weight

source	value	conditions	Ref.
Clostridium PA 3679	100,000-120,000	sucrose density gradient	(1)

Specificity and catalytic properties

source	substrate	K_m (mM)	conditions	Ref.
Clostridium PA 3679 (purified 37 x)	L-ornithine[(a)]	11.1	pH 8.0, Tris, 40°	(1)
	NAD[(b)]	6.1	pH 8.0, Tris, 40°	(1)

(a) the following were neither substrates nor inhibitors: L-lysine; L-alanine; L-leucine; L-isoleucine; L-proline or D-proline. D-Ornithine was not a substrate but it inhibited L-ornithine utilization.

(b) NAD (1.00), which was required by the enzyme, could be replaced by mercaptoethanol (0.50); NADP (0.34) or FAD (0.16) but not by FMN or NMN.

References

1. Costilow, R.N. & Laycock, L. (1971) JBC, 246, 6655.

S-(HYDROXYALKYL)GLUTATHIONE LYASE

(S-(2-Hydroxyalkyl)glutathione alkylepoxide-lyase)

S-(2-Hydroxyalkyl)glutathione = glutathione + alkyl-epoxide

Ref.

Molecular weight

source	value	conditions	
Rat liver	40,000 [2]	pH 6.5 (in the presence of EDTA and reduced glutathione); Sephadex G 100; gel electrophoresis (SDS).	(1,2)

Multiple forms have been observed.

Specific activity

Rat liver (2250 x) 29.2 1,2-epoxy-3(4-nitrophenoxy)propane (pH 6.5, Pi, 25°) (2)

Specificity and kinetic properties

source	substrate	K_m (mM)	conditions	
Rat liver (highly purified)	glutathione[a]	8.1	pH 6.5, Pi, 25°	(1,2)
	1,2-epoxy-3(4-nitrophenoxy)propane [b]	0.17	pH 6.5, Pi, 25°	(1)

(a) mercaptoethanol, cysteine, thioglycerol and dithiothreitol were inactive.

(b) 1,2-epoxy-3(4-nitrophenoxy)propane (1.00) could be replaced by 1,2-epoxy-3-phenoxypropane (1.45); 2,3-epoxypropyl methacrylate (1.14); 2,3-epoxypropyl acrylate (1.10); 1-allyloxy-2,3-epoxypropane (1.06); 3,3,3-trichloro-1,2-epoxypropane (0.97); 1,2-epoxybutane (0.66); 1,2-epoxyethylbenzene (0.61); epoxystyrene (0.36); 1,2-epoxy-3-(4-chlorophenoxy)propane (0.30); epichlorohydrin (0.26); glycidaldehyde (0.26); 1,2-epoxypropane (0.25); butadiene monoxide (0.08) or methyl-10,11-epoxyundecanoate (0.08) but not by benzene oxide; 1,4-dimethylbenzene oxide; naphthalene-1,2-oxide; phenanthrene-9,10-oxide; ethyl-2,3-epoxybutyrate; _cis_- or _trans_-epoxysuccinate; glycidol; _trans_-stilbene oxide; cyclohexene oxide; scopalamine, dieldrin or glycidyl-lauryl ether.

Light absorption data

The thiolysis of 1,2-epoxy-3-(4-nitrophenoxy)propane is accompanied by a change in absorption at 360 nm (molar extinction coefficient = 508 $M^{-1}cm^{-1}$). (1,2)

References

1. Jakoby, W.W. & Fjellstedt, T.A. (1972) The Enzymes, 7, 205.
2. Fjellstedt, T.A., Allen, R.H., Duncan, B.K. & Jakoby, W.B. (1973) JBC, 248, 3702.

β-CYANOALANINE SYNTHASE

(L-Cysteine hydrogen sulphide-lyase (adding HCN))

L-Cysteine + HCN = 3-cyanoalanine + H_2S

Ref.

Molecular weight

source	value	conditions	
Lupinus angustifolius (blue lupine seedlings)	53,000	Sephadex G 100, pH 8.5	(1,2)

The enzyme contains tightly bound pyridoxal phosphate. The purified enzyme had no cofactor requirement. (1,2)

Specificity and catalytic properties

source	substrate	K_m (mM)	conditions	
L. angustifolius (purified 1700 x)	L-cysteine[a]	2.5	pH 8.5, Tris, 30°	(1,2)
	cyanide[b]	0.55	pH 8.5, Tris, 30°	(1,2)

(a) L-cysteine (1.00) could be replaced by O-acetyl-L-serine (0.05; the products were 3-cyanoalanine and acetate) but very low activities resulted when L-cysteine was replaced by L-cystine; D-cysteine; O-succinyl-L-serine; N-acetyl-L-serine; S-methyl-L-serine; O-acetyl-DL-homoserine; DL-homocysteine; O-succinyl-DL-homoserine; L-threonine; DL-cystathionine; O-phospho-L-serine or L-serine.

(b) cyanide (1.00) could be replaced by methanethiol (0.17) in which case S-methylcysteine and H_2S were produced.

In addition to the 3-cyanoalanine synthase reaction (1.00), the enzyme also catalyzed an exchange reaction of sulphide into cysteine (0.50) and the following reaction (0.08; also see EC 4.2.99.8):

O-Acetylserine + H_2S = cysteine + acetate. (2)

The enzyme did not require mono- or di-valent metal ions for activity. It was not inhibited by EDTA. (2)

References

1. Hendrickson, H.R. & Conn, E.E. (1971) Methods in Enzymology, 17B, 233.
2. Hendrickson, H.R. & Conn, E.E. (1969) JBC, 244, 2632.

ADENYLATE CYCLASE

(ATP pyrophosphate-lyase (cyclizing))

ATP = 3':5'-cyclic AMP + PPi

Ref.

Equilibrium constant

$\frac{[\text{cyclic AMP}]\,[\text{PPi}]}{[\text{ATP}]}$ = 65 mM (pH 7.3, 25°. In the presence of Mg^{2+} and pyruvate) (3)

Molecular properties

Adenylate cyclase is usually associated with the cytoplasmic, mitochondrial or endoplasmic membrane of the cell. It has, however, been obtained from the cytosol of *Streptococcus salivarius* (purified 3200 x; three molecular forms have been obtained from this source) and *Brevibacterium liquefaciens* (purified 100 x). (1,4,5)

Catalytic properties

Catalytic studies on membrane-bound adenylate cyclase are difficult to interpret because of its low activity and because the membrane also has ATPase activity (EC 3.6.1.3).

The activity of adenylate cyclase may respond to a wide variety of hormones and other pharmacologically active compounds such as histamine, serotonin, batrachotoxin, veratridine, ouabain and fluoride. The mechanism of the hormonal activation of the cyclase may involve an allosteric interaction between the hormone and either the enzyme itself or the membrane with which it is associated. Thus, whereas solubilized preparations may be catalytically active they are in general unresponsive to hormonal regulation. (1)

The adenylate cyclase from *Neurospora crassa* (membrane bound) requires Mn^{2+} for activity (Mg^{2+} and Ca^{2+} were inactive). With this enzyme, substrate (ATP-Mn) dependence curves were sigmoidal. This enzyme was not activated by fluoride, by hormones, by amino acid precursors or catabolites or by oxo-acids. (6)

The enzyme from *S. salivarius* (cytosol) was Mg^{2+} (or Mn^{2+}) dependent. Fluoride activated the forward reaction but it inhibited the reverse reaction. (4)

The soluble adenylate cyclase from *B. liquefaciens* requires Mg^{2+} and pyruvate (or other 2-oxomonocarboxylic acid) for activity. It was highly specific for ATP (Km = 30 mM at pH 9.0 and 33°) which could be replaced by dATP (to form cyclic 3':5'-dATP) but not by any of the other nucleoside triphosphates or by AMP or ADP. AMP, dAMP, ADP and GTP were inhibitory. (5)

The physiology and pharmacology of cyclic AMP have been reviewed. (2,7)

References

1. Jost, J-P. & Rickenberg, H.V. (1971) Ann. Rev. Biochem, 40, 741.
2. Cyclic AMP and Cell Function. Annals N.Y. Acad. Sci., 185 (1971)
3. Hayaishi, O., Greengard, P. & Colowick, S.P. (1971) JBC, 246, 5840.
4. Khandelwal, R.L. & Hamilton, I.R. (1971) JBC, 246, 3297.
5. Hirata, M. & Hayaishi, O. (1967) BBA, 149, 1.
6. Flawiá, M.M. & Torres, H.N. (1972) JBC, 247, 6873; 6880.
7. Advances in Cyclic Nucleotide Research. Vol 1: Physiology and Pharmacology of Cyclic AMP (1972) Eds. Greengard, P., Robison, G.A. & Paoletti, P. Raven Press, New York.

Rubalcava, B. & Rodbell, M. (1973) JBC, 248, 3831.
Flawiá, M.M. & Torres, H.N. (1973) JBC, 248, 4517.

GUANYLATE CYCLASE

(GTPpyrophosphate-lyase (cyclizing))

GTP = 3':5'-cyclic GMP + PPi

Ref.

Specificity and catalytic properties

source	substrate	K_m (μM)	conditions	Ref.
Rat lung (cytosol; purified 20 x)	GTD[a]	20-100	pH 7.4, Tris, 30°	(1)
	Mn^{2+}[b]	500	pH 7.4, Tris, 30°	(1)

(a) ATP (which was inhibitory), GMP and GDP were inactive as substrates.

(b) Mn^{2+} (1.0) could be replaced by Mg^{2+} (less than 0.1). Very low activity was observed with Ca^{2+}.

The cyclase required the addition of dithiothreitol for activity but it was not stimulated by fluoride, glucagon or adrenaline. The enzyme was inhibited by oxaloacetate and phosphoenolpyruvate but not by pyruvate, malate, 2-oxoglutarate, succinate, aspartate or glucose. All nucleoside triphosphates tested and several di- and monophosphates were inhibitory. (1)

References

1. Hardman, J.G. & Sutherland, E.W. (1969) JBC, 244, 6363.

PROLINE RACEMASE

(Proline racemase)

L-Proline = D-proline

Ref.

Equilibrium constant

$\frac{[\text{D-proline}]}{[\text{L-proline}]}$ = 1.27 (pH 8.0, 37°) (1)

Molecular weight

source	value	
Clostridium sticklandii	$s_{20,w}$ = 3.13 S	(1)

The spectrum of the enzyme showed no absorption other than that attributable to the protein. (1)

Specific activity

C. sticklandii (286 x) 600 L-proline (pH 8.0, Tris, 37°) (1)

Specificity and catalytic properties

source	substrate	$\overline{V}$ relative	K_m (mM)	conditions	
C. sticklandii	L-proline[(a)]	1.00	2.3	pH 8.0, Tris, 37°	(1)
	D-proline	1.32	3.8	pH 8.0, Tris, 37°	(1)

(a) L-proline (1.00) could be replaced by allohydroxy-D-proline (0.05) or hydroxy-L-proline (0.02) and low activity was shown towards sarcosine but valine and alanine were inactive.

The enzyme required the addition of a reducing agent (e.g. mercapto-ethanol) for activity but there was no requirement for a cofactor such as pyridoxal phosphate or NAD. (1)

The following were inhibitory: pyrrole 2-carboxylate; maleate; 2-thiophenecarboxylate and sarcosine. (1)

The mechanism of the reaction has been investigated. (1)

References

1. Cardinale, G.J. & Abeles, R.H. (1968) B, 7, 3970.

DIAMINOPIMELATE EPIMERASE

(2,6-LL-Diaminopimelate 2-epimerase)

2,6-LL-Diaminopimelate = meso-diaminopimelate

Ref.

Equilibrium constant

$$\frac{[\text{meso-diaminopimelate}]}{[\text{2,6-LL-diaminopimelate}]} = 1.86 \quad (\text{pH } 7.0,\ 25\text{-}45°)$$ (1)

Specificity and Michaelis constants

source	substrate	V relative	K_m (mM)	conditions	
Bacillus megaterium (purified 25 x)	meso-diamino-pimelate	1.0	100.0	pH 7.0, 37°	(1)
	2,6-LL-diamino-pimelate	0.12	6.7	pH 7.0, 37°	(1)

2,6-DD-diaminopimelate is inactive as a substrate. This enzyme does not require pyridoxal phosphate or metal ions; it was not inhibited by EDTA. 2,3-Dimercaptopropan-1-ol was required for full activity. (1)

References

1. White, P.J., Lejeune, B. & Work, E. (1969) BJ, 113, 589.

ARGININE RACEMASE

(Arginine racemase)

L-Arginine = D-arginine

Ref.

Molecular weight

source	value	conditions	
Pseudomonas graveolens	167,000	pH 7.3	(1)

The enzyme contains one mole of pyridoxal phosphate per 42,000 daltons. The pyridoxal phosphate, which was essential for activity, could not be replaced by FAD, FMN or riboflavin. (1)

Specific activity

P. graveolens (5400 x) 1250 D-arginine (pH 10.0, glycine, 37°) (1)

Specificity and catalytic properties

source	substrate	K_m (M)	conditions	
P. graveolens	D-arginine	1.0×10^{-3}	pH 10, glycine, 37°	(1)
	pyridoxal phosphate	4.0×10^{-7}	pH 10, glycine, 37°	(1)

The enzyme could utilize the following: lysine (1.10); arginine (1.00); N6-acetyllysine (0.86); ornithine (0.44); 2,3-diaminopropionate (0.40); homoarginine (0.25), canavanine (0.19); 2,4-diaminobutyrate (0.18); ethionine (0.13); citrulline (0.13); homocitrulline (0.12); N5-acetylornithine (0.12); theanine (0.11); glutamine (0.07) and methionine (0.04). The following were inactive: alanine; asparagine; valine; leucine; isoleucine; histidine; phenylalanine; aspartate; glutamate; serine; threonine; proline; hydroxyproline; ω-N-nitroarginine; N2-acetyllysine and N2-acetylornithine.

The enzyme catalyzes a transamination reaction between L- or D-ornithine and pyruvate. The rate of transamination is very low compared with the racemization reaction. (2)

References

1. Yorifuji, T., Ogata, K. & Soda, K. (1971) JBC, 246, 5085.
2. Yorifuji, T., Misono, H. & Soda, K. (1971) JBC, 246, 5093.

AMINO-ACID RACEMASE

(Amino-acid racemase)

L-amino acid = D-amino acid

Ref.

Molecular weight

source	value	conditions	
Pseudomonas striata	110,000	ultracentrifugation	(1,2)

The enzyme contains 2 moles of pyridoxal phosphate per 110,000 daltons. (1)

Specific activity

P. striata (240 x) 708 L-methionine (pH 8.3, PPi, 37°) (2)

Specificity and Michaelis constants

source	substrate	relative velocity	K_m (mM)
P. striata	D-lysine[a]	1.00	30
	L-ethionine	0.76	35
	D-arginine	0.60	--
	L-methionine	0.48	38
	L-citrulline	0.16	--
	D-2,4-diaminobutyrate	0.10	--
	L-alanine	0.09	36
	L-serine	0.09	--
	L-2-aminobutyrate	0.05	33
	L-leucine	0.03	33
	L-histidine	0.02	--
	pyridoxal phosphate	-	0.2

(a) the following were inactive: D-glutamate, L-isoleucine, L-phenylalanine, L-threonine and L-valine. The conditions were: pH 8.3, PPi, 37°. (from Ref. 1).

References

1. Soda, K. & Osumi, T. (1971) Methods in Enzymology, 17B, 629.
2. Soda, K. & Osumi, T. (1969) BBRC, 35, 363.

PHENYLALANINE RACEMASE (ATP-HYDROLYSING)

(Phenylalanine racemase (ATP-hydrolysing))

ATP + L-phenylalanine + H_2O = AMP + PPi + D-phenylalanine

Ref.

Equilibrium constant

			Ref.
[D-phenylalanine] / [L-phenylalanine]	= 4	(pH 8.6, 37°)	(1)
	= 1	(pH 7.8, 37°)	(1)

Molecular weight

source	value	conditions	
Bacillus brevis	100,000	sucrose density gradient	(1)

Specific activity

B. brevis (250 x) 0.015 L-phenylalanine (pH 8.6, TEA, 37°) (1)

Specificity and Michaelis constants

source	substrate	K_m (µM)	conditions	
B. brevis	L-phenylalanine[a]	20	pH 8.6, TEA, 37°	(1)
	ATP[b]	150	pH 8.6, TEA, 37°	(1)

(a) L-tryptophan, L-tyrosine, L-valine and L-leucine were inactive.

(b) ATP (1.00) could be replaced by ADP (0.08) but not by CTP, UTP, dTTP, GTP or AMP.

The racemase was stimulated by PPi, AMP and dithiothreitol and it required Mg^{2+} or Mn^{2+} for activity. (1,2)

The mechanism of the reaction has been investigated. In addition to its racemase activity, the enzyme also catalyzes an L- or D-phenylalanine dependent ATP-PPi or ATP-AMP exchange reaction. (1)

References

1. Yamada, M. & Kurahashi, K. (1969) J. Biochem.(Tokyo), 66, 529.
2. Yamada, M. & Kurahashi, K. (1968) J. Biochem. (Tokyo), 63, 59.

MANDELATE RACEMASE

(Mandelate racemase)

L-Mandelate = D-mandelate

Ref.

Molecular weight

source	value	conditions	Ref.
Pseudomonas putida	200,000	gel filtration	(1,2)

The spectrum of the enzyme showed no absorption other than that attributable to the protein. (2)

Specific activity

P. putida (560 x) 195 D-mandelate (pH 7.5, Pi, 25°) (1,2)

Specificity and catalytic properties

source	substrate[(a)]	V relative	K_m (μM)	conditions	Ref.
P. putida	mandelate	1.00	93	pH 7.5, Pi, 25°	(2)
	4-chloromandelate	3.26	100	pH 7.5, Pi, 25°	(2)
	4-bromomandelate	3.76	256	pH 7.5, Pi, 25°	(2)
	4-hydroxymandelate	0.45	290	pH 7.5, Pi, 25°	(2)
	4-methoxymandelate	0.17	330	pH 7.5, Pi, 25°	(2)

(a) each substrate tested was the D-isomer. Mandelate (1.00) could also be replaced by 3-imidazoleacetate (0.05) but not by 3-indolelactate; 3-phenyllactate; lactate; 3,4-dihydroxyphenylglycollate or 4-hydroxy-3-methoxyphenylglycollate. The following were inhibitory: phenyloxyacetate; phenylmercaptoacetate; phenylacetate; DL-β-phenyllactate and certain other mandelate analogues.

The enzyme was activated by divalent metal ions (it was not completely inhibited by EDTA) but not by mono- or tri-valent metal ions. There was no cofactor requirement (e.g. pyridine nucleotide or flavin). (2)

The mechanism of the reaction has been investigated. (3)

References

1. Hegeman, G.D. (1970) Methods in Enzymology, 17A, 670.
2. Hegeman, G.D., Rosenberg, E.Y. & Kenyon, G.L. (1970) B, 9, 4029.
3. Kenyon, G.L. & Hegeman, G.D. (1970) B, 9, 4036.

CDP PARATOSE EPIMERASE

(CDP-3,6-dideoxy-D-glucose 2-epimerase)

CDP-3,6-dideoxy-D-glucose = CDP-3,6-dideoxy-D-mannose

Ref.

Equilibrium constant

$$\frac{[\text{CDPtyvelose}]}{[\text{CDPparatose}]} = 1.2\text{-}1.4 \text{ (pH 8.4, 37°)}$$ (1)

Specificity and catalytic properties

source	substrate	V relative	K_m (μM)	conditions	Ref.
Pasteurella pseudotuberculosis (purified 6 x)	CDPparatose[a]	1.0	850	pH 8.4, Tris, 37°	(1)
	CDPtyvelose[a]	0.4	520		
	NAD[b]	-	1		

(a) the enzyme could not utilize the following: CDP-D-glucose; CDPabequose; CDPascarylose; paratose 1-phosphate or tyvelose 1-phosphate.

(b) NAD, which was essential for the reaction, could not be replaced by reduced NAD; NADP or reduced NADP. Of these, reduced NAD was inhibitory.

Abbreviations

paratose	3,6-dideoxy-D-glucose
tyvelose	3,6-dideoxy-D-mannose
abequose	3,6-dideoxy-D-galactose
ascarylose	3,6-dideoxy-L-mannose
colitose	3,6-dideoxy-L-galactose

References

1. Matsuhashi, S. (1966) JBC, 241, 4275.

GLUCOSE 6-PHOSPHATE 1-EPIMERASE

(Glucose 6-phosphate 1-epimerase)

α-D-Glucose 6-phosphate = β-D-glucose 6-phosphate

Ref.

Equilibrium constant

$$\frac{[\beta\text{-D-glucose 6-phosphate}]}{[\alpha\text{-D-glucose 6-phosphate}]} = 1.70 \text{ (pH 7.6, 25°)}$$ (1)

Molecular weight

source	value	conditions	
Bakers' yeast	35,000	Sephadex G 100, pH 7.6	(1)

Specificity

The enzyme from bakers' yeast (purified 30 x) does not catalyze the anomerization of D-glucose. Glucose 6-phosphate isomerase (EC 5.3.1.9) also possesses glucose 6-phosphate 1-epimerase activity. (1)

References

1. Wurster, B. & Hess, B. (1972) FEBS lett, 23, 341.

MALEATE ISOMERASE

(Maleate *cis*-*trans*-isomerase)

Maleate = fumarate

Ref.

Equilibrium constant

The reaction was irreversible [F] with the enzyme from *Pseudomonas fluorescens* or *Alcaligenes faecalis* as the catalyst. (1,2)

Molecular weight

source	value	conditions	
P. fluorescens	74,000	pH 7.5; amino acid composition	(2)
A. faecalis	100,000	Sephadex G 100; ultracentrifugation	(1)

Specific activity

P. fluorescens 24.1 maleate (pH 8.4, Tris, 25°) (2)

Specificity and catalytic properties

source	substrate	K_m (mM)	conditions	
P. fluorescens	maleate[a]	0.3	pH 8.4, Tris, 25°	(2)

(a) the following compounds were inactive as substrates or inhibitors: maleamide; *cis*-aconitate; L-ascorbate; *cis*-2-butene-1,4-diol; citraconate, crotonate, diethylmaleate, *trans*-epoxysuccinate, glutaconate, isocrotonate, linolenate, maleurate, oxaloglycollate, tiglate and ferulate.

The isomerase has also been purified (100 x) from *A. faecalis*. This enzyme is also highly specific for maleate (Km = 2.8 mM) and it could not utilize citraconate or mesaconate. (1)

Maleate isomerase does not require any added cofactor or metal ion for activity. The enzyme from *A. faecalis* was not inhibited by EDTA. (1)

Light absorption data

The transformation of maleate to fumarate is accompanied by an increase in absorption at 290 nm (molar extinction coefficient = 103 $M^{-1}cm^{-1}$). (2)

References

1. Seltzer, S. (1972) *The Enzymes*, *6*, 382.
2. Scher, W. & Jakoby, W.B. (1969) *JBC*, *244*, 1878.

LINOLEATE ISOMERASE

(Linoleate Δ^{12}-cis-Δ^{11}-trans-isomerase)

9-cis-12-cis-Octadecadienoate = 9-cis-11-trans-octadecadienoate

Ref.

Equilibrium constant

$$\frac{[\text{9-cis-11-trans-octadecadienoate}]}{[\text{9-cis-12-cis-octadecadienoate}]} = 61$$

(pH 7.0, Pi, 35°) (1)

Specificity and catalytic properties

source	substrate	K_m (µM)	conditions	Ref.
Butyrivibrio fibrisolvens (particulate)	linoleate	12	pH 7.0, Pi, 35°	(1)
	linolenate	23	pH 7.0, Pi, 35°	(1)

The enzyme (B. fibrisolvens) is highly specific and only those compounds which possess a free carboxyl group and a 9-cis-12-cis double bond system are isomerized. The presence of additional double bonds are tolerated; thus both the $\Delta^{9,12,15}$ and $\Delta^{6,9,12}$ isomers of linolenate are isomerized. No cofactor or metal ion requirement has been found for this enzyme; it was, however, inhibited by EDTA. (1,2)

The isomerase was inhibited by unsaturated fatty acids. Thus, oleate and petroselinate were competitive inhibitors of linoleate utilization and high substrate (linoleate or linolenate) concentrations were inhibitory. (2)

Abbreviations

linoleate	9-cis-12-cis-octadecadienoate
linolenate	9-cis-12-cis-15-cis-octadecatrienoate

Light absorption properties

The conjugated diene system has an absorption band at 233 nm (molar extinction coefficient = 24,000 $M^{-1}cm^{-1}$). (1,2)

References

1. Kepler, C.R. & Tove, S.B. (1967) JBC, 242, 5686.
2. Kepler, C.R. & Tove, S.B. (1969) Methods in Enzymology, 14, 105.

N-(5'-PHOSPHO-D-RIBOSYLFORMIMINO)-5-AMINO-1-(5''-PHOSPHORIBOSYL)-4-IMIDAZOLECARBOXAMIDE ISOMERASE

(N-(5'-Phospho-D-ribosylformimino)-5-amino-1-(5''-phosphoribosyl)-4-imidazolecarboxamide ketol-isomerase)

N-(5'-phospho-D-ribosylformimino)-5-amino-1-(5''-phosphoribosyl)-4-imidazolecarboxamide = N-(5'-phospho-D-1'-ribulosylformimino)-5-amino-1-(5''-phosphoribosyl)-4-imidazolecarboxamide

Ref.

Molecular weight

source	value	conditions	
Salmonella typhimurium	29,000 [1]	ultracentrifugation at pH 8.0 with or without 8 M urea; amino acid composition; peptide mapping	(1,2)

The purified enzyme gives three enzymically active bands in the ratio 90 : 9 : 1 when subjected to disc gel electrophoresis. (1,2,4)

Specific activity

S. typhimurium (143 x) 7.9 BBM II (pH 8.6, TEA, 45°) (1,4)

Michaelis constant

source	substrate	K_m (µM)	conditions	
S. typhimurium (purified 20 x)	BBM II	57	pH 8.6, TEA, 37°	(3)

Light absorption data

The molar extinction coefficient of BBM II at 290 nm is 8,000 $M^{-1}cm^{-1}$. (1,3)

Abbreviations

BBM II — N-(5'-phospho-D-ribosylformimino)-5-amino-1-(5''-phosphoribosyl)-4-imidazolecarboxamide

BBM III — N-(5'-phospho-D-1'-ribulosylformimino)-5-amino-1-(5''-phosphoribosyl)-4-imidazolecarboxamide

References

1. Margolies, M.N. & Goldberger, R.F. (1966) JBC, 241, 3262.
2. Margolies, M.N. & Goldberger, R.F. (1967) JBC, 242, 256.
3. Smith, D.W.E. & Ames, B.N. (1964) JBC, 239, 1848.
4. Martin, R.G., Berberich, M.A., Ames, B.N., Davis, W.W., Goldberger, R.F. & Yourno, J.D. (1971) Methods in Enzymology, 17B, 24.

PHENYLPYRUVATE TAUTOMERASE

(Phenolpyruvate keto-enol isomerase)

keto-Phenylpyruvate = enol-phenylpyruvate

Ref.

Equilibrium constant

[enol-3,5-diiodohydroxyphenylpyruvate] / [keto-3,5-diiodohydroxyphenylpyruvate]	less than 0.1	(Pi)	(1)
	about 1	(borate)	(1)

Molecular properties

source	value	conditions	Ref.
Hog thyroid	44,000 [1]	pH 7.4; sucrose density gradient; amino acid composition	(1)

Specific activity

Hog thyroid (1000 x) 2.3 keto-hydroxyphenylpyruvate (pH 6.2, borate) (1)

Specificity and catalytic properties

source	substrate(a)	K_m (mM)	conditions	Ref.
Hog thyroid	hydroxyphenylpyruvate	1.2	pH 6.2, borate	(1)
	3,5-diiodohydroxyphenylpyruvate	1.2	pH 6.2, borate	(1)
	phenylpyruvate	8	pH 6.2, borate	(1)

(a) keto form.

The tautomerase has also been purified from hog kidney (45 x). This enzyme attacked the following (all in the keto-form): 4-hydroxyphenylpyruvate (1.00); phenylpyruvate (0.43); 3-hydroxyphenylpyruvate (0.12); 4-methoxyphenylpyruvate (0.14) and 2,5-dihydroxyphenylpyruvate (0.02). It was inactive with oxaloacetate. The enzyme required no cofactor or metal ion for activity; it was not inhibited by EDTA. (2)

The enzyme from lamb kidney (purified 1000 x) attacked 4-hydroxyphenylpyruvate (1.00); phenylpyruvate (0.65); 4-methoxyphenylpyruvate (0.56); 3-hydroxyphenylpyruvate (0.51); 4-nitrophenylpyruvate (0.45); vanilpyruvate (0.30) and indole 3-pyruvate (0.04). (3)

Light absorption data

The borate complex of enol-hydroxyphenylpyruvate has an absorption band at 330 nm (molar extinction coefficient = 6330 $M^{-1}cm^{-1}$). (1)

References

1. Blasi, F., Fragomele, F. & Covelli, I. (1969) JBC, 244, 4864.
2. Knox, W.E. & Pitt, B.M. (1957) JBC, 225, 675.
3. Constantsas, N.S. & Knox, W.E. (1966) ABB, 117, 59.

OXALOACETATE TAUTOMERASE

(Oxaloacetate keto-enol-isomerase)

keto-Oxaloacetate = enol-oxaloacetate

Ref.

Equilibrium constant

$$\frac{[\text{enol-oxaloacetate}]}{[\text{keto-oxaloacetate}]} = 0.1 \text{ (pH 5-10)}$$ (1)

Both the enolization and ketonization reactions are catalyzed by acid, base, buffers and metal ions. (1)

Specificity and Michaelis constants

source	substrate	K_m (μM)	conditions	Ref.
Porcine kidney (purified 56 x)	enol-oxaloacetate	66	pH 7.4, Pi	(2)

The enzyme was activated by Ca^{2+}, Mn^{2+} or Co^{2+}. (2)

Light absorption properties

The enol-form of oxaloacetate has an absorption band at 260 nm (molar extinction coefficient = 8,700 $M^{-1}cm^{-1}$). (1)

References

1. Annett, R.G. & Kosicki, G.W. (1965) Can. J. Biochem, 43, 1887.
2. Annett, R.G. & Kosicki, G.W. (1969) JBC, 244, 2059.

MUCONOLACTONE Δ-ISOMERASE

((+)-4-Hydroxy-4-carboxymethylisocrotonolactone Δ^2-Δ^3-isomerase)

(+)-4-Hydroxy-4-carboxymethylisocrotonolactone =
3-oxoadipate enol-lactone

Ref.

Molecular weight

source	value	conditions	
Pseudomonas putida	93,000	gel filtration	(1,2)

Specific activity

P. putida (560 x)	710	(+)-muconolactone (pH 8.0, Tris)	(1,2)

Specificity and catalytic properties

source	substrate	K_m (μM)	conditions	
P. putida	(+)-muconolactone[a]	80	pH 8.0, Tris	(1,2)

(a) (+)-muconolactone could not be replaced by (-)-muconolactone or 4-carboxymuconolactone.

Light absorption data

The transformation of muconolactone to 3-oxoadipate is accompanied by a decrease in absorbance at 230 nm (molar extinction coefficient = 1428 $M^{-1}cm^{-1}$). (1,2)

Abbreviations

(+)-muconolactone	(+)-4-hydroxy-4-carboxymethylisocrotonolactone; 4-carboxymethyl-Δ^2-butenolide
3-oxoadipate lactone	4-carboxymethyl-Δ^3-butenolide

References

1. Ornston, L.N. (1966) JBC, 241, 3795.
2. Ornston, L.N. (1970) Methods in Enzymology, 17A, 536.

METHYLITACONATE Δ-ISOMERASE

(Methylitaconate Δ^2-Δ^3-isomerase)

Methylitaconate = dimethylmaleate

Ref.

Equilibrium constant

$$\frac{[\text{dimethylmaleate}]}{[\text{methylitaconate}]} = 3.4 \text{ (pH 7.9, 34°)}$$ (1)

Molecular weight

source	value	conditions	Ref.
Clostridium barkeri	140,000	sucrose density gradient	(1)

Michaelis constants

substrate[a]	V relative	K_m (mM)	conditions	Ref.
methylitaconate	1.00	6.2	pH 7.9, Pi, 34°	(1)
dimethylmaleate	0.19	4.0	pH 7.9, Pi, 34°	(1)

(a) with the enzyme from C. barkeri (purified 70 x). The isomerase did not require the addition of any cofactor (e.g. cobamide coenzyme derivatives) or metal ion for activity.

References

1. Kung, H-F. & Stadtman, T.C. (1971) JBC, 246, 3378.

ACONITATE Δ-ISOMERASE

(Aconitate Δ^2-Δ^3-isomerase)

trans-Aconitate = *cis*-aconitate

Ref.

Molecular weight

source	value	conditions	
Pseudomonas putida	78,000 [2] [a]	Sephadex G 100, pH 7.0. The buffer contained EDTA and dithiothreitol.	(1)

(a) the monomeric form was inactive.

Catalytic properties

source	substrate	K_m (mM)	conditions	
P. putida (purified 45 x)	*cis*-aconitate	1.9-1.0	pH 8.0, Tris, 25°, + 20% glycerol	(1)
	cis-aconitate	25-9 [a]	pH 8.0, Tris, 25° no glycerol	(1)

(a) in the absence of glycerol, the Km for *cis*-aconitate decreased on increasing the enzyme concentration.

The mechanism of the reaction has been investigated. (1)

Light absorption data

At 260 nm the difference in molar extinction coefficient between *trans*-aconitate and *cis*-aconitate = 600 $M^{-1}cm^{-1}$. (1)

References

1. Klinman, J.P. & Rose, I.A. (1971) B, 10, 2253; 2259.

PROTEIN DISULPHIDE-ISOMERASE

(Protein disulphide-isomerase)

Catalyses the rearrangement of -S-S-bonds in proteins

Ref.

Molecular properties

source	value	conditions	Ref.
Beef liver microsomes	42,000 [1]	pH 7.5; amino acid composition; peptide mapping	(1,2)

Multiple forms have been observed. The visible and ultraviolet spectra of the enzyme showed no peaks other than one with a maximum at 278 nm. (1)

Catalytic properties

The isomerase catalyzes the rearrangement of random, "incorrect" pairs of half-cysteine residues to the native, "correct" disulphide bonds in several protein substrates. It also pairs up the correct cysteines in fully reduced proteins. It has been extensively purified (1800 x) from beef liver microsomes and the most purified preparations obtained catalyzed the reactivation of randomly cross-linked soybean trypsin inhibitor at an initial rate of 0.885 μmole per min per mg of enzyme protein. (1,2)

Protein disulphide-isomerase may be identical with protein-disulphide reductase (EC 1.8.4.2).

References

1. de Lorenzo, F., Goldberger, R.F., Steers, E., Givol, D. & Anfinsen, C.B. (1966) JBC, 241, 1562.
2. Fuchs, S., de Lorenzo, F. & Anfinsen, C.B. (1967) JBC, 242, 398.

LYSINE 2,3-AMINOMUTASE

(L-Lysine 2,3-aminomutase)

L-2,6-Diaminohexanoate = L-3,6-diaminohexanoate

Ref.

Equilibrium constant

$\frac{[\text{L-}\beta\text{-lysine}]}{[\text{L-lysine}]}$ = 6.7 (pH 7.8, 37°) (1)

Molecular properties

source	value	conditions	Ref.
Clostridium SB4	285,000 [6]	ultracentrifugation at pH 7.8 with or without SDS; gel electrophoresis (SDS); amino acid composition.	(1,2)

The enzyme contains 4.3 gm atoms of Fe^{2+}, 1.7 moles of pyridoxal phosphate and 4 moles of S-adenosyl-methionine per 285,000 daltons. It may consist of 2 catalytic subunits (which bind pyridoxal phosphate) and 4 regulatory subunits (which bind S-adenosyl-methionine). (1,2)

Specific activity

Clostridium SB4 (53 x) 6.4 L-lysine (pH 7.8, Tris, 30°) (1)

Specificity and catalytic properties

source	substrate	K_m (M)	conditions	Ref.
Clostridium SB4	L-lysine[a]	6.6×10^{-3}	pH 7.8, Tris, 30°	(1)
	S-adenosylmethionine[b]	2.8×10^{-8}		

(a) D-lysine was converted to β-lysine at less than 1% the rate with L-lysine. S-Aminoethylcysteine was a competitive inhibitor (Ki = 56 μM) but the following were neither substrates nor inhibitors: DL-ornithine; L-N^6-acetyllysine; DL-allo-S-hydroxylysine; L-arginine; citrulline or quinacrine.
(b) S-adenosylhomocysteine was a competitive inhibitor in the presence or absence of S-adenosylmethionine. The following neither stimulated nor inhibited the reaction in the presence or absence of S-adenosylmethionine: 5'-methylthioadenosine, L-methionine and S-methyl-DL-methionine.

Lysine 2,3-mutase requires Fe^{2+}, pyridoxal phosphate and S-adenosylmethionine for activity. It is inactivated by O_2; the inactivation is reversed under anaerobic conditions in the presence of dithiothreitol.(1)

Ref.

Abbreviations

lysine	2,6-diaminohexanoate
β-lysine	3,6-diaminohexanoate

References

1. Chirpich, T.P., Zappia, V., Costilow, R.N. & Barker, H.A. (1970) JBC, 245, 1778.
2. Zappia, V. & Barker, H.A. (1970) BBA, 207, 505.

β-LYSINE 5,6-AMINOMUTASE

(L-5,6-Diaminohexanoate aminomutase)

L-3,6-Diaminohexanoate = L-3,5-diaminohexanoate

Ref.

Equilibrium constant

The reaction is reversible. (1)

Molecular properties

Native β-lysine mutase (*Clostridium sticklandii*) is a tetramer (molecular weight = 170,000 daltons) composed of 2 different types of subunit: 2 of 32,000 and 2 of 52,000 daltons. The smaller protein (E_2 protein) may be identical with the E_2 protein of D-lysine 5,6-aminomutase (EC 5.4.3.4). The enzyme contains tightly bound 5'-deoxyadenosylcobalamin. During isolation, the cobamide coenzyme is degraded to hydroxy(adenyl)cobamide which is a strong inhibitor and which remains tightly bound to the protein (see below). (2)

Specific activity

Cl. sticklandii (84 x) 19.5 L-β-lysine (pH 7.5, triethanolamine 37°) (2)

Before activity determinations, the enzyme requires preincubation with pyridoxal 5'-phosphate, dithiothreitol, Mg^{2+} (or Mn^{2+}) and 5'-deoxyadenosylcobalamin to remove the strongly inhibitory hydroxy (adenyl) cobamide. In addition, it requires a monovalent cation (K^+, Li^+ and Rb^+ were about equally effective; NH_4^+, Na^+ and Cs^+ were less effective). (2)

Specificity and Michaelis constants

source	substrate[a]	K_m (M)
Cl. sticklandii	L-β-lysine	3×10^{-4}
	pyridoxal 5'-phosphate[b]	2.2×10^{-7}

(a) for conditions see 'specific activity'.
(b) pyridoxamine 5'-phosphate, pyridoxal, pyridoxol and pyridoxamine were inactive as activators.

Abbreviations

β-lysine 3,6-diaminohexanoate

References

1. Stadtman, T.C. & Grant, M.A. (1971) *Methods in Enzymology*, 17B, 206.
2. Baker, J.J., van der Drift, C. & Stadtman, T.C. (1973) B, 12, 1054. Stadtman, T.C. (1973) *Advances in Enzymology*, 38, 426.

D-LYSINE 5,6-AMINOMUTASE

(D-5,6-Diaminohexanoate aminomutase)

D-2,6-Diaminohexanoate = 2,5-diaminohexanoate

Ref.

Equilibrium constant

The reaction is reversible. The absolute configuration of 2,5-diaminohexanoate is not known. (1,2)

Molecular properties

D-Lysine 5,6-aminomutase (*Clostridium* ME) consists of two dissimilar protein moieties: a large acidic protein containing bound cobamide and pyridoxal phosphate (E_1, molecular weight = 170,000 daltons) and a smaller sulphydryl protein (E_2, molecular weight = 60,000 daltons). The molecular weight of the complex is 250,000 daltons. On its own, E_1 catalyzes a pyridoxal phosphate and Mg^{2+} dependent exchange of hydrogen at position 6 of D-lysine with water. Neither E_1 nor E_2 catalyzes the aminomutase reaction on its own. The E_2 protein may be a dimer and it may be identical with the E_2 protein of β-lysine 5,6-aminomutase (EC 5.4.3.3). (1,2,5)

Specific activity

Clostridium ME (215 x)	1.1	D-lysine (pH 8.5-9.0, Tris, 37°; in the absence of air. All the cofactors required were present, see below)	(5)

Specificity and catalytic properties

The aminomutase requires a mercaptan, ATP (which is not utilized during the reaction), pyridoxal phosphate (pyridoxamine phosphate, pyridoxol phosphate and pyridoxal were inactive), B_{12} coenzyme, Mg^{2+} and a monovalent cation for activity. It does not require pyruvate or FAD. (3,5)

ATP is an allosteric effector. In the absence of ATP, substrate dependence curves were sigmoidal (V = 1.0, Km with D-lysine = 20 mM) whereas in its presence nearly hyperbolic curves were obtained (V = 1.9; Km with D-lysine = 0.3 mM). ATP could be replaced by its phosphonic acid-, β,γ-methylene- or α,β-methylene-analogue. The following neither activated nor inhibited: 2'-dATP; AMP or GTP. CTP and ADP were poor activators and phospho*enol*pyruvate and *S*-adenosylmethionine did not activate. (5)

The mechanism of the reaction has been investigated. (3,4)

A number of amino acids and amines inhibit the aminomutase and of these the most effective were: L-β-lysine; L-ornithine; N^6-acetyl-DL-lysine; *S*-aminoethylcysteine and 1,4-diaminobutane. (1,5)

Abbreviations

lysine 2,6-diaminohexanoate.

References

1. Stadtman, T.C. (1973) Advances in Enzymology, 38, 430.
2. Stadtman, T.C. (1972) The Enzymes, 6, 539.
3. Morley, C.G.D. & Stadtman, T.C. (1972) B, 11, 600.
4. Stadtman, T.C. & Grant, M.A. (1971) Methods in Enzymology, 17B, 211.
5. Morley, C.G.D. & Stadtman, T.C. (1970) B, 9, 4890.

D-ORNITHINE 4,5-AMINOMUTASE

(D-Ornithine 4,5-aminomutase)

D-2,5-Diaminopentanoate = D-2,4-diaminopentanoate

Ref.

Equilibrium constant

The reaction is reversible with an equilibrium constant of about 1. (1)

Molecular weight

source	value	conditions	Ref.
Clostridium sticklandii	180,000 [2]	sucrose density gradient; gel electrophoresis (SDS)	(1)

The enzyme contains coenzyme B_{12} which is converted to the inactive hydroxycobalamin during the purification procedure. (1)

Specific activity

C. sticklandii (71 x) 4.4 D-ornithine (pH 9.0, Tris, 37°) (1)

Specificity and catalytic properties

source	substrate	K_m (M)	conditions	Ref.
C. sticklandii	D-ornithine[(a)]	6.7×10^{-3}	pH 9.0, Tris, 37°	(1)
	pyridoxal phosphate	3.6×10^{-7}	pH 9.0, Tris, 37°	(1)

(a) D-ornithine could not be replaced by the following inhibitors: L-ornithine; L-lysine; DL-lysine; β-L-lysine or 5-aminopentanoate. The following were neither inhibitors nor substrates: cadaverine, putrescine and N^2-acetyl-L-ornithine. D-Ornithine utilization was inhibited by D-2,4-diaminopentanoate.

The aminomutase requires the addition of coenzyme B_{12}, pyridoxal phosphate and dithiothreitol for activity. It was inhibited by O_2. (1)

Abbreviations

Ornithine 2,5-diaminopentanoate

References

1. Somack, R. & Costilow, R.N. (1973) B, 12, 2597.

2-METHYLENE-GLUTARATE MUTASE

(2-Methylene-glutarate carboxymethylenemethylmutase)

2-Methylene-glutarate = 2-methylene-3-methyl-succinate

Ref.

Equilibrium constant

$\frac{[\text{methylitaconate}]}{[\text{2-methylene-glutarate}]}$ = 0.23 (pH 7.9, Pi, 34°) (2)

Molecular weight

source	value	conditions	
Clostridium barkeri	170,000	sucrose density gradient	(2)

The mutase is readily separated from its B_{12} coenzyme content during purification.

Specificity and catalytic properties

substrate[a]	V relative	K_m (M)	
2-methylene-glutarate	-	7.1×10^{-3}	(2)
α-(5,6-dimethylbenzimidazolyl)-cobamide coenzyme[b]	1.00	7.3×10^{-8}	(2)
α-(benzimidazolyl)-cobamide coenzyme	1.03	3.0×10^{-7}	(2)
α-(adenyl)-cobamide coenzyme	0.54	1.25×10^{-6}	(2)

(a) with the enzyme from C. barkeri (purified 40 x). Conditions: pH 7.9, Pi, 34°.

(b) this is the natural cofactor.

The purified enzyme requires a B_{12} coenzyme for activity. There was no requirement for the addition of any other organic cofactor or for metal ions. The enzyme was not inhibited by EDTA. (2)

The mutase was inhibited by a number of substrate analogues (all with respect to 2-methylene-glutarate):- itaconate (C, Ki = 1.2 mM); mesaconate (C, Ki = 1.2 mM); succinate (C, Ki = 1.8 mM); 1-methyl-1,2-trans-cyclopropanedicarboxylate (C, Ki = 3.3 mM); L-malate (C, Ki = 7.0mM); maleate (C); glutaconate (NC, Ki = 4.5 mM); 1-methyl-1,2-cis-cyclopropanedicarboxylate (NC, Ki = 38 mM) and citrate (NC). (2)

Abbreviations

methylitaconate 2-methylene-3-methyl-succinate

References

1. Barker, H.A. (1972) Ann. Rev. Biochem, 41, 60.
2. Kung, H-F. & Stadtman, T.C. (1971) JBC, 246, 3378.

CHORISMATE MUTASE

(Chorismate pyruvatemutase)

Chorismate = prephenate

Ref.

Molecular properties

Chorismate mutase catalyzes the first specific reaction of the tyrosine and phenylalanine pathways:

```
                                PHENYLPYRUVATE -------→ PHENYLALANINE
                                      ↑
                                      |  prephenate
                                      |  dehydratase
                  chorismate
       CHORISMATE ─────────→   PREPHENATE
                    mutase
                                      |  prephenate
                                      |  dehydrogenase
                                      ↓
                                4-HYDROXYPHENYL ------→ TYROSINE
                                  PYRUVATE
```

Chorismate is also a precursor of tryptophan (see EC 4.1.3.27).

Chorismate mutase (EC 5.4.99.5) is associated with prephenate dehydratase (EC 4.2.1.51) or prephenate dehydrogenase (EC 1.3.1.12). The two types of complex have been highly purified from certain sources and in each case the two activities are carried out by a single protein complex.

source	activities	molecular weight	conditions	
Aerobacter aerogenes	mutase + dehydrogenase	76,000[2]	gel filtration; amino acid composition	(1,4)
Escherichia coli K 12	mutase + dehydrogenase	82,000[2]	gel filtration; amino acid composition; peptide mapping	(2)
E. coli K12	mutase + dehydratase	85,000[2]	gel filtration; amino acid composition	(3)
Salmonella typhimurium	mutase + dehydratase	100,000	sucrose density gradient	(5,6)

Ref.

Specific activity

		Ref.
A. aerogenes (100 x)	21 chorismate (mutase activity; pH 7.5, Tris, 37°)	(1)
	63 prephenate (dehydrogenase activity; pH 8.1, Tris, 37°)	(1)
E. coli K 12 (65 x)	20 chorismate (mutase activity; pH 7.5, Tris, 37°)	(2)
	67 prephenate (dehydrogenase activity; pH 8.1, Tris, 37°)	(2)
E. coli K 12 (240 x)	64 chorismate (mutase activity; pH 7.8, Tris, 37°)	(3)
	28 prephenate (dehydratase activity; pH 8.2, Tris, 37°)	(3)

Specificity and catalytic properties

source	substrate (Km,mM)	end product inhibitor	type of inhibition	
E. coli K 12	chorismate (0.045)	phenylalanine	partially competitive	(3)
	prephenate (1.0)	phenylalanine	sigmoidal	(3)
S. typhimurium	chorismate (0.08)	phenylalanine	sigmoidal	(3)
	prephenate (0.35)	phenylalanine	sigmoidal	(3)
E. coli K 12	chorismate (0.39)	tyrosine	no inhibition	(2)
	prephenate (0.37)	tyrosine	sigmoidal	(2)
	NAD (0.33)	-	-	(2)
A. aerogenes	chorismate (1.3)	tyrosine	no inhibition	(1)
	prephenate (0.35)	tyrosine	sigmoidal	(1)
	NAD (0.6)	-	-	(1)

References

1. Koch, G.L.E., Shaw, D.C. & Gibson, F. (1970) BBA, 212, 375; 387.
2. Koch, G.L.E., Shaw, D.C. & Gibson, F. (1971) BBA, 229, 795; 805.
3. Davidson, B.E., Blackburn, E.H. & Dopheide, T.A.A. (1972) JBC, 247, 4441; 4447.
4. Cotton, R.G.H. & Gibson, F. (1968) BBA, 160, 188.
5. Schmit, J.C., Artz, S.W. & Zalkin, H. (1970) JBC, 245, 4019.
6. Schmit, J.C. & Zalkin, H. (1969) B, 8, 174.

3-CARBOXY-*cis-cis*-MUCONATE CYCLOISOMERASE

(4-Carboxymuconolactone lyase (decyclizing))

4-Carboxymuconolactone = 3-carboxy-*cis-cis*-muconate

Ref.

Molecular weight

source	value	conditions	Ref.
Pseudomonas putida	190,000 [4]	Sephadex G 200, pH 7.1; gel electrophoresis (SDS); amino acid composition.	(1,2,3)

Specific activity

P. putida (848 x)	800	3-carboxymuconate (pH 8.0, Tris, 25°)	(3)

Specificity and catalytic properties

source	substrate	K_m (μM)	conditions	Ref.
P. putida (purified 198 x)	3-carboxymuconate[(a)]	75	pH 8.0, Tris, 25°	(1,2)

(a) *cis-cis*-muconate was inactive.

The isomerase did not require metal ions for activity. It was not inhibited by EDTA. (1,2)

Light absorption data

3-Carboxymuconate has an absorption band at 270 nm (molar extinction coefficient = 6390 $M^{-1}cm^{-1}$). (1,2)

Abbreviations

4-carboxymuconolactone	4-carboxymethyl-4-carboxy-Δ^2-butenolide
3-carboxymuconate	3-carboxy-*cis-cis*-muconate

References

1. Ornston, L.N. (1966) JBC, 241, 3787.
2. Ornston, L.N. (1970) Methods in Enzymology, 17A, 540.
3. Patel, R.N., Meagher, R.B. & Ornston, L.N. (1973) B, 12, 3531.

D-GLUCOSE 6-PHOSPHATE-*myo*-INOSITOL 1-PHOSPHATE SYNTHASE CYCLOISOMERASE

(1L-*myo*-Inositol-1-phosphate lyase (isomerizing))

D-Glucose 6-phosphate = 1L-*myo*-inositol 1-phosphate

Ref.

Molecular weight

source	value	conditions	
Rat testes	215,000	Sephadex G 200, pH 7.4	(1)

Specificity and Michaelis constants

source	substrate	K_m (mM)	conditions	
Rat testes (purified "several fold")	glucose 6-phosphate	0.75	pH 7.4, Tris, 37°	(2)
	NAD	0.5	pH 7.4, Tris, 37°	(1)
Yeast (110 x)[a]	glucose 6-phosphate	1.5	pH 8.0, Tris, 29°	(3)
	NAD[b]	0.1	pH 8.0, Tris, 29°	(3)

(a) the preparation obtained also contained inositol 1-phosphatase (EC 3.1.3.25). The two activities can be separated (Ref. 6).
(b) NAD could not be replaced by NADP, FAD, FMN, TPP, cyanocobalamin or pyridoxal phosphate.

The cycloisomerase has an absolute and specific requirement for NAD. Reduced NAD is not a product of the reaction. (1,3,4,5)

The enzyme isolated from yeast requires Mg^{2+} (Km = 4×10^{-4}M; Mn^{2+} was less active and Zn^{2+} and Co^{2+} were inactive) and NH_4^+ (K^+ was less effective) for full activity. (3)

The cycloisomerase from *Neurospora crassa* has been purified 360 x; its properties are described in Ref. 4.

Inhibitors

The enzyme (rat testes) was inhibited by 2-deoxy-D-glucose 6-phosphate (C(glucose 6-phosphate), Ki = 20 µM) and by D-glucitol 6-phosphate (mixed (glucose 6-phosphate)). (1,2)

The enzyme (yeast) was inhibited by a number of sugar phosphates. (3)

References

1. Barnett, J.E.G., Rasheed, A. & Corina, D.L. (1973) BJ, 131, 21.
2. Barnett, J.E.G., Brice, R.E. & Corina, D.L. (1970) BJ, 119, 183.
3. Charalampous, F. & Chen, I-W. (1966) Methods in Enzymology, 9, 698.
4. Pina, E. & Tatum, E.L. (1967) BBA, 136, 265.
5. Sherman, W.R., Stewart, M.A. & Zinbo, M. (1969) JBC, 244, 5703.
6. Chen, I-W. & Charalampous, F. (1966) ABB, 117, 154.

CYSTEINYL-tRNA SYNTHETASE

(L-Cysteine : $tRNA^{Cys}$ ligase (AMP-forming))

ATP + L-cysteine + $tRNA^{Cys}$ = AMP + PPi + L-cysteinyl-$tRNA^{Cys}$

Ref.

Molecular weight

source	value	conditions	
Bakers' yeast	160,000	Sephadex G 200, pH 7.25	(1)

Specificity and Michaelis constants

source	substrate[a]	K_m (mM)	conditions	
Bakers' yeast (purified 710 x)	L-cysteine[b]	0.57	pH 7.2, Tris, 37°	(1)
	ATP	1.1	pH 7.2, Tris, 37°	(1)
	PPi	1.0	pH 7.2, Tris, 37°	(1)
	Mg^{2+}[c]	1.04	pH 7.2, Tris, 37°	(1)

(a) in the ATP-PPi exchange reaction.

(b) L-cysteine could not be replaced by any of the other protein amino acids, by D-cysteine or by L-lanthionine. High concentrations of L-cysteine were inhibitory.

(c) Mg^{2+} (1.00), which was essential for the reaction, could be replaced by Co^{2+} (1.00), Mn^{2+} (1.00) or Zn^{2+} (0.25) but not by Ba^{2+}, Ca^{2+} or Ni^{2+}.

References

1. James, H.L. & Bucovaz, E.T. (1969) JBC, 244, 3210.

HISTIDYL-tRNA SYNTHETASE

(L-Histidine : tRNA ^{His} ligase (AMP-forming))

ATP + L-histidine + tRNA ^{His} = AMP + PPi + L-histidyl-tRNA ^{His}

Ref.

Molecular weight

source	value	conditions	
Salmonella typhimurium	100,000	pH 7.5, the buffer contained mercaptoethanol and EDTA.	(1)

The spectrum of the enzyme showed no light absorption other than that attributable to the protein. (1)

Specific activity

S. typhimurium (800 x) 5 L-histidine (pH 7.5, cacodylate, 37°) (1)

Michaelis constants

source	substrate	K_m (μM)	conditions	
S. typhimurium	L-histidine	25	pH 7.5, cacodylate, 37°	(1)
	ATP	140	pH 7.5, cacodylate, 37°	(1)
	S. typhimurium tRNA^{His}	0.11	pH 7.5, cacodylate, 37°	(1)

Histidyl-tRNA synthetase has also been purified (590 x) from Saccharomyces cerevisiae. (2)

References

1. De Lorenzo, F. & Ames, B.N. (1970) JBC, 245, 1710.
2. von Tigerstrom, M. & Tener, G.M. (1967) Can. J. Biochem, 45, 1067.

ASPARAGINYL-tRNA SYNTHETASE

(L-Asparagine : tRNA ligase (AMP-forming))

ATP + L-asparagine + tRNA =

AMP + PPi + L-asparaginyl-tRNA

Ref.

Molecular properties

source	value	conditions	
Rabbit liver	35,000	Sephadex G 100 and G 200, pH 7.5. The buffer contained mercaptoethanol and EDTA.	(1)

The fully active enzyme is composed of 2 components each of molecular weight 35,000 daltons. Each component on its own is enzymically inactive. (1)

Specificity

Aspartic acid was neither a substrate nor inhibitor of asparaginyl-tRNA synthetase. N-(4'-Aspartyl)-2-acetamido-2-deoxy-β-D-glucopyranosylamine was an inhibitor. (1)

References

1. Davies, M.R. & Marshall, R.D. (1972) BBRC, 47, 1386.

D-ALANINE : MEMBRANE-ACCEPTOR LIGASE

(D-Alanine : membrane-acceptor ligase (ADP-forming))

ATP + D-alanine + membrane-acceptor =

ADP + Pi + D-alanyl-O-membrane-acceptor

Ref.

Molecular weight

source	value	conditions	Ref.
Lactobacillus casei	39,000	Sephadex G 150, pH 6.5	(1)

Specificity and catalytic properties

The product of the reaction, D-alanyl-membrane-acceptor, has been characterized as a neutral hydroxylamine-labile ester which is not extractable into lipophilic solvents. There is evidence that the D-alanine is incorporated into the teichoic acid, alanyl polyglycerophosphate, of the cell membrane. (1)

source	substrate	relative velocity	K_m (μM)	conditions	Ref.
L. casei (cytosol)	D-alanine[a]	1.00	18	pH 6.5, maleate, 37°	(1)
	D-2-amino-n-butyrate	0.04	850	pH 6.5, maleate, 37°	(1)
	ATP[b]	-	3300	pH 6.5, maleate, 37°	(1)

(a) the following were neither substrates nor inhibitors: L-alanine; L-2-amino-n-butyrate; glycine; D- and L-threonine; D- and L-isoleucine and L-serine. D-And L-cycloserine; DL-alaninol and N-acetyl-D- and -L-alanine were poor inhibitors. The following were competitive inhibitors of D-alanine utilization: D-2-amino-n-butyrate (Ki = 0.6 mM); D-serine (34 mM); D-alanine hydroxamate (1.3 mM), D-2-amino-n-butyrate hydroxamate (7 mM), and D-alanine-amide (7.3 mM).

(b) ATP (1.00) could be replaced by UTP (1.26); CTP (1.07); GTP (0.84) or ADP (0.83) but not by AMP.

The enzyme is activated by Mg^{2+}, Ba^{2+}, Ca^{2+}, Mn^{2+} or Co^{2+}. It does not require RNA for activity. (1)

References

1. Reusch, V.M. & Neuhaus, F.C. (1971) JBC, 246, 6136.

PHOSPHOPANTOTHENOYL-CYSTEINE SYNTHETASE

(4'-Phospho-L-pantothenate : L-cysteine ligase (ADP-forming))

ATP + 4'-phospho-L-pantothenate + L-cysteine =

ADP + Pi + 4'-phospho-L-pantothenoyl-L-cysteine

Ref.

Molecular weight

source	value	conditions	
Rat liver (purified 120 x)	37,000	Sephadex G 75, pH 7.5	(1)

Specificity and Michaelis constants

source	substrate(a)	K_m (μM)	
Rat liver (purified 22 x)	4'-phospho-L-pantothenate	71-83	(2)

(a) conditions: pH 7.0, Tris, 37°.

The synthetase requires Mg^{2+} for activity and cannot utilize pantothenate in the place of 4'-phosphopantothenate. The enzyme in rat liver or kidney (neither was extensively purified) could utilize CTP, GTP or UTP in the place of ATP but that from *Proteus morganii* could only utilize CTP. (3)

References

1. Abiko, Y., Tomikawa, M. & Shimizu, M. (1968) J. Biochem. (Tokyo), 64, 115.
2. Abiko, Y. (1967) J. Biochem. (Tokyo), 61, 290.
3. Brown, G.M. (1959) JBC, 234, 370.

UDP-N-ACETYLMURAMOYL-L-ALA-D-GLU-meso-DAP SYNTHETASE

(UDP-*N*-acetylmuramoyl-L-Ala-D-Glu : *meso*-Dap ligase (ADP forming))

ATP + UDP-*N*-acetylmuramoyl-L-Ala-D-Glu + *meso*-Dap =

ADP + Pi + UDP-*N*-acetylmuramoyl-L-Ala-D-Glu-*meso*-Dap

Ref.

Equilibrium constant

The reversibility of the reaction with the enzyme from *Bacillus cereus* has been demonstrated. At pH 7.9 the forward reaction was strongly favoured. (1)

Specificity and Michaelis constants

source	substrate	K_m (mM)	conditions	
B. cereus (purified 260 x)	ATP	0.34	pH 8.4, Tris, 37°	(1)
	UDP-Mur NAc-dipeptide	0.32	pH 8.4, Tris, 37°	(1)
	meso-Dap	0.13	pH 8.4, Tris, 37°	(1)

The enzyme was highly specific for its substrates. The following uridine dinucleotides failed to react with *meso*-Dap:- UDP-Mur NAc; UDP-Mur NAc-L-Ala; UDP-Mur NAc-L-Ala-D-Glu-*meso*-Dap; UDP-Mur NAc-L-Ala-D-Glu-L-Lys; UDP-Mur NAc-L-Ala-D-Glu-*meso*-Dap-D-Ala-D-Ala and UDP-Mur NAc-L-Ala-D-Glu-L-Lys-D-Ala-D-Ala. *meso*-2,6-Diaminopimelate could not be replaced by LL-diaminopimelate; L-lysine; L-alanine; D-alanine; L-glutamate; D-glutamate; glycine; L-aspartate; L-valine; L-leucine; DL-isoleucine; DL-serine; DL-threonine; L-cysteine or L-histidine. UTP had about 16% the activity of ATP but GTP, CTP and nucleoside mono- and diphosphates were inactive. (1)

The enzyme requires Mg^{2+} (or Mn^{2+}, Co^{2+}, Zn^{2+} or Cd^{2+}) for activity. Considerable activation was observed at high concentrations of NH_4^+ and K^+ but not with Na^+. (1)

The enzymes purified from *Corynebacterium xerosis* and *Staphylococcus aureus* were also highly specific for their substrates. (2)

Abbreviations

Dap	2,6-diaminopimelate
UDP-Mur NAc-dipeptide	UDP-*N*-acetylmuramoyl-L-Ala-D-Glu

References

1. Mizuno, Y. & Ito, E. (1968) *JBC*, *243*, 2665.
2. Ito, E., Nathenson, S.G., Dietzler, D.N., Anderson, J.S. & Strominger, J.L. (1966) *Methods in Enzymology*, *8*, 324.

UDP-N-ACETYLMURAMOYL-L-ALA-D-GLU-meso-DAP-D-ALA-D-ALA SYNTHETASE

(UDP-N-acetylmuramoyl-L-Ala-D-Glu-meso-Dap:D-Ala-D-Ala ligase (ADP-forming))

ATP + UDP-N-acetylmuramoyl-L-Ala-D-Glu-meso-Dap + D-Ala-D-Ala = UDP-N-acetylmuramoyl-L-Ala-D-Glu-meso-Dap-D-Ala-D-Ala + ADP + Pi

Ref.

Catalytic properties

The synthetase has been purified from *Bacillus subtilis* (148 x). It catalyzes the forward reaction (F) in the presence of ATP and Mg^{2+} and the reverse, hydrolytic, reaction (R) in the presence of ADP, Pi and Co^{2+}. ATP was not formed in the reverse reaction. (1)

The synthetase has also been purified from *Escherichia coli* (400 x). This enzyme could not utilize D-alanine or L-alanine in the place of D-ala-D-ala. (2)

source	substrate[a]	reaction	K_m (μM)	metal ion	
B. subtilis	UDP-Mur NAc pentapeptide[b]	R	8	Co^{2+} [c]	(1)
	ADP	R	7	Co^{2+}	(1)
	UDP-Mur NAc tripeptide	F	30	Mg^{2+} [d]	(1)
	ATP	F	50	Mg^{2+}	(1)
	D-Ala-D-Ala	F	200	Mg^{2+}	(1)

(a) in cacodylate buffer, pH 7.3, and 30°.

(b) UDP-Mur NAc pentapeptide with meso-Dap replaced by L-Lys was hydrolyzed at 25% the rate. UDP-Mur NAc-L-Ala-D-Glu-meso-Dap-D-Ala was not hydrolyzed.

(c) when Co^{2+} was replaced with Mg^{2+} or Mn^{2+}, little activity was observed.

(d) at pH 7; Co^{2+}, Mn^{2+} and Mg^{2+} were about equally active, but at the pH optimum (pH 8-9) Co^{2+} had little activity.

Abbreviations

Dap	2,6-diaminopimelate
UDP-Mur NAc-pentapeptide	UDP-N-acetylmuramoyl-L-Ala-D-Glu-meso-Dap-D-Ala-D-Ala
UDP-Mur NAc-tripeptide	UDP-N-acetylmuramoyl-L-Ala-D-Glu-meso-Dap

References

1. Egan, A., Lawrence, P. & Strominger, J.L. (1973) JBC, 248, 3122.
2. Comb, D.G. (1962) JBC, 237, 1601.

DETHIOBIOTIN SYNTHETASE

(7,8-Diaminopelargonate : carbon-dioxide cyclo-ligase (ADP-forming))

ATP + 7,8-diaminopelargonate + CO_2 = ADP + Pi + dethiobiotin

Ref.

Molecular properties

source	value	conditions	Ref.
Escherichia coli[a]	42,000 [2]	Bio-Gel P 150, pH 7.5; gel electrophoresis (SDS)	(2)
Pseudomonas graveolens	$S_{20,w}$ = 3.5 S		(1)

(a) the enzyme does not contain bound biotin. It was not inhibited by avidin.

Specificity and catalytic properties

source	substrate	K_m (μM)	conditions	Ref.
E. coli (purified 190 x)	ATP[a]	5.0	pH 7.7, Tris, 37°	(2)
	DAPA[b]	1.3	pH 7.7, Tris, 37°	(2)
	$NaHCO_3$[c]	3400	pH 7.7, Tris, 37°	(2)
	Mg^{2+}[d]	600	pH 7.7, Tris, 37°	(2)

(a) ATP (1.0) could be replaced by CTP (0.5) but little activity (less than 0.1) occurred when any of the following replaced ATP: ITP, UTP or GTP. ADP was a competitive inhibitor of ATP (Ki = 0.23 mM). ATP + $NaHCO_3$ could not be replaced by carbamoyl phosphate.
(b) DAPA (1.00) could be replaced by diaminobiotin (0.37) to produce biotin.
(c) the actual substrate of the reaction is CO_2.
(d) Mg^{2+} (1.00) could be replaced by Mn^{2+} (0.35).

Dethiobiotin synthetase from E. coli was not inhibited by biotin or by low concentrations of dethiobiotin: concentrations of dethiobiotin above 10 mM inhibited. (2)

The enzyme from P. graveolens (purified 2000 x) was highly specific for DAPA which could not be replaced by 7-oxo-8-aminopelargonate; 7-amino-8-oxopelargonate or biotin diaminocarboxylate. (3)

Dethiobiotin synthetase requires Mg^{2+} for activity. (2,3)

Abbreviations

DAPA 7,8-diaminopelargonate

References

1. Eisenberg, M.A. (1973) Advances in Enzymology, 38, 317.
2. Krell, K. & Eisenberg, M.A. (1970) JBC, 245, 6558.
3. Yang, H., Tani, Y. & Ogata, K. (1970) Agr. Biol. Chem. (Tokyo), 34, 1748.

UREA CARBOXYLASE

(Urea: carbon-dioxide ligase (ADP-forming))

ATP + urea + CO_2 = ADP + Pi + allophanate

Ref.

Urea carboxylase (*Saccharomyces cerevisiae*) is a biotinyl protein and it is a component of a system which converts urea to CO_2 and NH_3. This system consists of 2 enzymes, namely urea carboxylase and allophanate amidohydrolase. The hydrolase catalyzes the following reaction:

allophanate + H_2O = $2NH_3$ + $2CO_2$

The overall reaction is irreversible [F] and it requires CO_2, Mg^{2+} and K^+. Allophanate hydrolase activity occurs in a 4-fold excess over the urea carboxylase activity and since the Km of the hydrolase for allophanate is low (0.37 mM) allophanate has a transitory existence. The isolation of an enzyme - $^{14}CO_2$ complex has been reported. (1,2,3)

Urea carboxylase and allophanate hydrolase from *Chlorella vulgaris* have been separated but no separation technique has been successful with the complexes from *S. cerevisiae* (purified 250 x) or *Candida utilis* (purified 150 x). (2,4,5)

The kinetics of the overall reaction have been studied with the enzyme system from *C. utilis* (conditions: pH 8.0, Tris). The following Michaelis constants were obtained: urea (0.1 mM); HCO_3^- (0.5 mM); ATP (0.25 mM) and K^+ (2.5 mM). The system was highly specific for urea, for bicarbonate (formate, acetate and a variety of other organic or inorganic anions tested were inactive) and ATP (1.0) or dATP (0.3) (ADP, AMP and all the other common nucleoside mono-, di- or tri-phosphates were inactive). Either Mg^{2+} or Mn^{2+} satisfied the divalent cation requirement of the system; there was also a requirement for K^+ which could be replaced by NH_4^+, Rb^+ or Cs^+ but not by Li^+ or Na^+. The system was strongly inhibited by ADP (C(ATP)) but not by any of the other common nucleoside mono-, di- or tri-phosphates. It was also inhibited by avidin. (5,6)

References

1. Whitney, P.A. & Cooper, T.G. (1973) JBC, 248, 325.
2. Whitney, P.A. & Cooper, T.G. (1972) BBRC, 49, 45.
3. Whitney, P.A. & Cooper, T.G. (1972) JBC, 247, 1349.
4. Thompson, J.F. & Muenster, A.M.E. (1971) BBRC, 43, 1049.
5. Roon, R.J. & Levenberg, B. (1972) JBC, 247, 4107; 7539.
6. Roon, R.J. & Levenberg, B. (1970) Methods in Enzymology, 17A, 317.
   Whitney, P.A., Cooper, T.G. & Magasanik, B. (1973) JBC, 248, 6203.

5'-PHOSPHORIBOSYLAMINE SYNTHETASE

(Ribose-5-phosphate : ammonia ligase (ADP-forming))

ATP + ribose 5-phosphate + NH_3 = ADP + Pi + 5'-phosphoribosylamine

Ref.

Specificity and catalytic properties

source	substrate	K_m (mM)	conditions	
Chicken liver (purified 93 x)	NH_4Cl [(a)]	6.3	pH 9.0, Tris	(1)
	ribose 5-phosphate [(b)]	1.8	pH 9.0, Tris	(1)

(a) glutamine and asparagine were less active and carbamoyl phosphate was inactive.

(b) 5-phospho-α-D-ribose 1-diphosphate was inactive.

The enzyme requires Mg^{2+}. EDTA was inhibitory. (1)

The enzyme was inhibited by certain purines and their derivatives (e.g. adenine, AMP and GMP). (1)

References

1. Reem, G.H. (1968) JBC, 243, 5695.

BIOTIN-[PROPIONYL-CoA-CARBOXYLASE (ATP-HYDROLYSING)] SYNTHETASE

(Biotin : apo-[propionyl-CoA :
carbon-dioxide ligase (ADP-forming)] ligase (AMP-forming))

ATP + biotin + apo-[propionyl-CoA : carbon-dioxide ligase (ADP-forming)] =
AMP + PPi + [propionyl-CoA : carbon-dioxide ligase (ADP-forming)]

Ref.

Specificity and catalytic properties

The synthetase has been purified (25 x) from rabbit liver. It is highly specific for propionyl-CoA apocarboxylase (EC 6.4.1.3): acetyl-CoA apocarboxylase (EC 6.4.1.2) and methylcrotonoyl-CoA apocarboxylase (EC 6.4.1.4) were inactive. *d*-Biotinyl 5'-adenylate (free or enzyme bound) is an intermediate in the reaction:

(*a*) ATP + biotin = *d*-biotinyl 5'-adenylate + PPi

(*b*) *d*-biotinyl 5'-adenylate + propionyl-CoA apocarboxylase =
AMP + propionyl-CoA holocarboxylase

Reaction (*a*) requires Mg^{2+}.

source	substrate	K_m (M)	conditions	
Rabbit liver	d-biotin	4.7×10^{-9}	pH 7.6, Tris, 37°	(1)
	ATP (a)	1.3×10^{-6}	pH 7.6, Tris, 37°	(1)
	d-biotinyl 5'-adenylate	5.6×10^{-6}	pH 7.6, Tris, 37°	(1)

(a) ATP (1.00) could be replaced by UTP (0.24); CTP (0.19); GTP (0.11); dTTP (0.07) or ITP (0.04).

References

1. Siegel, L., Foote, J.L. & Coon, M.J. (1965) JBC, 240, 1025.

BIOTIN CARBOXYLASE

(d-Biotin : carbon-dioxide ligase (ADP-forming))

ATP + d-biotin + CO_2 = ADP + Pi + carboxy-d-biotin

Ref.

Molecular properties

source	value	conditions	
Escherichia coli	100,000 [2]	pH 7 (EDTA + dithiothreitol); Sephadex G 200, pH 7.0; gel electrophoresis (SDS).	(1)

E. coli acetyl-CoA carboxylase (EC 6.4.1.2) can be resolved (reversibly) into three protein components:

(i) Biotin carboxylase (EC 6.3.4.aa)
(ii) Carboxyl carrier protein which contains a covalently-bound biotin prosthetic group. Carboxyl carrier protein rather than free d-biotin is the natural substrate for biotin carboxylase.
(iii) Malonyl-CoA: d-biotin carboxyl transferase.

Dissociation of acetyl-CoA carboxylase from animal tissues requires drastic conditions and gives rise to catalytically inactive components. (1,3)

Specific activity

E. coli (2000 x) 1 d-biotin (pH 7.0, TEA, 30°. In the absence of ethanol). (1,2)

Specificity and catalytic properties

source	substrate	V relative	K_m (mM)	conditions	
E. coli	d-biotin[a]	1.0	170	pH 7.0, TEA, 30°	(1)
	d-biotin[b]	6.5	170	pH 7.0, TEA, 30°	(1)
	ATP-Mg	-	0.1	pH 7.0, TEA, 30°	(1)
	HCO_3^-	-	2.9	pH 7.0, TEA, 30°	(1)

(a) d-biotin (1.00) could be replaced by d,l-O-heterobiotin (0.21); biocytin (0.19); d-homobiotin (0.10); d-biotin methyl ester (0.09); d-norbiotin (0.08) or d,l-dethiobiotin (0.07). The following were inactive: l-biotin, d-2-thiobiotin; 1'-N-carboxy-d-biotin dimethyl ester; 2-imidazolidone and urea.
(b) in the presence of 10% ethanol. The ethanol had no effect on the kinetic constants of ATP-Mg or HCO_3 or on the molecular properties of the enzyme. Acetyl CoA carboxylase was inhibited rather than activated by ethanol.

Biotin carboxylase requires Mg^{2+} (or Mn^{2+}) for activity. (1)

The carboxyl carrier protein component of E. coli acetyl-CoA carboxylase has been extensively purified. In its native form it has molecular weight 45,000 daltons but since it is very susceptible to proteolysis, fragments are often obtained (molecular weights down to 9,100 daltons). The fragments are catalytically active. (4)

Ref.

References

1. Dimroth, P., Guchhait, R.B., Stoll, E. & Lane, M.D. (1970) PNAS, 67, 1353.
2. Dimroth, P., Guchhait, R.B. & Lane, M.D. (1971) Hoppe-Seylers Z. Physiol. Chem, 352, 351.
3. Guchhait, R.B., Moss, J., Sokolski, W. & Lane, M.D. (1971) PNAS, 68, 653.
4. Fall, R.R. & Vagelos, P.R. (1973) JBC, 248, 2078.

DNA LIGASE

(Poly(deoxyribonucleotide):
poly(deoxyribonucleotide) ligase (AMP-forming))

$$ATP + (deoxyribonucleotide)_n + (deoxyribonucleotide)_m = AMP + PPi + (deoxyribonucleotide)_{n+m}$$

Ref.

Molecular weight

source	value	conditions	Ref.
Rabbit spleen or bone marrow	95,000	Sephadex G 200, pH 7.1. The buffer contained EDTA and mercaptoethanol.	(1,2)

Specificity and catalytic properties

DNA ligase catalyzes the repair of phosphodiester bond scissions in bihelical DNA. It also acts on polydeoxyribonucleotide substrates as small as pentanucleotides. The enzyme requires a 5'-phosphomonoester and an adjacent 3'-hydroxyl group at the site of the single-strand break. At the site of joining, a 3'-5' phosphodiester bond is formed. The two end groups must not be separated by a gap caused by missing nucleotides. The enzyme therefore, repairs DNA damaged by, for example, pancreatic DNase (EC 3.1.4.5). (1,3,5,6)

At high enzyme concentrations, DNA ligase joins deoxyribo-oligonucleotides on ribonucleotide templates and ribo-oligonucleotides on deoxyribonucleotide templates. (4)

DNA ligase catalyzes the ATP-PPi exchange reaction. (1,5)

source	substrate	K_m(M)	conditions	Ref.
Bacteriophage T-4 (host = Escherichia coli. Purified 100 x)	DNA[a]	1.5×10^{-9}	pH 7.6, Tris, 37°	(5,6)
	ATP[b]	1.4×10^{-5}	pH 7.5, Tris, 37°	(5,6)
Rabbit spleen (purified 145 x)	ATP[c]	2×10^{-7}	pH 7.5, Tris, 37°	(1)
Rabbit bone marrow (purified 220 x)	ATP	2×10^{-7}	pH 7.6, Tris, 37°	(2)
	dATP	4×10^{-5}	pH 7.6, Tris, 37°	(2)

(a) the Km given is in terms of phosphomonoesters at single-strand breaks.

(b) other triphosphates and NAD were inactive. dATP was a competitive inhibitor (Ki = 3.5×10^{-5}M).

(c) GTP, dTTP and NAD were inactive. dATP did not inhibit.

The enzyme requires Mg^{2+} (partially replaced by Mn^{2+}) and dithiothreitol (partially replaced by mercaptoethanol). (1,5)

Ref.

References

1. Lindahl, T. (1971) Methods in Enzymology, 21D, 333.
2. Lindahl, T. & Edelman, G.M. (1968) PNAS, 61, 680.
3. Gupta, N.K., Ohtsuka, E., Weber, H., Chang, S.H. & Khorana, H.G. (1968) PNAS, 60, 285.
4. Kleppe, K., van de Sande, J.H. & Khorana, H.G. (1970) PNAS, 67, 68.
5. Weiss, B. (1971) Methods in Enzymology, 21D, 319.
6. Weiss, B., Jacquemin-Sablon, A., Live, T.R., Fareed, G.C. & Richardson, C.C. (1968) JBC, 243, 4543.

POLYNUCLEOTIDE SYNTHETASE (NAD)

(Poly(deoxyribonucleotide) : poly(deoxyribonucleotide) ligase (AMP-forming, NMN-forming))

$$NAD + (deoxyribonucleotide)_n + (deoxyribonucleotide)_m = AMP + NMN + (deoxyribonucleotide)_{n+m}$$

Ref.

Specificity and catalytic properties

The synthetase (*E. coli*, purified 210 x) joins DNA strands in a 3'-5'-phosphodiester bond. It requires a 5'-phosphoryl terminus on the DNA which must be in hydrogen-bonding juxtaposition with a 3'-hydroxyl terminus (also see EC 6.5.1.1). Thus, the enzyme was able to join single strand breaks in duplex DNA caused by pancreatic DNase (EC 3.1.4.5). (1)

The enzyme is highly specific for NAD (Km = 3×10^{-8}M). The following were much less effective: reduced NAD; thionicotinamide-NAD and 3-acetylpyridine-NAD. The following were neither substrates nor inhibitors: NADP; reduced NADP; α-NAD; desamino-NAD; NMN; ADP-ribose; FMN; FAD; CoA; ADP; ATP; ATP + GTP + CTP + UTP or dATP + dGTP + dCTP + dTTP. (1)

The enzyme requires Mn^{2+} or Mg^{2+} for activity; Ca^{2+}, Co^{2+} or Ni^{2+} were inactive. (1)

References

1. Zimmerman, S.B., Little, J.W., Oshinsky, C.K. & Gellert, M. (1967) PNAS, 57, 1841.

INDEX

Enzymes in the Enzyme Handbook (1969) and systematic names have not been included. The following trivial names often occur in the literature: *carboxykinase* and *carboxytransphosphorylase* (see sub-subgroup 4.1.1.); *dehydrase* (dehydratase or hydratase : 4.2.1); *kinase* (phosphotransferase : 2.7.1-6); *kinosynthetase* (synthetase : 6.3.2, 6.3.4); *methylpherase* (methyltransferase : 2.1.1); *oxidase* (often dehydrogenase - see group 1); *phosphorylase* (2.4.1, 2.4.2, 2.7.7); *pyrophosphorylase* (2.4.2, 2.7.7); *synthetase* (often synthase or lyase : 2.1.1, 2.3.1, 2.4.1, 2.7.7, 4.1.2, 4.1.3) and *thiokinase* (synthetase : 6.2.1). The names amidinotransferase, formyltransferase, glucosyltransferase etc. are often replaced by transamidinase, transformylase, transglucosylase etc. The Index includes certain enzyme complexes and systems (e.g. fatty acid synthetase, tryptophan pathway); substances such as ferredoxin and acyl carrier protein and sources for the light absorption data (LAD) of certain dyes (e.g. dichlorophenol-indophenol) or cofactors (e.g. NAD(P)).

Adenylylsulphatase 3.6.2.1
ADPribose phosphorylase 2.7.7.35
ADPsulphurylase 2.7.7.5
Adrenodoxin reductase 1.6.7.1
Aeromonas proteolytica aminopeptidase 3.4.11.10
D-Alanine carboxypeptidase 3.4.12.6
D-Alanine: membrane acceptor ligase 6.1.1.bb
Alanine racemase see 4.1.99.2
L-Alanyl-tRNA UDP-*N*-acetylmuramoyl pentapeptide transferase 2.3.2.10
Alcohol dehydrogenase (acceptor) 1.1.99.8
Alcohol dehydrogenase (NAD(P)) 1.1.1.71
Alcohol oxidase 1.1.3.13
Aldohexose dehydrogenase 1.1.1.119
L-3-Aldonate dehydrogenase 1.1.1.45
Aldose dehydrogenase 1.1.1.121
Alkane 1-hydroxylase 1.14.15.3
Alkane 1-monooxygenase 1.14.15.3
Alkyl amidase 3.5.1.bb
Allantoicase 3.5.3.4
Allophanate amidohydrolase see 6.3.4.6
ω-Amidase 3.5.1.3
N-Amidino-L-aspartate hydrolase 3.5.3.aa
Aminoacetone synthase 2.3.1.29
D-Amino acid dehydrogenase 1.4.99.1
Amino acid racemase 5.1.1.10
Aminoacyl-lysine aminopeptidase 3.4.13.4
Aminoacyl-lysine dipeptidase 3.4.13.4
Aminoacylproline aminopeptidase 3.4.11.9
Aminoacyl-tRNA hydrolase 3.1.1.29
2-Aminoadipate aminotransferase 2.6.1.39
Aminoadipate semialdehyde dehydrogenase 1.2.1.31
Aminoadipate semialdehyde-glutamate reductase 1.5.1.10
*N*²-(4-Aminobutyryl)-L-lysine hydrolase 3.4.13.4
Aminocarboxymuconate-semialdehyde decarboxylase 4.1.1.45
2-Amino-4-hydroxy-6-hydroxymethyldihydropteridine pyrophosphokinase 2.7.6.3
2-Aminoisobutyrate decarboxylase 4.1.1.64
δ-Aminolaevulinate synthase 2.3.1.37
Aminopeptidase B 3.4.11.6
Aminopeptidase P 3.4.11.9
2-Aminophenol oxidase 1.10.3.4
D-Aminopropanol dehydrogenase see 1.1.1.75
L-Aminopropanol dehydrogenase 1.1.1.75
Aminopropyltransferase 2.5.1.16
5-Aminovaleramidase 3.5.1.aa
Ammonia kinase 2.7.3.8
Angiotensin I and II see 3.4.15.aa
Angiotensin-converting enzyme 3.4.15.aa
Angiotensinogen see 3.4.15.aa
Anthranilate 1,2-dioxygenase (deaminating, decarboxylating) see 1.14.12.2
Anthranilate 2,3-dioxygenase (deaminating) 1.14.12.2
Anthranilate hydroxylase 1.14.12.2
Anthranilate phosphoribosyltransferase 2.4.2.18
Anthranilate synthase 4.1.3.27
D-Apiitol dehydrogenase 1.1.1.114
D-Apiose reductase 1.1.1.114
Apiosyltransferase 2.4.1.cc
Apo-pyridoxal enzyme hydrolase 3.4.99.aa
Aquocob(I)alamin adenosyltransferase 2.5.1.17
α-L-Arabinofuranosidase 3.2.1.55
Arabinonate dehydratase 4.2.1.5
D-Arabinose dehydrogenase (NAD(P)) 1.1.1.117
Arabinosidase 3.2.1.55
Arabonate dehydratase 4.2.1.5
Arene-oxide hydratase 4.2.1.64

Enzyme	EC
L-Fucose dehydrogenase	1.1.1.122
Fucose 1-phosphate guanylyltransferase	2.7.7.30
α-L-Fucosidase	3.2.1.51
1,2-α-L-Fucosidase	3.2.1.63
4-Fucosyltransferase	2.4.1.65
2-Furoyl-CoA hydroxylase	1.2.99.aa
Galactinol hydrolase	see 2.4.1.ff
Galactinol-raffinose galactosyltransferase	see 2.4.1.ff
Galactinol-stachyose galactosyltransferase	see 2.4.1.ff
Galactinol-sucrose galactosyltransferase	2.4.1.ff
Galactolipase	3.1.1.26
D-Galactose dehydrogenase (NADP)	1.1.1.120
Galactose 6-phosphate dehydrogenase	1.1.1.ee
Gastricsin	3.4.23.3
GDP-6-deoxy-D-talose dehydrogenase	see 1.1.1.aa
GDPfucose pyrophosphorylase	2.7.7.30
GDPglucose glucohydrolase	3.2.1.42
GDPglucosidase	3.2.1.42
GDPhexose pyrophosphorylase	2.7.7.28
GDPmannose 4,6-dehydratase	4.2.1.47
GDPmannose oxidoreductase	4.2.1.47
GDP-4-oxo-D-rhamnose reductase	1.1.1.aa
GDP-D-talomethylose dehydrogenase	see 1.1.1.aa
Geranyltransferase	2.5.1.10
β1,3-Glucan hydrolase	3.2.1.39
1,3-β-Glucan synthase	2.4.1.34
1,3-β-Glucan-UDPglucosyltransferase	2.4.1.34
Gluconate dehydratase	4.2.1.39
Glucose dehydrogenase (Aspergillus)	1.1.99.10
Glucose dehydrogenase (NADP)	1.1.1.119
Glucose 6-phosphate cycloisomerase	5.5.1.4
Glucose 6-phosphate 1-epimerase	5.1.3.aa
D-Glucose 6-phosphate-*myo*-inositol 1-phosphate synthase cycloisomerase	5.5.1.4
D-Glucoside 3-dehydrogenase	1.1.99.13
β-Glucoside kinase	2.7.1.aa
Glucuronate 1-phosphate uridylyltransferase	2.7.7.aa
Glutamate acetyltransferase	2.3.1.35
Glutamate carboxypeptidase	3.4.12.10
D-Glutamate cyclase	4.2.1.48
Glutamate methyltransferase	2.1.1.21
Glutamate-2-oxo-adipate transaminase	2.6.1.39
Glutamate synthase	2.6.1.53
Glutaminase II	2.6.1.15
Glutamine-oxoacid aminotransferase	2.6.1.15
Glutamine-synthetase adenylyltransferase	2.7.7.42
Glutamine transaminase	2.6.1.15
γ-Glutamyl carboxypeptidase	3.4.12.aa
γ-Glutamylcyclotransferase	2.3.2.4
γ-Glutamyl lactamase	2.3.2.4
γ-Glutamyltransferase	2.3.2.2
γ-Glutamyl transpeptidase	2.3.2.2
Glutarate-semialdehyde dehydrogenase	1.2.1.20
Glutathione	see 2.3.2.2
Glutathione *S*-alkyltransferase	2.5.1.12
Glutathione CoAS-SG transhydrogenase	1.8.4.3
Glutathione cystine transhydrogenase	1.8.4.4
Glutathione epoxidase	4.4.1.7
Glutathione *S*-epoxidetransferase	4.4.1.7
Glutathione peroxidase	1.11.1.9
Glycerol dehydratase	4.2.1.30

Glycerol dehydrogenase 1.1.1.72
Glycerol dehydrogenase (NADP, dihydroxyacetone forming) 1.1.1.gg
Glycine acetyltransferase 2.3.1.29
Glycine cleavage system 2.1.2.10
Glycine-cytochrome c reductase 1.4.99.aa
Glycine decarboxylase 2.1.2.10
Glycine methyltransferase 2.1.1.20
Glycine-oxaloacetate aminotransferase 2.6.1.35
Glycine synthase 2.1.2.10
Glycocholase 3.5.1.24
Glycocholate hydrolase 3.5.1.24
Glycogen 6-glucanohydrolase 3.2.1.68
Glycollate dehydrogenase 1.1.1.79
Glyoxylate reductase (NADP) 1.1.1.79
GTP cyclohydrolase 3.5.4.16
Guanidinobutyrase 3.5.3.7
Guanidinosuccinate hydrolase 3.5.3.aa
Guanosine aminase 3.5.4.15
Guanosine deaminase 3.5.4.15
Guanyl cyclase 4.6.1.2
Guanylate cyclase 4.6.1.2
Guanylyl cyclase 4.6.1.2
L-Gulonate dehydrogenase 1.1.1.45

Haem oxygenase (decyclizing) 1.14.99.3
Halogenated tyrosine transaminase 2.6.1.24
Hexopyranoside: cytochrome c oxidoreductase 1.1.99.13
Hexose oxidase 1.1.3.5
Hexose 1-phosphate nucleotidyltransferase 2.7.7.28
Histidine methyltransferase 2.1.1.bb
Histidine pathway see 3.5.4.19
Histidyl-leucine dipeptidase see 3.4.15.aa
Histidyl t-RNA synthetase 6.1.1.21
Holoacyl carrier protein synthase 2.7.8.7
Homocysteine methyltransferase 2.1.1.13 and 2.1.1.14
L-Homocysteine synthase 4.2.99.10
Homoisocitrate dehydrogenase 1.1.1.dd
Homoserine acetyltransferase 2.3.1.31
Hydrogen carrier protein see 2.1.2.10
D-3-Hydroxyacyl carrier protein dehydratase see 4.2.1.58
S-(Hydroxyalkyl) glutathione lyase 4.4.1.7
4-Hydroxybenzoate hydroxylase 1.14.13.2
4-Hydroxybenzoate 3-monooxygenase 1.14.13.2
3-Hydroxybenzyl-alcohol dehydrogenase 1.1.1.97
D-3-Hydroxybutyryl acyl carrier protein dehydratase 4.2.1.58
3-Hydroxybutyryl CoA dehydrogenase (NADP) 1.1.1.ff
2-Hydroxycinnamate reductase 1.3.1.11
10-Hydroxydecanoate dehydrogenase 1.1.1.66
ω-Hydroxydecanoate dehydrogenase 1.1.1.66
3-Hydroxydecanoyl acyl carrier protein dehydratase 4.2.1.60
ω-Hydroxylase 1.14.15.3
5'-Hydroxyl polynucleotide kinase 2.7.1.78
Hydroxylysine kinase 2.7.1.81
Hydroxymethyldihydropteridine pyrophosphokinase 2.7.6.3
5-Hydroxy-*N*-methylpyroglutamate synthase 3.5.1.36
D-6-Hydroxynicotine dehydrogenase 1.5.3.6
L-6-Hydroxynicotine dehydrogenase 1.5.3.5
3-Hydroxyoctanoyl acyl carrier protein dehydratase see 4.2.1.58 and 4.2.1.60

3-Hydroxy-2-oxoacid reductioisomerase 1.1.1.89
3-Hydroxypalmitoyl-acyl carrier protein dehydratase see 4.2.1.58 and 4.2.1.60
2-Hydroxyphenylpropionate hydroxylase 1.14.13.4
15-Hydroxy-prostaglandin dehydrogenase 1.1.1.141
16α-Hydroxysteroid dehydrogenase 1.1.1.147
17β-Hydroxysteroid dehydrogenase 1.1.1.64
20α-Hydroxysteroid dehydrogenase 1.1.1.149
21-Hydroxysteroid dehydrogenase 1.1.1.151
3-β-Hydroxysteroid sulphotransferase 2.8.2.2

Imidazoleacetate hydroxylase 1.14.13.5
Imidazoieacetate 4-monooxygenase 1.14.13.5
Indanol dehydrogenase 1.1.1.112
Indoleacetaldoxime dehydratase 4.2.1.29
Indole 3-glycerolphosphate synthase 4.1.1.48
Indolelactate dehydrogenase 1.1.1.110
Inositol cyclase 5.5.1.4
Inositol 1-phosphatase 3.1.3.25
Inositol 1-phosphate synthase 5.5.1.4
Insulinase 3.4.22.aa
Iodophenol methyltransferase 2.1.1.26
Iodotyrosine aminotransferase 2.6.1.24
Iron-cytochrome c reductase 1.9.99.1
Isoamylase 3.2.1.68
Isomaltase 3.2.1.10
Isophenoxazine synthase 1.10.3.4
Isoprenoid-alcohol kinase 2.7.1.66

Kallidins I and II see 3.4.21.8
Kallikrein 3.4.21.8
3-Ketoacyl acyl carrier protein reductase 1.1.1.100
3-Ketoacyl acyl carrier protein synthase 2.3.1.41
2-Keto-3-deoxy-L-arabonate aldolase 4.1.2.18
2-Keto-3-deoxy-L-arabonate dehydratase 4.2.1.43
2-Keto-3-deoxy-D-fuconate aldolase 4.1.2.18
5-Keto-D-fructose reductase 1.1.1.123 and 1.1.1.124
3-Keto-L-gulonate reductase 1.1.1.45
Kinin see 3.4.21.8
Kininogen see 3.4.21.8
Kininogenase 3.4.21.8
Kininogenin 3.4.21.8

Lactaldehyde dehydrogenase 1.2.1.22
β-Lactamase II 3.5.2.8
Lactate-malate transhydrogenase 1.1.99.7
Laminaribiose phosphorylase see 2.4.1.30
Laminarinase 3.2.1.39
L-Leucine dehydrogenase 1.4.1.9
Leucyltransferase see 2.3.2.8
Limit dextrinase 3.2.1.10
Linoleate isomerase 5.2.1.5
Lysine 2,3-aminomutase 5.4.3.2
β-Lysine 5,6-aminomutase 5.4.3.3
D-Lysine 5,6-aminomutase 5.4.3.4
Lysine 6-aminotransferase 2.6.1.36
Lysine 2-monooxygenase 1.13.12.2
Lysine 6-monooxygenase 1.13.12.aa
D-α-Lysine mutase 5.4.3.4
β-Lysine mutase 5.4.3.3
Lysine oxidase 1.13.12.aa
Lysine racemase see 5.1.1.9

Malate dehydrogenase (decarboxylating) 1.1.1.39
Malate dehydrogenase (NADP) 1.1.1.82
Malate-lactate transhydrogenase 1.1.99.7
Maleate isomerase 5.2.1.1
Malic enzyme 1.1.1.39
Malonyl acyl carrier protein decarboxylase see 2.3.1.41
Malonyl-CoA acyl carrier protein transferase 2.3.1.39
Malonyl CoA: *d*-biotin carboxyl transferase see 6.3.4.aa
Mandelate racemase 5.1.2.2
Mannanase 3.2.1.25
Mannase 3.2.1.25
β-Mannosidase 3.2.1.25
Melilotate dehydrogenase 1.3.1.11
Melilotate hydroxylase 1.14.13.4
Melilotate 3-monooxygenase 1.14.13.4
Methionine synthase 4.2.99.10
Methionyl dipeptidase 3.4.13.aa
Methionyl-tRNA formyl-transferase 2.1.2.9
Methylamine-glutamate methyltransferase 2.1.1.21
N-Methylaminoacid dehydrogenase 1.5.99.bb
Methylene blue (LAD) see 1.2.99.aa
2-Methylene-glutarate mutase 5.4.99.4
5,10-Methylenetetra-hydrofolate reductase 1.1.1.68
N-Methylglutamate dehydrogenase 1.5.99.bb
N-Methylglutamate synthase 2.1.1.21
Methylglyoxal synthase 4.2.99.11
Methylhydroxypyridine-carboxylate dioxygenase 1.14.12.4
Methylitaconate Δ-isomerase 5.3.3.6
Methylmalonate semialdehyde dehydrogenase (acylating) 1.2.1.27
Methylmalonyl-CoA decarboxylase 4.1.1.41
N-Methyl-2-oxoglutaramate hydrolase 3.5.1.36
6-Methylsalicylate decarboxylase 4.1.1.52
Methylviologen (LAD) see 1.8.99.1
Microsomal haem oxygenase 1.14.99.3
Monodehydroascorbate reductase (reduced NAD) 1.6.5.4
Monophenol monooxygenase see 3.4.21.cc
Muconolactone-Δ-isomerase 5.3.3.4
Muramoyl-pentapeptide carboxypeptidase 3.4.12.6
Muramoyl-tetrapeptide carboxypeptidase see 3.4.12.6
Mycodextranase 3.2.1.61
Myrosinase 3.2.3.1

NAD(P) (LAD) see 1.1.1.ee
reduced NADP-cytochrome c_2 reductase 1.6.2.5
NDPhexose pyrophosphorylase 2.7.7.28
Nicotinate dehydrogenase 1.5.1.13
Nicotinate hydroxylase 1.5.1.13
Nicotinatenucleotide-dimethylbenzimidazole phosphoribosyl-transferase 2.4.2.21
Nitrite reductase 1.7.7.1 and 1.7.99.3
Nitrite reductase (cytochrome) 1.7.2.1
Noradrenaline *N*-methyl-transferase 2.1.1.28
Nucleosidase 3.2.2.8
Nucleoside phosphoacyl-hydrolase 3.6.1.24
Nucleoside phospho-transferase 2.7.1.77
Nucleoside tetraphosphate hydrolase 3.6.1.14

Enzyme	EC number
Nucleosidetriphosphate-adenylate kinase	2.7.4.10
Nucleosidetriphosphate pyrophosphatase	3.6.1.19
Nucleotidase	3.1.3.31
Oestradiol dehydrogenase	1.1.1.64
17β-Oestradiol-UDP-glucuronyltransferase	2.4.1.59
Oestriol-UDP(16α-glucuronyl) transferase	2.4.1.61
1,4-β-Oligoglucan: orthophosphate glucosyltransferase	2.4.1.49
β-1,3-Oligoglucan phosphorylase	2.4.1.30
Oligo-1,3-glucosidase	3.2.1.39
Oligo-1,6-glucosidase	3.2.1.10
Orcinol hydroxylase	1.14.13.6
Orcinol 2-monooxygenase	1.14.13.6
Ornithine acetyltransferase	2.3.1.35
D-Ornithine 4,5-aminomutase	5.4.3.aa
Ornithine cyclase	4.3.1.aa
Ornithine cyclodeaminase	4.3.1.aa
Ornithine decarboxylase	4.1.1.17
Ornithine oxidase	1.13.12.2
Orsellinate decarboxylase	4.1.1.58
Orsellinate depside esterase	3.1.1.aa
Oxaloacetate (LAD)	see 1.1.99.7
Oxaloacetate tautomerase	5.3.2.2
Oxaloglycollate reductase (decarboxylating)	1.1.1.92
Oxalyl-CoA decarboxylase	4.1.1.8
N-Oxide dealkylase	4.1.2.24
3-Oxoacyl acyl carrier protein reductase	1.1.1.100
3-Oxoacyl acyl carrier protein synthase	2.3.1.41
3-Oxoadipate enol-lactonase	3.1.1.24
2-Oxoaldehyde dehydrogenase	1.2.1.23
2-Oxoglutarate semialdehyde dehydrogenase	1.2.1.26
2-Oxoglutarate synthase	1.2.7.3
4-Oxoproline reductase	1.1.1.104
Palatinase	3.2.1.10
Pantetheinephosphate adenylyltransferase	see 2.7.1.24
Pantoate dehydrogenase	1.1.1.106
Pantothenate kinase	2.7.1.33
Parapepsin II	3.4.23.3
Patulin pathway	see 1.1.1.97
Penicillin amidase	3.5.1.11
Pepsin C	3.4.23.3
Peptidyllysine oxidase	1.13.12.aa
Phaseolain	3.4.12.1
Phenol hydroxylase	1.14.13.7
Phenol 2-monooxygenase	1.14.13.7
Phenylacetaldehyde dehydrogenase	1.2.1.aa
Phenylalanine pathway	see 5.4.99.5
Phenylalanine racemase (ATP-hydrolysing)	5.1.1.11
Phenylalanyltransferase	see 2.3.2.8
Phenylethanolamine *N*-methyltransferase	2.1.1.28
Phenylpyruvate decarboxylase	4.1.1.43
Phenylpyruvate-glutamine aminotransferase	2.6.1.15
Phenylpyruvate tautomerase	5.3.2.1
Phosphatidate cytidylyltransferase	2.7.7.41
Phospho*enol*pyruvate protein phosphotransferase	see 2.7.1.69
Phospho*enol*pyruvate synthase	2.7.1.dd
1-Phosphofructokinase	2.7.1.56
6-Phospho-β-D-galactosidase	3.2.1.bb
6-Phospho-β-glucosidase	3.2.1.cc
3-Phosphoglycerate phosphatase	3.1.3.aa
Phosphohistidinoprotein hexose phosphotransferase	2.7.1.69
Phosphomevalonate kinase	2.7.4.2
Phosphonatase	3.11.1.1
Phosphonoacetaldehyde hydrolase	3.11.1.1

Phosphopantothenoyl-cysteine decarboxylase 4.1.1.36
Phosphopantothenoyl-cysteine synthetase 6.3.2.5
Phosphoprotamine phosphatase see 2.7.1.70
Phosphoramidate-ADP phosphotransferase 2.7.3.8
Phosphoribokinase 2.7.1.18
5'-Phosphoribosylamine synthetase 6.3.4.7
Phosphoribosyl-AMP cyclohydrolase 3.5.4.19
N-(5'-Phosphoribosyl)-anthranilate isomerase see 4.1.1.48
Phosphoribosylanthranilate pyrophosphorylase 2.4.2.18
N-(5'-Phospho-D-ribosyl-formimino)-5-amino-1-(5''-phosphoribosyl)-4-imidazolecarboxamide isomerase 5.3.1.16
O-Phosphorylethanolamine phospho-lyase 4.2.99.7
Phosphoserine amino-transferase 2.6.1.52
Picolinic carboxylase 4.1.1.45
Pinitol dehydrogenase see 1.1.1.143
Plant carboxypeptidase 3.4.12.1
Plant RNase 3.1.4.23
Polynucleotide 5'-hydroxyl kinase 2.7.1.78
Polynucleotide ligase 6.5.1.1 and 6.5.1.2
Polynucleotide synthetase (ATP) 6.5.1.1
Polynucleotide synthetase (NAD) 6.5.1.2
Polyol dehydrogenase(NADP) 1.1.1.139
Porphobilinogen deaminase 4.3.1.8
Premonophenol monooxygenase activating enzyme 3.4.21.cc
Prephenate dehydratase see 5.4.99.5
Prephenate dehydrogenase see 5.4.99.5
Presqualene reductase see 2.5.1.bb
Presqualene synthase 2.5.1.bb
Proinsulin see 3.4.21.aa
Proinsulinase 3.4.21.aa
Proline hydroxylase 1.14.11.2
Proline,2-oxoglutarate dioxygenase 1.14.11.2
Proline racemase 5.1.1.4
Propionyl-CoA carboxylase 4.1.1.41
Propionyl CoA holo-carboxylase synthetase 6.3.4.10
Propylamine transferase 2.5.1.16
Prostaglandin dehydrogenase 1.1.1.141
Protamine kinase 2.7.1.70
Protaminephosphate phosphatase see 2.7.1.70
Protein-acetylgalact-osaminyl transferase 2.4.1.41
Protein (arginine) methyltransferase see 2.1.1.24
Protein disulphide isomerase 5.3.4.1
Protein methylase I see 2.1.1.24
Protein methylase II 2.1.1.24
Protein *O*-methyl-transferase 2.1.1.24
Protein-UDPacetylgalactos-aminyltransferase 2.4.1.41
Protocatechuate 4,5-dioxygenase 1.13.11.8
Protocollagen proline hydroxylase 1.14.11.2
Pseudouridine kinase 2.7.1.bb
N-Pteroyl-L-glutamate hydrolase 3.4.12.10
Pteroyl-oligoglutamyl conjugase 3.4.12.aa
Putrescine oxidase 1.4.3.aa
Pyloruspepsin 3.4.23.3
Pyridoxal dehydrogenase 1.1.1.107
Pyridoxamine-phosphate aminotransferase 2.6.1.aa
Pyridoxin 4-oxidase 1.1.3.12
Pyridoxol 5'-dehydrogenase 1.1.99.9
Pyridoxol 4-oxidase 1.1.3.12
Pyrimidine-nucleoside phosphorylase 2.4.2.2